T0096969

An Industrial Heritage

ALLIS-CHALMERS CORPORATION

by

Walter F. Peterson, Ph. D.
President, University of Dubuque

with an epilogue by

C. Edward Weber, Ph. D.
Dean, College of Business Administration,
University of Wisconsin — Milwaukee

MILWAUKEE COUNTY HISTORICAL SOCIETY

Library of Congress Card No. 76-57456

Copyright 1976, 1977, 1978 by the

Milwaukee County Historical Society
910 North Third Street
Milwaukee, Wisconsin 53203

Produced in the United States of America by
NAPCO Graphic Arts, Inc.
Milwaukee, Wisconsin

ALLIS-CHALMERS

FOREWORD

For many years people have gathered historical information and pictures about Allis-Chalmers Corporation. Many employees were interviewed for their firsthand knowledge about events and persons who played roles in those events; company documents were closely scrutinized; publications, including newspapers and magazines, were gone over carefully, some dating back to before 1847 when a predecessor company of Allis-Chalmers began to do business.

The late Alberta Price Johnson, M.A., who was an instructor at Wauwatosa High School in Wauwatosa, Wisconsin, broke ground for this present volume with her research, interviews, and writing before and during World War II. After the war, her investigations culminated in five typed volumes entitled *Mill Stones to Atom Smashers*, detailing the origins and development of Allis-Chalmers during the periods 1847–1870, 1870–1900, and 1941–1945. Dr. Walter F. Peterson, now President, University of Dubuque, while chairman of the Social Sciences Department at Milwaukee-Downer College, Milwaukee, Wisconsin, subsequently wrote a volume covering the period 1901–1941.

Then Dr. Peterson, while teaching at Downer, the University of Wisconsin-Milwaukee, and Lawrence University, synthesized this material into a single narrative which appears as the first ten chapters of this book.

Celebration of the Bicentennial of the United States in 1976 became the compelling factor in the decision to publish a history of Allis-Chalmers and its growth through 129 years.

The Milwaukee County Historical Society, which through the years has provided invaluable advice and active help in composing this company history, provided the final motivating thrust when it offered to publish the book as a bicentennial project.

Ten years had passed, however, since Dr. Peterson had completed his portion of the book. Much had occurred during that decade, so much that it was felt an additional chapter was warranted to include the Bicentennial year.

The Society was able to secure the services of Dr. C. Edward Weber, Dean of the Business Administration School of the University of Wisconsin-Milwaukee, to look at the past decade and analyze what had happened to Allis-Chalmers during that period when the American economy became the first in the world to pass the trillion dollar mark in 1971. Thus came about the final chapter of this saga of free enterprise.

No one knows better than I the sobering truth that business histories rarely produce bestsellers. In part, this attitude is due to an impression most people have that business history is boring. The truth of the matter happens to be just the opposite. For business history is filled with the same drama as the tragedies and comedies of Shakespeare and makes for fascinating reading. Even though this type of history seldom, if ever, achieves the Olympian heights of classical literature, nevertheless, the personalities, events, and achievements described in a corporate history are essential to an understanding of the making of the United States what it is today. We hope our corporate history will offer the reader a deeper, truer, more authentic understanding of industry and the part industry plays in the American pursuit of life, liberty, happiness, and a more perfect country.

So, to the many people who contributed their time and effort to making this volume possible go our special thanks. This includes Mrs. Johnson, Dr. Peterson and Dr. Weber; Sharon Mallman, who edited the manuscript for publication, and the staff and the editorial committee of the Milwaukee County Historical Society; Quentin J. O'Sullivan, Manager of the Allis-Chalmers News Bureau, who was editorial director for the project. We also acknowledge that this history of Allis-Chalmers was made possible only through the efforts of thousands of employes who, since 1847, contributed their work and ideas to help make the company grow to what it is today. We also acknowledge contributions made in the early days by the owners and later by the managers, officers and directors of Allis-Chalmers. Obviously all of these people, supported by the confidence of the shareowners through the years, understandably could not be named in this book, so we take this opportunity to recognize their efforts.

To them and all the others whose efforts have contributed to making this book a reality, I offer my deep appreciation and gratitude.

DAVID C. SCOTT
chairman of the board,
chief executive officer
and president

Milwaukee, Wisconsin
March, 1978

CONTENTS

Edward P. Allis

CHAPTER ONE

THE BEGINNING

THE WEST HELD LIMITLESS OPPORTUNITY for able and ambitious men, and in 1846 Wisconsin was part of that West. The area had raw materials in profusion and in happy combination. The soil produced not only an ever greater amount of wheat for milling but also food in abundance for the working man. Wisconsin's forests contained timber of all types. Northern Wisconsin and Upper Michigan would soon yield iron for machinery and implements. The numerous excellent streams meant ample power.

Milwaukee was Wisconsin's principal city, located where the meandering Milwaukee River joins Lake Michigan. The enterprising man could look beyond the marshy lowlands to the steep bluffs and looming forests in the distance. Milwaukee's first daily paper was established in 1844, and through public agitation an ordinance was passed the next year preventing cattle from running freely in the streets. After ten years as a village, Milwaukee became a city on January 31, 1846. Her commanding position on Lake Michigan attracted growing numbers of Easterners, particularly New Yorkers, and immigrants, especially Germans. The population of 8,000 in 1845 had grown to almost 10,000 in 1846 and was to be 12,000 by the close of 1847. In 1846 the thriving city exported 213,448 bushels of wheat, 1,770,650 pounds of lead, 295 dozen pails and 10,562 pounds of wool. An optimistic and enterprising Milwaukee was building plank roads to the west and was agitating for a railroad. Because Milwaukee had easy access to plentiful supplies of tan bark and had facilities for collecting hides and pelts from the interior, it provided an excellent base for the development of this growing market. In fact, Milwaukee was to become the world's largest tanning center by 1870. In 1845 Rufus King, in his first visit to Milwaukee, reported that several merchants had done retail business in the previous two years of "forty to fifty thousand." Edward Phelps Allis hoped to become at least that successful by exploiting the resources and opportunities of Wisconsin.[1]

Edward P. Allis was only twenty-one when he arrived in Milwaukee in the spring of 1846. Born in Cazenovia, New York on May

12, 1824, the son of Jere and Mary Allis, he had been educated at Cazenovia Academy, Geneva College, and Union College (B.A., 1845). In Milwaukee, Allis went into partnership with his friend and college classmate, William Allen. This association with Allen was profitable because it provided direction, experience, and financial backing. Allen's family, which had been in the tanning business in Cazenovia, decided to relocate in 1846 when their supply of tan bark in New York was exhausted. The father, Rufus, and his sons William and George came to Milwaukee where both hides and hemlock were readily available. Financial support from Rufus Allen enabled William Allen and Edward Allis to open a leather goods store, announced in the *Milwaukee Sentinel* on May 2, 1846. The two New Yorkers named it the "Empire Leather Store."[2]

Milwaukee offered the opportunity Allis wanted, to develop a business, build a home, and rear a family. Established in the leather business, he returned to Geneva, New York, to marry Margaret Marie Watson on September 12, 1848. On February 22, 1849 he drew $2,972.07 out of his business account to cover the initial cost of a home and furnishings. William, the first of twelve children, was born on November 14, 1849.[3]

A man of only medium stature, Allis soon grew a beard to appear older. Quiet and capable, he made an immediate impact on the civic and business leaders of Milwaukee. E. P. Allis was a many-faceted man. A successful businessman, he had an active interest in science, art and literature as well as politics and civic affairs. He considered himself a friend of labor, and astounded the conservative business community by running as the Greenback candidate for governor of Wisconsin in 1877 and 1881. Save for this

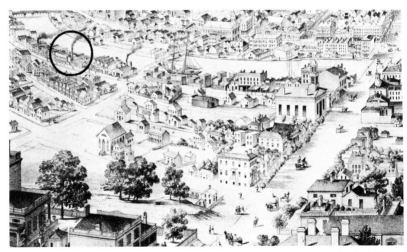

Milwaukee in 1853; the Reliance Works on Second Street north of Wells Street (shown on the left side of the drawing) was the beginning of the Allis-Chalmers Corporation.

brief venture into politics, his public activity diminished in his early forties as he became deaf. Workers at Reliance remembered his riding to work on his horse each morning and striding rapidly through the shops, ear trumpet in hand, before returning home in the evening. As he became more successful, he spent a greater amount of time with his books and his growing art collection. Allis was respected rather than loved; his contemporaries have left little in the way of description or comment about him. In fact, perhaps no man of his time who made such an impact on Milwaukee and Wisconsin left so few source materials which might explain his attitudes and motivations. Edward P. Allis moved through his life an enigma, known and remembered for his canny business sense, his development of a special business technique, and his exploitation of every business opportunity.[4]

Allis methodically set about utilizing business opportunities in Milwaukee. With the leather store prospering, about 1848 he and Allen bought a large tract of hemlock timber at Two Rivers, Wisconsin. On this site they built the "Mammoth Tannery" of the Wisconsin Leather Company, which they organized in 1850. On November 25, 1851, the editor of the *Sentinel* acknowledged a present of a "very handsome piece of calf-skin leather manufactured by the Wisconsin Leather Company at their large establishment at Two Rivers." The letter sent with the gift chronicled the facts about the new enterprise.

Milwaukee
November 25, 1851

Dear Sir:

Accompanying this is a piece of calf-skin leather, together with an order for its "making up." It is from our Tannery, and being among the "first fruits," gives an earnest of what we can do. In a little more time, we do not mean to be excelled. It may not be uninteresting to you to know that a little more than a year ago the site of our settlement at Two Rivers was a wilderness. Now our Mammoth Tannery is in full blast, and we have, in addition to the various Tannery Buildings, ten dwelling houses, all fully occupied. We have fewer men in our employment just now than at any time since the commencement of the works, but there are, at this time, 90 souls in our settlement, all dependent for their means of support upon the Tannery; which number will have to be increased from time to time, as we progress. We have expended in this enterprise, for land, buildings, hides, labor, etc., upwards of $60,000, and are now just commencing to receive returns.

We are yours,
"WISCONSIN LEATHER COMPANY"

Allis's and Allen's store remained the Milwaukee "depot" for the Wisconsin Leather Company.[5]

The "Mammoth Tannery" at Two Rivers was important in the

career of Edward Allis, who served as its managing director. At the tannery he had his first experience with management and direction of a sizable work force—about 100 men in 1852—and the handling of a large volume of materials. On July 27, 1852, the *Sentinel* published a description of the Two Rivers operation. The tannery, seven large buildings dominated by a 100-foot chimney of Milwaukee brick, worked cattle hides into leather at the rate of 800 a week, or about 40,000 a year. By late July, large shipments of rough leather had been sent East, and the partners estimated that until the close of navigation they would ship weekly twelve to fourteen thousand pounds.[6]

Although the profitable leather business of Allis and Allen had expanded into larger quarters in Milwaukee, Edward Allis wanted to operate alone rather than in partnership. In 1856 he sold his interest in the Wisconsin Leather Company. Whether he anticipated the depression of 1857 or not, he disposed of his investment at a good time and a good price, and did not reinvest or engage in business until after the panic.[7]

Edward Allis was looking ahead, however, rather than removing himself from business life. Some time after selling his interest in the tannery, he purchased for $15,000 twenty acres of land on the road to South Point in Milwaukee's Fifth Ward. Although eighteen years earlier this land had sold for $1,000, the *Sentinel* in 1856 remarked, "Even at the largely enhanced price now given for it, it will be a bargain to the purchaser." And a bargain it was to be: it would become the Florida and Clinton Street site of the Reliance Works in 1867.[8]

Milwaukee riverfront near the Reliance Works about 1860.

Very early in his business career, Allis began contributing his time, talents, and money to the community welfare. On December 8, 1847, he joined other business and civic leaders of Milwaukee— S. Osgood Putnam, John A. Van Dyke, Edward D. Holton, Henry W. Tenney, Garret Vliet and Isaac N. Mason—in forming The Young Men's Association. This organization, made reasonably select through annual dues of $12.50, led the cultural life of Milwaukee for the next three decades. At the first meeting, Edward Allis was elected secretary. Within a year 120 members enjoyed the privileges of the club rooms and the 810-volume private lending library. The association grew in membership and number of books, as well as in scope. In the fall of 1850 it sponsored a series of public lectures (family tickets cost one dollar) by some of the most well-known men then appearing before American audiences. Famous lecturers included Horace Mann, Bayard Taylor, Horace Greeley, Ralph Waldo Emerson, James Russell Lowell, Mark Hopkins, Wendell Phillips, and Henry Ward Beecher. The Young Men's Association provided many personal contacts for Allis, and his leadership in this venture brought him recognition in the Milwaukee community.[9]

Mrs. Allis was as energetic and able in community affairs as her husband. They were responsible for the establishment of the Unitarian Church in Milwaukee and offered it continuing financial support. After eleven of the twelve children she bore survived infancy, her concern for the unfortunate in the community was evidenced in her assistance in founding the Visiting Nurses Asso-

The Reliance Works before its purchase by Allis.

ciation, her leadership in the establishment of the Wisconsin Industrial School for Girls and the Wisconsin Training School for Boys, and her participation in University Settlement work in Milwaukee.[10]

Having developed into a capable young man with leadership and organizational abilities, an interest in civic betterment, and an aptitude for business, E. P. Allis was increasingly drawn into the power structure of early Milwaukee. In 1852 and again in 1854 he was elected treasurer of the Milwaukee Volunteer Fire Department, perhaps a modest beginning. But in 1855, when the Milwaukee Board of Trade increased its Board of Directors to fifteen, Allis and John Plankinton, later of meat packing fame, were among those elected. During the same year Allis was elected a trustee of the Merchants' Mutual Insurance Company; he thus became associated with Harrison Ludington, Daniel Newhall, and George D. Dousman, all prominent men in wheat trading, which was then showing a steady and promising growth. On February 11, 1858, he became a director and vice president of the Fox River Valley Railroad, then being constructed by a force of 400 men. When the Board of Trade and the Milwaukee Corn Exchange merged to form the Chamber of Commerce later that same year, Allis became a member of the new organization.[11]

Opportunities to improve his business connections and add to his practical experience kept coming to Allis. While still in his thirties he was elected a director and treasurer of the Northern Illinois Railroad Company. During his forties he was elected a director of the National Exchange Bank and the Northwestern

The Reliance Works after its removal to Florida and Clinton Streets in 1867.

Mutual Life Insurance Company. As a member of all these boards, Allis had an overall view of business and industrial activities and prospects in the Milwaukee area.[12]

On May 27, 1861, an advertisement in the *Milwaukee Sentinel* began: "RELIANCE WORKS of EDWARD P. ALLIS & CO. (late Decker and Seville)." Thus Allis announced to Milwaukee that he had actively re-entered business. Since he was always keenly alert to business developments, it is probable that he had been aware of the Reliance Works since its founding in 1847. Under the direction of James Seville, financed by Charles Decker of Dayton, Ohio, the Reliance Works had become important in the manufacture of flour milling equipment and had begun to produce sawmills and various cast iron products. By early 1857 the Reliance Works employed about seventy-five men and was one of the largest of the eighteen iron manufacture shops in Milwaukee. But the fortunes of Decker and Seville had declined after the panic of 1857 until, on February 7, 1861, a notice of a sheriff's sale to satisfy a judgment against the firm was published.[13]

Through the purchase of Decker and Seville's Reliance Works, E. P. Allis became a manufacturer of sawmills, flour mills and iron products. At first glance the Reliance Works did not appear to be a great bargain. Because of the panic of 1857 and subsequent financial limitations, the works had declined to little more than a shop for odd jobs and repairs, with about twenty workers. Allis, however, had his eye not on the existing operations but on the potential.[14]

The Reliance Works was considerably enlarged at the new location.

The Reliance Works was not large or impressive, but it provided Allis with a base for current operations and future expansion. The factory was 150 feet square, with the Milwaukee River at the rear. A smaller building, 21 by 37 feet, housed the mill shop on the first floor and Allis's office on the second. The third and smallest building was a mill repair shop. The first floor of the main shop contained an engine room, tumbling barrel room, foundry, blacksmith shop, and machine shop. The cupola for nesting the iron could cast about four tons a day. The works had eight lathes in the machine shop, which also included one planer, one bolt machine, one punch machine, two grindstones and a number of smaller machines. An old steam engine connected by a long belt to the line shaft provided power. By the early sixties, E. P. Allis had doubled the work force to forty men, all of whom worked from 7 A.M. until 6 P.M., (except Saturday when they stopped at 5 P.M.) with one hour off for lunch.[15]

Allis's business increased steadily throughout the Civil War, the works reaching capacity production in 1865–1866. The principal product of the works continued to be mill machinery and French burr mill stones, "many of the best mills in Wisconsin and Minnesota having their outfit from it." Allis also continued the manufacture of iron products, particularly specialized work for the growing industries of Milwaukee. Because he believed in exploring any new business opportunity compatible with the general pattern of his works, he had begun the manufacture of steam and water heating apparatuses, which were gaining in popularity for homes and offices. With his trade extending over the Northwest, he employed more than seventy-five men and through much of 1866 operated night and day with a total production of about $150,000.[16]

By the time Edward P. Allis reached the age of forty, he had become a prominent Milwaukee businessman. He had built a sound reputation and had won the respect of men in business and finance. But all this activity had not made him a rich man. When a list of forty-eight Milwaukee men whose annual incomes ranged from $10,000 to $77,000 was published in August of 1866, E. P. Allis ranked forty-seventh with an income of $10,000.[17]

In 1866 E. P. Allis assessed his business position and prospects. Although he had not profited from government contracts during the Civil War, still this had been a prosperous period, and there was every indication that this prosperity would continue. His extensive contacts with Milwaukee's leaders in business and finance gave him confidence in his actions and offered sources of financial support. But Allis could not really profit from the expanding economy to the fullest possible extent because his Reliance Works was already operating at capacity. The immediate problem that he faced was the need for expansion.[18]

The twenty-acre site on Florida and Clinton Streets, purchased in 1856, solved the problem. Although really no more than a tamarack swamp, it could be cleared and filled. Moreover, because both old and new sites were near the Milwaukee River, existing buildings and equipment could be moved easily and cheaply. Grading and filling on the new site began in August of 1866, and by December the work on the foundry, the largest in the Northwest, was underway. The Reliance Works moved down river in the spring of 1867. The buildings on the old site were cut into sections of fifty feet, loaded on flatboats and moved to the new site "piece by piece, department by department, and erected and used as a millstone shop, as milling work was very heavy at that time." By late summer of 1867, the Reliance Works was in operation at the Clinton and Florida Streets location.[19]

Allis had been producing water-powered mills, the pattern for generations of flour milling. But the age of the steam engine had arrived, and Milwaukee manufacturers had ample reason for an interest in steam: the city's primary dam had partially washed out in 1866, and inadequate repairs caused the remainder to wash out in the next year. As industries along the Milwaukee River converted to steam engines, the new Reliance Works began to experiment with their production. The first Allis steam engine blew its whistle on New Year's Eve, 1868. For the next eight years the company featured in its catalogs a simple, basic steam engine.[20]

The leading Milwaukee competitors of E. P. Allis in 1867 were the Bay State Iron Manufacturing Company, the Badger Iron Works, Kelsey Langworthy and Company, the Novelty Iron Works, the Milwaukee Foundry, the Chestnut Street Foundry, and Portable Saw Mills. Allis removed from rivalry the largest of his competitors, the Bay State Iron Manufacturing Company, when he bought it in 1869 from owner William Goodnow. Only two blocks away at the corner of Lake and Barclay Streets, this eight-block site was operated as a branch of the Reliance Works. Allis had not absorbed a smaller rival but rather had purchased a concern almost twice the size of the Reliance Works. In 1867 Bay State had employed about 175 men and produced steam engines, boilers, portable saw mills, gearing, shaftings, pipes, upright and portable engines and castings of all types. The acquisition of Bay State's skilled and experienced personnel, along with its buildings and new product lines, in a single stroke raised E. P. Allis and Company to leadership in Milwaukee industry.[21]

Following the move to the new site and the purchase of the Bay State Works, E. P. Allis was faced with consolidating production facilities, personnel and product lines. He moved machinery from the Bay State buildings to the Reliance Works to concentrate plant facilities and increase efficiency. He increased sales of all basic product lines until by 1871 $350,000 worth of machinery was produced by 200 workers.[22]

However successful he was, E. P. Allis had not done anything dramatic enough to capture the attention of Milwaukee, nor had he found the product that would give his works a national reputation. In the period from 1871 to 1874 he achieved the first of these goals, for the attention of Milwaukeeans was turned toward E. P. Allis and the Reliance Works as they undertook to provide the city with its first modern water system.

Through the Civil War Milwaukee had depended on springs and wells for its water, despite the fact that Lake Michigan was readily at hand. By the close of the war the population, now approaching 60,000, would no longer tolerate the situation. In 1868 the Common Council engaged E. S. Chesborough, city engineer of Chicago, to examine Milwaukee's needs and to provide a plan for adequate water supply. His report was in hand when the state legislature in 1871 authorized the city to readjust its debt limit to construct a water system. Guided by Chesborough's plans, the Board of Water Commissioners proposed to build a pumping station two and one-half miles north of the harbor entrance, a reservoir capable of storing twenty-one million gallons of water, and fifty miles of distributing mains.[23]

At that time no one in Milwaukee expected E. P. Allis to submit a bid on any part of this equipment, particularly not on the iron pipes since he had no foundry to produce them. But six months before the bids were to be opened, Allis astounded the community by beginning to build Milwaukee's first pipe works. Because neither he nor his men had experience in casting pipe, he hoped to insure the success of his venture by hiring two experts: John Pennycook, formerly of the Shoots Works, Edinburgh, Scotland, and more recently of a pipe works in Chicago, and William Wall, a found-ryman. Milwaukee was proud of this brave enterprise, and the *Sentinel* carried a question which had the force of an affirmation, "Why should not the iron gas pipe and water pipe of all dimensions needed by our city be made here?"[24]

Meanwhile the $100,000 pipe foundry began to take shape. The slate-roofed building was ninety-six feet long, seventy feet wide and forty-nine feet high. Because the pipe would be cast in a vertical position in a pit, special care had to be taken to prevent water from rising in the pits. After a pit twelve feet deep and forty feet square was dug, a puddling of one foot of blue clay was added. This was covered with timbers ten inches square, laid close to-gether. The timbers, in turn, were covered with another puddling of clay, and then another layer of massive timbers. Over all this, brick was laid in cement. A railway to carry large ladles of molten iron from the three cupolas of the foundry to the pits was installed, as were three cranes to move the iron anywhere in the pits.[25]

E. P. Allis and Company took the greatest share of the pipe contract for the Milwaukee water system when it was awarded in March of 1872, with a total of 2,600 tons. Two Philadelphia firms were commissioned 500 tons each and a Burlington, New Jersey firm was awarded 200 tons. Delivery was to begin on May 7, 1872, but work progressed so rapidly that by the end of March, Allis announced the first pipe casting. A large number of Milwaukee's most influential citizens witnessed the entire operation as the first pipe, twelve feet five inches long with a six-inch inside diameter, was cast. The *Sentinel* recorded a hearty sigh of relief and pleasure when the spectators heard the official announcement that it was a complete success. By the end of the year Milwaukeeans were jubilant over the casting of nineteen miles of pipe. The Reliance Works thus became the friend of every household and the pride of the entire city. "Milwaukee has good reason to be proud of her manufacturing interests, for they are her hope and life, but of none should she be more proud than the Reliance Works."[26]

The prestige of E. P. Allis was further enhanced when the *Sentinel* learned that, prior to the bidding, Allis had been contacted by a "cunning agent" of an Eastern industrial combine and had refused to cooperate with them in maintaining high prices. His action saved Milwaukee between eighteen and twenty dollars a ton. Praise for Allis was almost overwhelming. "Wisconsin-like, . . . no inducement can lead him or persuade him to enter into a combination against the city he has helped to build up. This is another evidence that Wisconsin may trust her sons when her interests are at stake."[27]

There was really nothing very dramatic about cast iron pipe laid underground, however useful and sanitary it was. But the plans for the Milwaukee water system called for more than pipe alone, and

These Allis compound condensing beam and flywheel engines were Milwaukee's first pumping engines. Installed in the North Point Pumping Station, they were started up on September 14, 1874.

two giant pumping engines could excite the imagination. Engines of this size had never been produced by the Reliance Works, nor for that matter by anyone in the West. The popularity of E. P. Allis rose even higher when he announced the hiring of Robert W. Hamilton, a renowned mechanical engineer of Hartford, Connecticut, to draw up plans for the pumping engine so that Allis could submit a bid. The public's unrestrained enthusiasm caused the *Sentinel* to write, ". . . there is no good reason why it should not be done here, and if any house can do it, the Reliance Works can."[28]

In this undertaking, Allis bid against some of the leading engine-builders of America who furnished plans and proposals in a lively competition to erect the two "compound condensing beam and fly-wheel pumping engines, both coupled to one fly-wheel and so arranged that they could be run together or separately." The specifications required each engine to have a "capacity sufficient to deliver 8 million gallons of water in 24 hours under a vertical lift of 150 feet and through a fort [sic] main of 6480 feet—1600 feet of which will be 30 inches in diameter and the remainder 24 inches in diameter."[29]

The proposals were opened by the Commissioners on April 22, 1872, and after about two weeks' consideration the award went to Allis's Milwaukee firm even though its bid of $168,000 was the highest. The article announcing the award of the pumping engine contract was headed "Another Triumph," and it closed with a hope that the "enterprise of the proprietors" would meet with a "generous return." Hamilton's drawings presented a "design of handsome proportions." It was of great importance to the citizens that Milwaukee would be a step ahead of Chicago, whose pumping engines had been built in the East.[30]

The contract called for completion of the engines by September 1, 1873. Work was begun immediately, first on a new building to accommodate the engine work and then on the engines themselves. The work was beset by difficulties and delays, and not until the end of July, 1873, did real progress on the engines begin. Then within a month the company regretfully announced that "some unavoidable delays have rendered more time necessary." The press was tolerant of the delay, assuring the public that Reliance would spare no effort to complete them at an early date. Allis held out hope that the water works, allowing ten weeks for installation of the engines, would be in operation by January 1, 1874.[31]

January came and went and the engines were still in production, with the Allis Company working a twenty-four hour day. So great was the pride in the Reliance Works and such was the confidence in E. P. Allis that the press remained uncommonly patient during the next seven months, reporting only that by mid-July the engines were "looming up into handsome proportions."[32]

The momentous day when the massive engines of the water works were set in motion was September 14, 1874. The first report

RELIANCE WORKS

MILWAUKEE, WISCONSIN.

E. P. ALLIS & CO., PROPRIETORS,

Machinists and Founders, Mill Builders and Furnishers,

Thankful for past favors, and being determined to continue in the front-rank in progressive Milling, we have made new arrangements by which our customers may be benefited through us, and solicit correspondence from all who contemplate building or improving their mills.

Our **Burr Blocks** are selected by careful men, from the very choicest stock of the French quarries, and put up by the most skilful workmen. Where parties desire it, we furnish them ready for grinding, dressed with the

Emery Mill Stone Dresser.

We make **Portable Mills** in great variety, complete in every respect, and ready for use.

In the line of **New Process Flouring,** we are prepared with the

Downton Purifiers,
Reliance Purifiers,
Affleck Purifiers,
Peerless Middlings Dusters,
Rolls For Crushing Middlings,

And invite special attention of progressive Millers to this branch of our trade.

Orders received for all the standard and popular articles of

Cleaning Machinery.

We have for several years sold and warranted

Dufour Bolting Cloth

only, but intend hereafter to keep in stock low price Cloths for Millers who want them.

We are agents for all of the best

Water Wheels,

and manufacture superior

Steam Engines,

to give Millers reliable power.

Competent Millwrights and Draughtsmen employed and sent to examine and locate mills and prepare plans and estimates.

This early advertisement in the Milwaukee Sentinel *lists products of the Reliance Works.*

was merely that the engines worked "very satisfactorily," but a week later the praise began to mount. A "distinguished" engineer pronounced them the best in the world, tests indicated that they operated with "singular economy" and that their speed could be regulated with more facility than other engines of their class. Moreover, they could easily throw twenty million gallons of water every twenty-four hours, although the contract called for only sixteen million gallons, and they supplied enough water for the city's needs by running only six to eight hours a day. The Commissioners seemed to feel that they were vindicated in giving the contract to Allis even though his bid had been the highest and despite a delay of a full year beyond the contract date. All Milwaukee regarded the engines as an enduring monument to Allis's skill as a builder of first-class engines.[33]

The Reliance Works had an enormous variety in its products. Although the interest of Milwaukee focused on the manufacture of water pipe and the two large pumping engines, equipment for flour mills remained the basic product line. At the state fair in 1872, Reliance displayed five different models of portable grist mills. The company was proud that Japan had ordered one of its models. It also produced several types of water wheels of its own design. Pile drivers, dredges, hoisting machines, steam pumps, pulleys, shaftings, fifty varieties of fuel-saving grates, and all sizes of Whittier's steam radiators extended Reliance services to a wide variety of industries. It also made the "celebrated warming apparatus of the Union Paul Water Heating Company," and supplied hundreds of tons of castings to "parties who control the rights in Indiana, Illinois, Iowa, and Missouri" and "eastern parties who are under contract to supply heating apparatus for the various public buildings in the west." With the Reliance Works going full strength, the number of workers increased from 200 in 1871 to 350 in 1872. In the latter year the value of the products was estimated at one million dollars.[34]

In managing the growing business, Allis was joined by his three older sons during the early and middle seventies. William Watson Allis, born in 1849, Edward Phelps Allis, Jr., born in 1851 and Charles Allis, born in 1853, entered the business as soon as they had completed their schooling. Save for Louis, born in 1866, the other four sons never took an active part in the business. William, Edward and Charles worked steadily and hard, although without enthusiasm, at helping to manage the business.

Although a nationwide depression struck in 1873, neither the Reliance Works nor Milwaukee was much affected at first. In fact, between 1873 and 1875 there was some expansion of the works and employment increased, although by no means as rapidly as in the previous three years. By 1875 the works covered nearly six acres and employed about 400 men. Wages were average for the times. Apprentices worked their first three months without pay and for the next nine months received five cents an hour. Skilled workers received $1.25 to $1.50 a day for a fifty-nine hour week, and the weekly payroll came to about $4,000.[35]

By 1875 materials consumed annually amounted to 5,000 tons of coke, 3,000 tons of coal, 10,000 to 12,000 tons of pig iron, 6,000 to 8,000 bushels of charcoal, 250 tons of millstones, 150 tons of hay, 30 to 40 carloads of fire-brick, clay and fine sand, 200 loads of lake sand and some thousands of loads of foundry loam and clay. In August of 1875 the Reliance Works was "one of the busiest manufactories .... No kind of hard times there." The following year the *Sentinel* reported that "the Reliance Works of E. P. Allis & Company are now running more full of work than at any time within two or three years. The firm is running both the Reliance and Bay State Works with a double set of hands."[36]

The prosperity of the Reliance Works which seemed so apparent to the outside observer in 1876 did not, in fact, exist. Events of the summer and fall of that year indicated that E. P. Allis had simply overextended himself. He had relied too much on the expertise of Pennycook and Wall in the pipe foundry and Hamilton in the engine works to make the profits while he handled the financing. The ventures into pipe and engines had paid handsomely in prestige but not in profits. He had borne the expense and overhead of a full year beyond the contract period for the pumping engines. Although there was no question that his pipes were of excellent quality, which once again bolstered his image in the community, for all the investment involved in this venture there were no profits. Not only was he basically a small pipe manufacturer as compared with the Eastern companies, but he was pricing his product as much as twenty dollars a ton below Eastern prices. As Irving Reynolds, later chief engineer for the Allis-Chalmers Company, once remarked, "thousands of tons of pipes wrecked Allis— he was competing with Cincinnati pipe."[37]

Having survived the initial panic of 1873, Allis, with the rest of the country, was caught in a prolonged depression which prevented the kind of expansion of plant and profits which might have saved him. The number of business failures grew, more than doubling between 1873 and 1878, the year described in Dun's annual report as "the fifth year of a depression unparalleled in extent, character and duration."[38]

The business community then as now was bound together by a complex maze of financial arrangements. In a prolonged depression the failure of one firm could embarrass other firms that might have survived unscathed despite their shaky foundations. Allis's financial difficulties, when they came to light in September of 1876, were linked with the insolvent Milwaukee Iron Works. Allis was in debt to the Iron Works to the extent of about $56,000. A reporter interviewing Allis found "an honestness characteristic of him," when he gave assurance that "the firm would certainly meet every expense of indebtedness, besides assuming its obligation on the paper of the Iron Co." His statement was more optimistic than forthright, for E. P. Allis and Company was insolvent.[39]

Creditors of E. P. Allis and Company met in the office of Finches, Lynde and Miller, and found the liabilities to be $437,728.98, of which $162,000 was secured by real estate and $275,729.89 unsecured. Assets totalled $176,833.31 exclusive of the real estate security. Taking into account the remaining property of the firm, the creditors estimated arrearage at about $160,000. In a letter to his creditors Allis explained that "he had made quite an investment to increase the facilities of his Works and that this, with the loss on low bid contract work, had hastened the present state of affairs." The creditors appointed a committee of five, headed by Marcus A. Hanna, to "obtain a true list of all said Allis's liabilities and also a true inventory of all assets and their values and to report to the creditors on October 4." Before the committee reported, E. P. Allis felt compelled "by the pressure of the times" to reduce the wages of his employees by twenty-five percent. Although some of the men refused to accept this initially, most of them soon resumed work.[40]

The committee recommended a compromise at the rate of fifty cents on the dollar, payable in eighteen months, two, three and four years, and secured by a trust mortgage on the debtor's property. To carry out the arrangement Edward H. Brodhead, a trustee, brought proceedings in bankruptcy and secured a court order sanctioning the compromise. The unsecured liabilities amounted to $276,000, and the assets, independent of a prior mortgage, $120,000.[41]

The prestige that Allis had developed among Milwaukee's business and financial leaders proved to be an invaluable asset at this point. The confidence of his creditors in Allis's ability was evident in the relatively easy terms extended him. Allis responded to this situation by proposing to expand his works, products and personnel, all on borrowed money, so that he could make money to pay off his arrearages.

The business career of Edward Allis through 1876 clearly indicated that he was the entrepreneur whose primary concern was financing. Since he entered business with no personal or family resources, he operated on borrowed money and the profits which accrued to his enterprise. Because he was sole owner he had no responsibility to stockholders or partners for his actions and followed a constant policy of expansion of facilities on borrowed money. This policy was designed for prosperity, not for depression, as he learned in 1876. But even as he surveyed the financial wreckage of the Allis Company in 1876, he was developing a business technique which was to make him the most successful Wisconsin industrialist in the 1880s.

NOTES TO CHAPTER ONE

Unless otherwise specified, all unpublished materials cited in the notes are from the files of the Allis-Chalmers Manufacturing Company.

1 Bayrd Still, *Milwaukee, The History of a City* (Madison, 1948), pp. 54, 61, 70–72, 109; John G. Gregory, *History of Milwaukee, Wisconsin* (Chicago, 1931), 1:240, 348. The population figures given are approximate. All the sources are speculative and offer no basis for agreement.

2 *Dictionary of American Biography* (New York, 1928), pp. 219–220; *Milwaukee Sentinel*, May 2, 1846, October 21, 1847, April 2, 1889; Gregory, 1:550–551; Still, p. 188; Allis & Allen *Journal*, p. 1. Under the date March 10, 1847, there is the following entry, "To R. Allen & Son (Am't this day due them) $10,703.40." Julius P. Bolivar MacCabe, *Directory of the City of Milwaukee, 1847–48* (Milwaukee, 1847), p. 79.

3 Charles Allis, *Allis Genealogy* (Milwaukee, 1893), Louis Allis Scrapbook, vol. 2; Jerome A. Watrous, *Memoirs of Milwaukee County* (Madison, 1909), 2:502; Allis & Allen *Journal*, p. 376.

4 *Milwaukee Sentinel*, April 2, 1889.

5 *Milwaukee Sentinel*, November 26, 1851, July 27, 1852.

6 *Milwaukee Sentinel*, July 27, 1852, February 9, 1883, April 2, 1889; *Proceedings of the American Society of Civil Engineers*, 1889, Louis Allis Scrapbook, vol. 1.

7 Watrous, 1:575; *Milwaukee Sentinel*, January 5, 1853. *The Milwaukee City Directory* for 1854–55 lists Allis as being connected with "Wis. Leather Co." (p. 20); the *Directory* for 1856–57 does not (p. 29). *United States Biographical Dictionary* (Chicago, 1877), pp. 462–465.

8 *Milwaukee County Deeds*, Vol. 54, June 27, 1856, to February 2, 1857, p. 312; *Milwaukee Sentinel*, April 28, 1856.

9 Gregory, 2:1077–1078; Still, pp. 214–215, quoting Lillian Krueger, "Social Life in Wisconsin," *Wisconsin Magazine of History*, 22:421–425 (June, 1939); Allis & Allen *Journal*, p. 295.

10 *Evening Wisconsin*, December 20, 1909; *Milwaukee Journal*, December 20, 1909; *Milwaukee Sentinel*, December 21, 1909; *Unity*, December 30, 1909.

11 *Milwaukee Sentinel*, January 14, 1852, January 10, 1854, February 6, March 12, 1855, February 15, 1858.

12 *Milwaukee Sentinel*, April 8, 1859, January 31, 1861, January 6, 10, 1863; *The United States Biographical Dictionary*, pp. 462–465. The author has been unable to identify the source or sources of loans to E. P. Allis in this early period. Allis was a director of the Northwestern Mutual Life Insurance Company from 1866 to 1874 and at his death in 1889 owed that company $243,000. No mention of support from Northwestern Mutual to Allis appears in either of the two studies of Northwestern Mutual Life. *Semi-Centennial History of the Northwestern Mutual Life Insurance Company of Milwaukee, Wisconsin, 1859–1908* (Milwaukee, 1908); Harold F. Williamson and Orange A. Smalley, *Northwestern Mutual Life; A Century of Trusteeship* (Evanston, 1957).

13 *Milwaukee Sentinel*, May 27, 1861; James Seville, "Milwaukee's First Railway," *Early Milwaukee: Papers From the Archives of the Old Settlers' Club of Milwaukee County* (Milwaukee, 1916), p. 89.

14 *Milwaukee Sentinel*, May 21, 1861; Seville, p. 92.

15 Statement on July 15, 1919, by James Laherty at the age of eighty to Mr. Porter, regarding the Edward P. Allis plant, Louis Allis Scrapbook, vol. 2.

16 *Milwaukee Sentinel*, March 12, 1870, February 1, 1867.

17 *Milwaukee Sentinel*, August 1, 1866.

18 *Milwaukee Sentinel*, February 1, 1867.

19 *Milwaukee Sentinel*, August 17, December 7, 1866, February 1, 1867; James Laherty to Mr. Porter, July 15, 1919, Louis Allis Scrapbook, vol 2; Henry Riemenschneider to Louis Allis, October 9, 1903, Louis Allis Scrapbook, vol. 2.

20 Henry Riemenschneider to Louis Allis, October 9, 1903, Louis Allis Scrapbook, vol 2. The brass whistle is now on display in the Allis-Chalmers exhibit area.

21 *Milwaukee Sentinel*, February 1, 1867, March 12, 1870; *Evening Wisconsin*, April 2, 1889. On May 5, 1869, the Bay State Manufacturing Company, the buildings and the eight lots on which they were situated were sold at auction to William Goodnow, former president of the company, for $59,335.00. In view of the almost immediate transfer of this property to E. P. Allis, it is very likely that

Goodnow purchased the Bay State works under an agreement with and support from E. P. Allis. *Milwaukee Sentinel*, May 5, 1869.

22 *Milwaukee Sentinel*, November 18, 1871.

23 Still, p. 248; H. P. Bohmann, *Milwaukee Water Works* (Milwaukee, 1935), pp. 1–4.

24 *Milwaukee Sentinel*, September 20, 1871, April 1, May 16, 1872.

25 *Milwaukee Sentinel*, September 20, November 18, 1871, April 1, 1872.

26 *Milwaukee Sentinel*, March 8, 9, April 1, May 13, 16, 1872, January 28, 1873; *The Directory of the City of Milwaukee, 1872–73*, p. 13. The advertisement of the Reliance Works included "cast iron water and gas pipe" for the first time in their list of products.

27 *Milwaukee Sentinel*, January 20, 1873.

28 *Milwaukee Sentinel*, February 24, May 13, 1872.

29 Bohmann, p. 1; *Milwaukee Sentinel*, May 13, 1872.

30 *Milwaukee Sentinel*, May 13, 1872.

31 *Milwaukee Sentinel*, July 23, 28, August 29, 1873; Bohmann, pp. 1–4.

32 *Milwaukee Sentinel*, July 11, September 12, 1874.

33 *Milwaukee Sentinel*, September 15, 21, 1874. There is no record of the other bids on the Milwaukee pumping engines. It may, however, have been quite a bit higher than the other bonds, for the *Sentinel* on September 19, 1874 noted: "E. P. Allis and Co., of this city, offer to furnish the engines for the West Side Water Works, of Chicago, for $340,000, but the work was let to the American Bridge Company, of that city, for $188,000."

34 *Milwaukee Sentinel*, March 23, 1871, May 16, October 1, 1872.

35 *The United States Biographical Dictionary*, pp. 462–465; J. M. J. Keogh, "Retirement Address," *Power Review*, September, 1937; *Milwaukee Sentinel*, July 12, 1876.

36 *The United States Biographical Dictionary*, pp. 462–465; *Milwaukee Sentinel*, August 12, 1875, July 12, 22, 1876.

37 Irving Reynolds to Alberta J. Price, July 31, 1945.

38 Victor S. Clark, *History of Manufactures in the United States* (New York, 1929), 2:158.

39 *Milwaukee Sentinel*, September 21, 1876. The first public indication that the Milwaukee Iron Works was in financial difficulty was on September 8, 1876. By September 11 the management promised an early settlement of the difficulties. When this proved impossible, the creditors and directors authorized proceedings in bankruptcy, and the Milwaukee Iron Works was declared bankrupt on October 19, 1876. It was first estimated that Allis owed the iron company between $60,000 and $70,000. It was finally determined that the extent of his debt was $56,000. *Milwaukee Sentinel*, September 8, 9, 11, 21, October 4, 5, 6, 19, 23, 1876.

40 *Milwaukee Sentinel*, September 22, 23, 27, 1876.

41 *Milwaukee Sentinel*, October 6, 1876. According to Irving Reynolds, Allis paid his obligations with personal notes. A little later, two banker friends of Edward Allis, Charles Nash and W. G. Fitch, who were in a position to know that his business affairs were improving, went around to the creditors and bought up the Allis notes for ten cents on the dollar. Later, when Allis paid off at face value they made a handsome profit, much to the chagrin of the original holders. Irving Reynolds to Alberta J. Price, July 31, 1945.

George Madison Hinkley

CHAPTER TWO

E. P. ALLIS DEVELOPS A BUSINESS TECHNIQUE

DURING THE EARLY 1870s, Edward P. Allis was developing a technique of management that was ultimately to lead him to success. An observer described his method: "It has been Mr. Allis' policy to secure the assistance of the best specialists in the different lines of machinery manufacture, and thus turn out the best machinery made, to which is due in large measure his great success." Allis brought together engineering talent for the production of goods and financial support for the constant expansion of his works. The Reliance Works was a jumble of buildings erected with minimal planning; these were, in a sense, representative of the Allis philosophy, for they had been erected with profit in mind. All were frame structures which bordered on the flimsy. But no expense was spared on equipment or the securing of the best available personnel, because these factors were basic to the production of profit.

The Allis engineers were responsible for the excellence of product and efficiency in production that would result in profits. The first application of this policy with John Pennycook in the pipe works and Robert Hamilton in pumping engines led, in part, to the financial embarrassment of Allis in the fall of 1876. The disaster was not a reflection on the technique, however, but rather on Pennycook and Hamilton, whose names dropped rapidly from sight. Following his plan of seeking out the best specialists available, Allis persisted until it was successful.

As early as 1873 Allis had hired the first of the engineering triumvirate which would lead him and the Reliance Works to international fame and financial success. This man was George Madison Hinkley, who became head of the Reliance Works sawmill department. The second major appointment was that of William Dixon Gray in 1876 to head the flour milling department. Allis rounded out his staff of brilliant engineers by engaging a year later the services of the greatest genius of them all, Edwin Reynolds, who became the master builder of steam engines in the late nineteenth century. The American Society of Civil Engineers, which invited Edward P. Allis to become a Fellow in 1883, published this appraisal of his successful business technique. "Mr. Allis was not an engineer, not an inventor, not a mechanic, but he had in full measure that rare talent for bringing together the work of the engineer, the inventor, the mechanic, that it might come to full fruition, and the world at large be the gainer thereby." The success of Edward P. Allis is found in the success of Hinkley's sawmill equipment, Gray's flour milling inventions, and Reynolds's giant steam engines.[1]

As it turned out, E. P. Allis could not have picked better men than Hinkley, Gray and Reynolds, nor could Allis and his engineers have lived at a better time. After the wreckage of the depression of 1873 had been cleared away, the United States very rapidly developed to maturity as an industrial nation. From an economic point of view the period from 1873 to 1893 was the golden age of American history. During this time the public debt was rapidly reduced, even though taxes remained low. The federal government was usually more concerned with a surplus than a deficit. Gradually, the spiritual unity of the nation was restored after the violent shock of civil war. Manpower resources were unlimited, with young and ambitious Europeans settling in cities and on farms. Inventions of all kinds added greater comfort and convenience to daily life. But most of all, a consciousness of progress, development and growth made possible an optimism in American life that has, perhaps, never been so great.

The mind of E. P. Allis was not so preoccupied with iron pipe and pumping engines in the early seventies that he did not note opportunity and change. A remarkable opportunity was to be found in his own back yard in the lumber industry. Given the enormous stands of accessible timber, Allis might almost have anticipated that between the Civil War and 1890 the principal center of the lumber industry would be the Great Lakes region. In fact, during that period, Michigan and Wisconsin accounted for nearly thirty percent of the national lumber production. During the decade following Hinkley's appointment as manager of the sawmill department, the quantity of white pine sawed annually in the Great Lakes area was to double, increasing from roughly four million to eight million feet. Moreover, the industry was soon to develop in the West and in the South. During the eighties the total value of the product was to increase from $210,000,000 to $404,000,000. Supplying the rapidly expanding lumber industry with equipment represented enormous opportunity.[2]

The element of change that Allis must have noted was that sawmill methods during the previous decade had been undergoing rapid development. Introduction of the circular saw increased cutting capacity more than ten times, although early circular saws were exceedingly wasteful, sawing out at each cut a half inch of kerf. The movements of the log carriage had been accelerated, and the double edger and later the gang edger had been introduced. At the close of the sixties steam replaced manual labor in handling logs. These and many other lesser improvements were accompanied by the increasing efficiency and power of the driving engines. In short, the good sawmill of 1870 bore little resemblance to the sawmill of 1860, and the process of development was proceeding apace.

Until 1873 the sawmill department of the Reliance Works had been no more than a sideline of the milling business, but here was an obvious opportunity if the right man could be found. George Madison Hinkley was the right man. E. P. Allis appointed him head of the sawmill department in October, 1873.

Hinkley was born in Seneca, New York, on May 24, 1830. After mastering the carpenter's trade, he became a millwright and from 1851 to 1861 worked on a succession of mill jobs in Michigan. In 1862 he enlisted as a corporal in Company I, Sixth Michigan Cavalry. Captured by the Confederates at the Battle of Trevilian Station, Virginia, on June 11, 1864, he was sent for a time to Andersonville prison until he was paroled in December, 1864. After his discharge in January, 1865, he built saw mills in Michigan until 1866, when he came to Milwaukee to build and operate a shingle mill. In 1870 Hinkley decided to develop his ideas for improving sawmill machinery and to establish his own business. He invented and sold a saw swage, a mill lathe and other devices which Filer and Stowell, sawmill manufacturers in Milwaukee,

Northern Wisconsin logging scene in the 1870s.

produced for him. His worth and potential as inventor and engineer moved E. P. Allis to hire him for the Reliance Works.[3]

When he joined the Allis organization, Hinkley contributed his valuable patents. Actually his productivity in new sawmill devices had just begun for, during the thirty-two years that he was head of the sawmill department, his patented inventions totalled thirty-five. Hinkley, and later Gray and Reynolds, were induced to come to the Reliance Works because Allis allowed them to keep all or part of their patents. Allis then paid his managers for the use of their patented devices. Moreover, the name plates on machines or in company catalogs frequently featured the name of the department head, thus giving him international recognition.[4]

When George Madison Hinkley came to the Reliance Works, annual sales of sawmilling equipment had not reached $1,000. Hinkley poured all his talents and energies into his job. At the outset he did all of the drafting, traveled, and carried on the correspondence as well. His patents covered most of the machinery turned out, and his genius was such that some of the mill appliances invented by him were used in mills for two to three decades afterward without marked changes. As the reputation of the Reliance Works and of Hinkley's inventions grew after 1873, so did the sales of sawmill equipment. E. P. Allis had the satisfaction of seeing sawmill sales reach nearly $400,000 per year by the time of his death in 1889.[5]

Logging was a rough business in the late nineteenth century, and the sawmill owners were a hard-bitten lot. Only a particular type of person could sell effectively to them regardless of the quality of the product. Hinkley could do it. He was known for his commanding bearing, his ample beard and his forceful manner.

Before the development of the band mill, circular saws were used; a double saw arrangement like this was needed to cut larger logs.

One sawmill man remembered his "highly scientific and gifted knowledge of picturesque language." Once when something had gone wrong, an Allis boy rushed out of his office to suggest moderating the language, only to give up when Hinkley furiously expanded his original statement and added that he would "kow-tow to nobody!" But he understood the logger and sawmill owner and could speak their language. Here was a man who knew what he wanted and had the courage and ability to go after it. Hinkley employed no tricks of salesmanship but sold the products of the Reliance Works solely on their merits, "recommending them for the value that was in them, and of that value and its most minute details no man ever had more intimate and thorough knowledge."[6]

When George M. Hinkley assumed management of the sawmill department, the Allis catalog featured only a circular saw, described as a fast-running disc "with teeth on its periphery." Within a year the catalog of the sawmill department was increased to a substantial seventy pages. Hinkley's patents, together with his ingenuity and energy, had made the difference.[7]

At the very time when E. P. Allis was in financial difficulties in 1876, the sawmill department was sending a complete mill to Japan and filling many larger orders as the reputation of the

Hinkley was one of the first to develop a practical band mill. It was announced to the trade in 1885.

23

department grew. In the spring of 1878, it sent ten carloads of sawmilling equipment to Texas, including two large double sawmills, setworks, engines, and boilers—everything necessary for a complete outfit. Later the same year the *Sentinel* reported that "in the matter of sawmills the reputation of Messrs. Allis & Co. stands alone."[8]

From 1880 to 1883 Allis's leadership was tested in the courts. In the hard-fisted and free-wheeling sawmill business, a less energetic man than Hinkley and a smaller concern than the Edward P. Allis Company would have had difficulty maintaining the identity and integrity of its patents. Following the development of rapidly operating log carriages, the mechanical "dog"—the device which held the log in place—was of utmost importance. As a consequence, in 1880, the Allis Company sued Filer, Stowell and Company for infringement of their patented dog. Allis and Hinkley sought to recover royalties from all firms that had manufactured or were using their patented device, to the extent of $600 to $800 for use of the dog during past years, and recognition of rights in the future. When the courts sustained the Allis position, the lumbermen of Oshkosh formed the Northwestern Sawmill Protective Association to defend themselves against an additional Allis claim of twenty-five cents per thousand feet of lumber cut by mills using its devices if not manufactured by the Reliance Works. When the courts decided this claim in favor of the Allis Company, they appointed a referee to determine extent of damages. Allis continued to press his claims against a growing list of firms and lumbermen. The case against Filer and Stowell of Milwaukee was

Sawmills were "big business" in the late 1800s.

settled in 1883, with their agreement to pay for past infringement and to take out a license from E. P. Allis and Company covering future use of the patent. This action provided the principle for settlement of the remaining cases.[9]

At the fairs and exhibitions popular after the Civil War manufacturers of all types entered their products in competition for prizes, hoping to widen their markets through the education of the public. Hinkley supervised elaborate displays of Allis sawmill equipment all over the country, and the entire Hinkley family was usually present at important fairs, promoting the interests of the company. Hinkley's progress in developing a first-class sawmill department can be seen in the impressive collection of prizes awarded his sawmill equipment at the New Orleans World's Fair of 1885. For a circular sawmill in practical operation, he was awarded a medal of second class; headblocks in operation with circular sawmill, medal of second class; collective display of sawmill machinery, medal of second class; gang edger, medal of first class; automatic lumber trimmer, medal of first class; two-saw lumber trimmer, honorable mention; lathe and picket machine and bolter, honorable mention; flooring machine, medal of first class; and for the Reliance mill dogs, operated in connection with circular saw mills, medal of first class. This record becomes more impressive when it is compared with two other Milwaukee manufacturers who also entered their equipment at the New Orleans fair. Filer, Stowell and Company received honorable mention for its display of mill machinery, and the T. H. Wilkin Company won a medal of first class for its saw

Interior of a typical sawmill in the late nineteenth century.

stretcher. G. M. Hinkley's sawmill department was obviously helping to establish the national and international reputation of the Allis Company.[10]

Although Hinkley did not invent the band saw, he is given credit for its perfection. The band saw was made from an endless steel band with teeth on one edge which ran over two flat-faced wheels, one above and one below the level at which the log was sawed. The great advantage was that, since the steel band was one half the thickness of the old circular saw, it reduced the waste from sawdust proportionally at every cut. Hinkley had closely observed the initial experiments in devising a band saw. When he was convinced it could work a great advantage, he proceeded to perfect it even though the first blades had to be rolled in France. In 1884 and 1885, no one at the Reliance Works or in the United States for that matter, could roll an endless steel band.[11]

With his characteristic skill and energy, Hinkley pushed the development of the band mill. His first one was announced on December 6, 1885, in a notice entitled "TO THE ATTENTION OF LUMBERMEN."

> We have just completed our new band saw mill, which is without question, the best machine of its kind ever offered to the market. One of these mills is now set up at our works, corner of Florida and Clinton Streets where it will remain on exhibition until December 15. It will then be removed to Dorchester, Wisconsin, and placed in active operation about January 1 in the mill of the Jump River Lumber Company. We make this announcement in order that parties interested in band saw mills may have an opportunity to inspect our machines.

This was a nine-foot mill for saws ten inches wide. The lower wheel had a cast iron rim on the outside of which a hardwood rim was bolted. The weight of this lower wheel was about 3000 pounds. The top wheel was constructed almost entirely of the best seasoned hardwood to make it as light as possible, and at the same time perfectly rigid. After the new band mill was placed in operation at the Jump River Lumber Company of Prentice, Wisconsin, the E. P. Allis Company received the following undated letter:

> Your combined Band and Rotary Mill put in for us was started up about the first of February last. It started off perfectly and our satisfaction has been constantly increasing. We are cutting from mixed logs, knotty, frozen, shaky and sound, at the rate of 3,000 feet per hour, of measured lumber, requiring no more care than a circular mill. We expect with a little more familiarity with operating the mill, to saw 35,000 feet per day. We have examined other mills in operation and unhesitatingly say we have seen none that compare favorably with this one. We cordially recommend anyone desiring a mill to examine this one in operation.
>
> Jump River Lumber Company

Although the later development of the Hinkley Automatic Power Swage and the Hinkley Power Guide, along with numerous other inventions, rounded out Hinkley's contributions to the sawmill industry, the *American Lumberman* regarded the perfected band saw as "the monument of his rare genius and mechanical ability."[12]

Hinkley lived and produced his equipment during the greatest lumber expansion, when every manufacturer of sawmilling equipment was pushed to the utmost to meet both the great demand and the intense competition. At his death in 1905 the *American Lumberman* paid him tribute:

> Mr. Hinkley was as great a man in his line of business as Carnegie in his. He has been as useful in his day and generation, in view of the circumstances which surrounded him, as any great inventor whose name could be mentioned. His relation to the improvement of saw mill machinery was almost akin to that of Edison to electrical development or of Ericsson to the evolution of naval construction. Had he so elected his name would have been as eligible to enrollment in a national hall of fame as any of those cited. But he chose—if he gave that matter a thought—that his works should be his monument.[13]

But Hinkley distinguished himself within the company not only as an inventor and machinist. His business ability was prized

William Dixon Gray at his desk in the Reliance Works.

equally highly by E. P. Allis. The management and sales of the department were wholly in his charge. By 1889 he raised the status of his department to the first rank, and its annual sales to $400,000. He vindicated the business technique by which E. P. Allis operated and, in a measure, helped create his fortune.

Hinkley's contribution to the success of E. P. Allis and his company must, however, be placed in perspective. Although Hinkley was the first eminently successful appointment to the post of departmental manager, still the sawmill department was the least important of the company's three major departments, both in volume of sales and in profits.

From the days of Decker and Seville the Reliance Works had been known for the excellence of its flour mills, and that department had been long most important in the company. Given his business philosophy, Allis needed a man with the happy combination of imagination, mechanical ability, and a good head for profits. He found that man in June of 1876. His name was William Dixon Gray.

Born in Lauder, Scotland, on July 22, 1843, Gray was eleven when he emigrated with his family to a Canadian farm. In 1861 he was apprenticed to a millwright, and in 1865 he moved to Minneapolis attracted by its promise of becoming a grain and milling center. Interested in the theoretical as well as the practical aspects of milling, he mastered drafting and then rose to the position of mill engineer. In 1866 he joined the firm of Pray and Webster, then erecting a large mill in Minneapolis for Cadwallader C. Washburn. After a decade of excellent experience, W. D. Gray, tall, dignified and aloof, came to E. P. Allis and Company in January of 1877, as superintendent of the milling machinery department.[14]

In his new position, Gray had something to work with. During its ten years at Florida and Clinton Streets, the Reliance Works had grown until it covered "two whole squares," together with the eight full lots of the adjoining Bay State Works. Annually it made thousands of mill-picks of the best English steel and sent them to all parts of the country. The millstone shop, forty-one by eighty-two feet, employed about fifty men as builders and dressers of millstones. They produced approximately 300 run of French burr stones in 1876, and the company sold about 20,000 yards of silk bolting cloth for sifting flour. All railroads in Milwaukee had tracks running to the Reliance Works. The main offices of the plant on Florida Street were connected by telegraph with the central office of the Western Union Telegraph Company so that messages could be handled speedily. During the fall of 1876 Allis employed about 600 men.[15]

With Gray in charge of the efficient flour milling department, orders for mills continued to pour in. Early in 1878 he had a contract to build a thirty-run mill at Niagara Falls. The *Sentinel* reported that, "For size and beauty of design the mill will be without equal in the country. The plans of the firm were considered so superior that they were able to command their own price for the Works." Because of this and other orders, Allis was obliged to add one hundred feet of space to the machine shop, and before that addition was completed the pressure of orders compelled him to begin another addition of equal size. The *Chicago Journal of Commerce* referred to the Reliance Works in May, 1878 as "The Largest Complete Mill Building Establishment in the World." It described Gray's flour milling department as the company's largest department, one that had "proven by the extent and quality of its production, to be greatly in advance of all similar institutions." It noted that during 1877 Reliance had constructed seventy mills and had placed them in operation in all parts of the United States, Austria, Hungary, New Zealand, Japan and other countries. The account concluded that this record "only bespeaks the excellent management and vast facilities of this department of the Reliance Works." W. D. Gray was obviously doing well for himself and for E. P. Allis.[16]

The ingenious Oliver Evans of Delaware had invented the American automatic flour mill in the late eighteenth century. Since that time a number of notable improvements had been made, such as the use of French burr instead of granite as millstones, and the substitution of silk for wool in bolting cloths. But basically the milling process dated back to ancient times, and Gray had only brought this process to near perfection. The principal portion of the mill was, of course, the grinding mechanism, which consisted of two flat, circular stones, the upper one delicately balanced on the lower. The gristing surfaces of both were grooved or dressed to provide a simultaneous cutting, squeezing, and crushing action. The upper stone, called the runner, was set so that, when in

Wheat was ground with burr stones in flour mills of the 1870s, before the advent of the roller mill.

motion, nearly its entire weight rested upon the kernels of wheat. The success of the milling operation depended on the accuracy with which this mechanism was adjusted.

In even the best mills one basic problem still existed; there was a fundamental disregard for the nature of the grain. A grain of wheat is composed of three parts: a horny outer husk which becomes bran in the milling process, an inner starchy kernel which is converted into flour, and an oily germ or embryo which enters the discarded middlings. The accepted milling process crushed all these parts into one mass. Clear white flour, however, could be obtained only by the gradual reduction process, as developed in Hungary by mid-century. But the Hungarian system was too expensive for general use and proved quite impractical when millers tried to introduce it in the United States.[17]

Gray was not satisfied with perfecting an age-old system. The kind of mind that eventually introduced fifty-seven patented inventions into the milling industry also believed in revolution:

I believe it better to step out of the well-worn rut and try the experiment of doing better, at the risk of making a mistake, or even a total failure; for he who never makes mistakes never does much of anything! So it is good with individuals and corporations to have someone come forward occasionally to say that this system of things has continued long enough, and we will have a change, even at the expense of being considered a crank, a revolutionist or even a rebel if by so doing conditions are changed for the better.

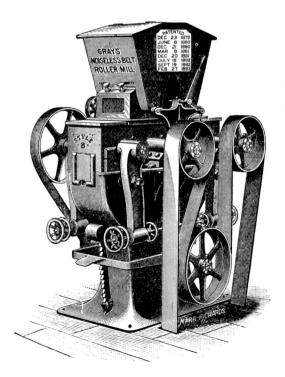

One of Gray's early roller mills. The first installation was in June, 1878.

This revolution in milling did take place. In 1899 Gray wrote that "there is hardly one thing in a modern flour mill in common with the mill of twenty-five years ago. The historic millstone that was considered by all millers of former days as being the life and heart of the mill, has given way to rolls, an entirely new creation."[18]

The principle of rolls, familiar to washer women and tinsmiths, had been used in flour milling in Europe, particularly in Hungary. In 1873 Friederich Wegmann of Wegmann, Bodmer & Co. of Zurich, Switzerland, invented a roller assembly that was markedly improved over any before that time, and it was introduced into Budapest in 1874. As early as 1876 the Allis company exhibited cast iron rollers at the Centennial Exposition at Philadelphia. Although these rolls received an award, they were rather crude compared to the European ones. In 1877 Oscar Oexle was invited to the United States by Cadwallader C. Washburn to represent the Wegmann interests, and Edward P. Allis and Company later became the importers and distributors of the entire Wegmann machine. After observing the performance of the Wegmann machines and then experimenting with them, Gray came to these conclusions:

After having had considerable experience under these rolls, I decided that they would do better work and make better flour than millstones. I had faith that they would become generally used if they could be put in better shape; that is if they could be mounted in a frame with the proper adjustment and drive where the wear and tear would not be so great and where the miller could control them so that they would be less noisy, and have a larger capacity. So, by way of self protection, I started to see if I could produce the desired results. By the cut of the Wegmann Machine it may be seen that the frame was light and flimsy, with nothing rigid about it. I constructed my frame of one solid casting, so that there could be no rack or tremble, to it. The stationary roll was held rigidly in place as the bed-stone in a constructed hurst frame, and then, with the proper adjustments for the movable roll there was no trouble to keep the rolls in good grinding position. I decided, also, that a belt was needed to do away with the train of spur wheels on the Wegmann Machine. I may say that the first Gray roller made by me was not as perfect as it was a few years later, but it was a success from the start, and I believe a very decided improvement on anything that had gone before it.

Gray installed the first Gray belt roller mills about June, 1878, in the Eagle Mill of John B. A. Kern in Milwaukee. Gray then installed eight roller mills in the mill he was building for Schoellkopf and Matthews at Niagara Falls. But in both these mills the rollers were used in combination with millstones.[19]

So firm was the traditional pattern that, for some years, many millers installed stones in addition to rolls in case the rolls would not do the job. Other millers felt that they could not get along without a few stones to grind fine middlings. Resistance to innovation made it difficult to gain permission to build an all-roller mill.

In fact, when the opportunity arrived, it came about by accident.[20]

In the spring of 1878 while Cadwallader C. Washburn in Minneapolis was building a new B Mill, the Washburn A Mill exploded. The new foundation for the rebuilding of the A Mill turned out to be eighteen feet longer than necessary for the equipment to be installed by the Allis Company. Gray persuaded Washburn to let him set up in the extra space a small experimental mill, completely roller, of a capacity of about 100 barrels per day. Gray also persuaded Washburn to drive half the roller assemblies by belts instead of by gears as was the European custom. After trial, the belt method of power transmission proved so superior that it became the standard for American mills. The secrecy of the experiments in the spring of 1879 stirred rumors of failure, while quite the opposite was true. The roller system proved so superior that Washburn converted both the A and B mills to rolls.[21]

Having succeeded in proving to Washburn the superiority of rolls, Gray approached Charles A. Pillsbury. He knew that if he could convince these two important Minneapolis millers, countless others would follow. Gray, in the late seventies, had built for Dorilus Morrison the conventional Excelsior Mill which the Pillsbury Company had leased and was running on shares for Morrison. Gray suggested to Pillsbury that he be allowed to convert it into a

This view of Minneapolis, an important flour milling center, shows the Washburn Crosby mills of the late nineteenth century. Washburn and Pillsbury installed Gray roller mills soon after they were developed.

roller mill. After a good deal of discussion, Pillsbury reluctantly told Gray to go ahead and change the mill by saying, "Go quick for fear I'll change my mind." In mid-afternoon Gray ordered his men, who had been servicing the equipment, to begin tearing out all the milling machinery. He kept them at work all night so that next morning when Pillsbury arrived, all the old equipment stood on a platform at the front door. To reinstall the equipment, had he changed his mind, would have cost too much by that time. When the Excelsior started up as a roller mill, Charles Pillsbury was in Europe and his brother Fred C. Pillsbury was in charge. Not believing that good flour could be made from a complete roller mill and fearing that it would not measure up to "Pillsbury's Best," he followed the new flour to the bakeries of New York and Boston. There the roller-milled flour passed all tests.[22]

Gray and the Reliance salesmen could point out that one advantage of the roller mill was that it gave the miller far greater control over the grinding process than was possible with millstones. Not only could the rate of feed be better regulated, but the grain could be spread evenly over the rolls. Moreover, the rolls could be set with great accuracy so that the miller could count on a precise spacing between them. When corrugated, non-touching rollers were set to run at slightly different speeds, the old crushing action was eliminated, and instead the grain was opened by the corrugation. Through successive grindings in the gradual reduction process, the grain remained in granule form rather than being crushed. From the first introduction of rollers, millers found that they could produce a larger proportion of better grade flour. Rollers also had a great economic advantage over millstones. The cost of

The assembly of Gray roller mills in the Reliance Works.

dressing or regrooving the stones, estimated at about seventy-five dollars per pair annually, was excessive, especially in view of the stones' low productivity. Gray could prove that replacing mill-stones with rollers cost nothing because the expense saved in stone dressing was sufficient to pay the interest on the capital required for the installations of rollers.

Gray's roller mill was first exhibited during June of 1880 at a miller's convention in Cincinnati. Of 170 exhibitors, including 14 from Europe, only E. P. Allis and Company and the Downton Manufacturing Company showed roller mills. Gray's roller mills were operational, and millers from all parts of the country brought sacks of grain to pass through the rollers. This demonstration, along with the excellent reports from the Washburn and Pillsbury mills, convinced those men, who placed a good number of orders at the convention.[23]

The demonstration at Cincinnati was so successful that in September, E. P. Allis bought the old Kilbourn Mill in Milwaukee, completely remodeled it, and equipped it as a model roller mill to exhibit the superiority of the system. Within a year this mill, renamed the Daisy Mill, was open for the inspection of sightseers as well as of prospective customers, and trained attendants carefully explained the most minute details. When the Daisy Mill burned in December of 1885, it was not rebuilt at the same site because it lacked accessibility to the docks. Instead, Allis bought another mill on the south side of Milwaukee which did have access

This improved roller mill, developed by the late 1880s, was used with little change for fifty years.

to shipping. When refitted as a roller mill, it had a daily capacity of from 1,000 to 1,100 barrels. The new Daisy Roller Mill was incorporated by Allis, Lou R. Hurd, Edwin Reynolds and Gray with a capital of $101,000.[24]

In addition to being profitable, the Daisy Mill provided excellent entertainment and instruction for prospective customers. Allis, alert to the need for public relations, always entertained men who could bring him business. For example, on September 15, 1886, seventeen members of the Pennsylvania Millers' Association received the usual hospitality from Allis as they were returning from an excursion to the Northwest. Allis and others of the company met them at the station in the morning to escort them through the Reliance Works. After lunch at the Plankinton House, he took them by carriage to the Daisy Mill for most of the afternoon, then to the breweries and Allis's own residence before returning to the Plankinton House for dinner. They completed the day at a performance of the *Private Secretary* at the opera house.[25]

The installation of roller mills for Washburn and Pillsbury brought the endorsement that Gray had wanted. The Daisy Mill and the exhibition at Cincinnati provided places for observation by and demonstration to the public. Demand steadily increased until by February, 1883, employment at the Reliance Works mounted to between 800 and 900 men, many of them producing milling equipment. Gray's Noiseless Roller Mill had become one of the principal manufactures, with production increased to about 100 pairs of rolls per month.[26]

The Allis salesmen also worked hard to promote the new product throughout the country. One salesman wrote to the company from Painted Post, New York:

> I intended to of went to Waterloo, but thought would wait and get this order as was afraid they might get out of notion if didant close while they were hot. Send me more contract blanks.

Salesmen may have been unschooled, but they were dedicated, if this one was in any way typical; he managed to take another order the same day for a mill at Corning, New York.

> I tried hard to get shorter terms, but was afraid I might ruffle the old man. Accepted these terms rather than nothing although he agreed before to purchase with the other party but he is sutch a conservative old fellow was afraid he might back out.

Reliance salesmen knew that Allis did not insist on cash. One of the first mills installed after Washburn's was for Archibald, Schumeier & Smith of St. Paul, Minnesota. They paid $18,000 down on a $27,000 order, and promised to pay "cash as work progresses and as wanted by the party of the first part." A note was appended to this contract in Allis's own hand, "the above unit is payable in 3, 6, & 9 months from completion of contract, 8% interest."[27]

Because roller mills offered economy, efficiency, and in the early days, prestige, orders came in faster than Reliance could build and install the mills. While Gray and the Allis Company did their very best to satisfy demand, only a few qualified millwrights could supervise proper installation. On March 17, 1882, Upham & Sons Co. of Blue Rapids, Iowa, wrote the Allis Company, "Gents: We have bought this mill with express purpose of changing to the Roller System and want it in running order shure by the last day of June." On May 17, Upham and Sons complained that "we are very

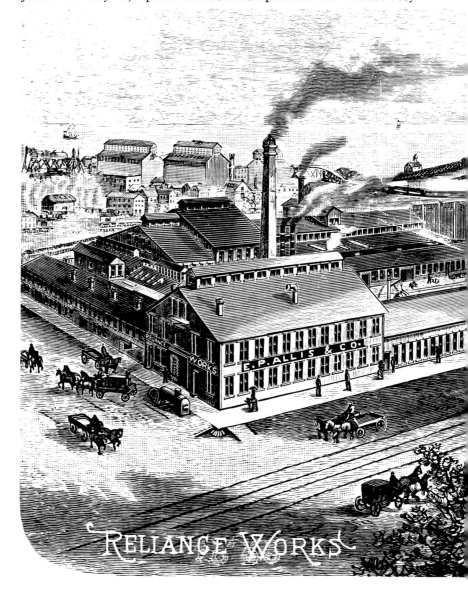

The Reliance Works in the 1880s. The facilities had expanded greatly since 1865 and at this time extended from Florida Street to National Avenue.

sorrow that you did not notify us that you could not ship the machinery according to agreement." When Upham and Sons wrote Allis on July 1, they had some of the machinery but were dissatisfied with Davidson, the millwright, who had "no push to him," and they feared frost before the mill was in operation. The company replied that "Millwrights are very scarce. We will try and get one for you as soon as possible." On August 8, Davidson was still there, some of the machinery had still not been sent, and a flurry of telegrams continued charges of the millwright's incompetence. The mill was finally in operation by mid-September, and arrange-

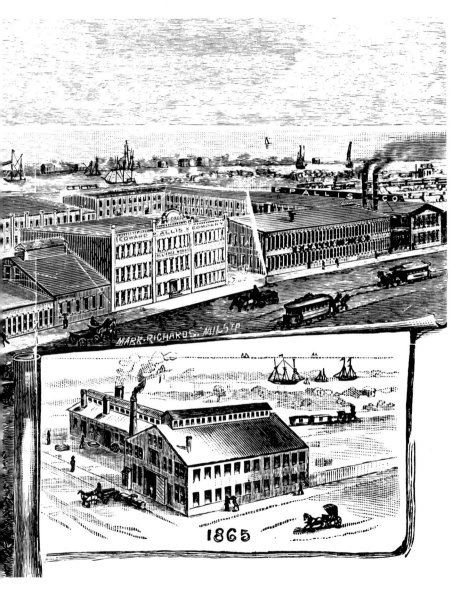

ment for payment was concluded by September 25 with a note "$8,000 paid, Notes $1,522—6 months due, $2,000 @ 6% year from date. Settled all accounts and claims." Such correspondence, sometimes amusing, certainly reveals how the roller boom hit E. P. Allis and Company, and temporarily overwhelmed production, shipping and installation facilities.[28]

W. D. Gray played a large part in the revolution of the milling business during the early eighties. In commenting on a court decision regarding his roller machines, Gray provided some assessment of his contribution:

> He (Judge Blodgett) did not seem to grasp the idea that I did not claim to have invented rolls broadly, that I did not claim to have invented rolls with journals, set screws and levers, but that I had invented a new and useful and successful combination that made the complete whole new and useful, while all the pieces and parts and materials that these parts were made of were old.

Gray had Americanized a foreign product. Wegmann had regarded his roller machine as the performer of a specific function, grinding middlings to flour. But Gray saw it as a means to mass production. He recognized that in our society "a machine or process can be fully useful only if it is able to handle enormous quantities of raw material successfully, without undue wear or expense or loss of effectiveness, and also if it can itself be manufactured by quantity-production methods."[29]

A farming revolution also resulted from the roller process. Winter wheat, which is relatively deficient in gluten, is also soft in texture; spring wheat is rich in gluten and therefore hard to reduce to flour. Because millers had always had difficulty producing satisfactory flour from the hard spring wheat, St. Louis, in the center of winter wheat culture had become the principal milling center of the continent after mid-century. Although the great plains of the American Northwest had the greatest wheat fields in the world, they were secondary because they grew hard spring wheat and their product commanded a lower price. The introduction of the Hungarian process in Minneapolis in the early seventies began to adjust this imbalance, but it was the roller mill that wrested the milling crown from Budapest in Europe and St. Louis in the United States and gave it to Minneapolis. The rapid growth of Minneapolis to a principal milling center can be seen in the following figures:

1860	30,000 barrels
1865	98,000
1869	250,000
1875	843,000
1880	2,152,000
1885	5,973,000
1890	7,434,000
1895	12,577,000

1900	14,863,000
1910	15,813,000
1915	20,443,000

The roller process, which could economically produce excellent flour from spring wheat, was thus responsible for an agricultural as well as an industrial revolution.[30]

From the days of Decker and Seville, the Reliance Works had steadily developed as a producer of milling equipment, until under E. P. Allis and W. D. Gray, it had become the leading manufacturer of roller mills in the United States. By 1884 the Allis Company could say: "Indeed we are within the truth when we state that more of the Gray Patent Roller mills have been sold and are at the present in use than all other kinds together." Twenty-five thousand pairs of Gray rollers were in use by 1888, and by 1894 eighty-five percent of the combined capacity of the Minneapolis, Duluth, and Milwaukee mills was milled by Gray rolls. This worked a great advantage to the Allis Company since it was at that time the only firm in the United States prepared to totally equip a mill of any size or type. A roller mill was rather expensive for the miller, for to be really useful the roller process had to be flanked and supported by innumerable reels, purifiers, aspirators and other devices to separate and dress the flour after it left the machine.[31]

No figures are available on sales and profits on milling equipment during the eighties. But one fact is obvious: E. P. Allis had found the kind of man he needed to head the flour milling department. By producing a milling revolution with the roller mill, Gray had helped Allis pay off the debts incurred in 1876 and had provided for the continuing expansion of the Reliance Works. As did George M. Hinkley, William Dixon Gray helped to validate the technique of business success developed by Allis.

NOTES TO CHAPTER TWO

1   *Proceedings of American Society of Civil Engineers*, 1889, Louis Allis Scrapbook, vol. 1.
2   Victor S. Clark, *History of Manufactures in the United States* (New York, 1929), 2:482–483.
3   Allis-Chalmers Company, *Sales Bulletin*, December, 1905, p. 1; *American Lumberman*, December 23, 1905, p. 1.
4   President W. H. Whiteside, *Circular Letter No. 62*, December 20, 1905. The *Milwaukee Sentinel*, February 29, 1888, notes that a patent was granted on a sawmill carriage, one half to George M. Hinkley and one half to E. P. Allis and Co. The contract, in Allis-Chalmers files, between William W. Allis, president of Edward P. Allis Company and Edwin Reynolds, April 9, 1890, reaffirming his previous contract, gave him full right to his patents. Mr. Ernest C. Shaw, who knew G. M. Hinkley well, understood that Mr. Hinkley also held some patents of departmental co-workers. Ernest C. Shaw to Alberta J. Price, August 23, 1954; Edward P. Allis and Company *Catalog*, 1885.
5   *The American Lumberman*, December 23, 1905, pp. 1, 37.
6   Ernest C. Shaw to Alberta J. Price, August 23, 1954; Axel Soderling to Alberta J. Price, August 3, 1954.
7   Edward P. Allis and Company *Catalogs*, 1871, 1875. These catalogs no longer exist. The information is from notes on the catalogs in the files of the company.

8 *Milwaukee Sentinel,* October 9, 1876, March 19, May 15, 1878.

9 *Milwaukee Sentinel,* August 16, September 27, 1880, January 29, 1881, October 4, 1882, March 22, September 2, 1883.

10 Earl Langdon Hinkley to Alberta J. Price, August 17, 1954; *Milwaukee Sentinel,* May 23, 1885.

11 It was characteristic of all the sawmill developments of the sixties to increase output or to save labor. Little effort was made to save timber, which was both cheap and abundant. Ernest C. Shaw to Alberta J. Price, August 23, 1954.

12 *Milwaukee Sentinel,* December 6, 1885. in the *Southern Lumberman,* December 15, 1931, p. 82, Allan Hall, then manager of the milling machinery department of Allis-Chalmers, provided details on construction of the mill. Jump River Lumber Company letter; *American Lumberman,* December 23, 1905, p. 1.

13 *American Lumberman,* December 23, 1905, p. 37.

14 Frank A. Flower, ed. *History of Milwaukee* (Chicago, 1881), p. 1288; Jerome A. Watrous, *Memoirs of Milwaukee County* (Madison, 1909), 2:951.

15 *The United States Milling and Manufacturing Journal,* January, 1877, p. 1.

16 *Milwaukee Sentinel,* January 28, 1878; The *Chicago Journal of Commerce* article was reprinted in the *Sentinel,* May 15, 1878.

17 John Storck and Walter Dorwin Teague, *Flour for Man's Bread; A History of Milling* (Minneapolis, 1952), pp. 161–199; Frederick Merk, *Economic History of Wisconsin During the Civil War Decade* (Madison, 1916), pp. 134–136.

18 Data on Gray's patents from Allis-Chalmers patent department; W. D. Gray, "A Quarter Century of Milling," Part XIV, The *Weekly Northwestern Miller,* January 24, 1900, p. 169; Gray, Part I, October 18, 1899, p. 740.

19 Storck and Teague, pp. 234–236; *Milwaukee Sentinel,* September 28, 1876. Before Gray, the Allis Company had tried marble rolls for middlings but these had proven too soft to be practicable. *The United States Milling and Manufacturing Journal,* July, 1877, p. 1, carried this notice from E. P. Allis Co.: "We are happy to inform the milling public that we have secured the entire manufacture and sale of the WEGMANN PATENT PORCELAIN ROLLER MILLS for America, and are prepared to supply all orders for the same." Gray, Part VII, December 6, 1899, p. 1092. In his twenty-part account of "A Quarter Century of Milling" in the *Weekly Northwestern Miller,* Gray consistently used the spelling Weggmann.

20 Gray, Part VII, December 6, 1899, pp. 1092–1093.

21 Gray, Part VII, December 13, 1899, p. 1145; Gray, Part X, December 27, 1899, pp. 1241–1243; Storck and Teague, pp. 246–251.

22 Gray, Part X, December 27, 1899, p. 1243; Storck and Teague, p. 254.

23 Storck and Teague, p. 252; Gray, Part X, December 27, 1899, p. 1243. Gray used 1881 in his account of the Cincinnati Exhibition. The year actually was 1880.

24 *Milwaukee Sentinel,* September 29, 1880, July 30, 1881, December 12, 24, 25, 1885, January 5, 1886. On February 16, 1886, the *Sentinel* noted that Allis was building a 35,000-bushel storage elevator for the New Daisy Mill.

25 *Milwaukee Sentinel,* September 15, 1886. For the Milwaukee Industrial Exposition of 1887, Allis showed a model of the New Daisy Mill which was the hit of the show. "One of the most perfect, unique and expensive representations is to be seen in the Daisy model roller flour mill, exhibited and constructed by E. P. Allis and Co. It is complete in every detail, engine, boiler, scale, rollers, scourer and polisher, flour packer, middlings purifier and in fact a complete mill throughout. It is lighted by miniature one candle power incandescent electric lights, and when in operation does not fail to attract the attention of every passer." *Milwaukee Sentinel,* September 8, 1887. This model roller mill was on display in the main office of the Reliance Works through the nineties. Axel Soderling to Alberta J. Price, August 3, 1954.

26 *Milwaukee Sentinel,* February 3, 1883.

27 J. H. Miller to E. P. Allis & Co., May 19, 1882, regarding contract with W. S. Hodgman & Co., Painted Post, New York; J. H. Miller to E. P. Allis & Co., May 19, 1882 regarding contract with Archibald, Schumeier & Smith, St. Paul, Minnesota, July 20, 1880.

28 Exchange of letters and telegrams between Upham & Sons Co., Blue Rapids, Iowa, and E. P. Allis & Company from March 17, 1882 to September 25, 1882.

29 Gray, Part XIX, March 14, 1900, p. 521; Storck and Teague, p. 236.

30 Merk, pp. 136–7; Storck and Teague, p. 210; Gray, Part II, October 25, 1899, p. 799.

31 E. P. Allis & Company *Catalogue,* 1884; Storck and Teague, p. 255.

Edwin Reynolds

CHAPTER THREE

THE STEAM ENGINES OF EDWIN REYNOLDS

BY 1877 ALLIS, as "Head of the Mill-Furnishing House of Edward P. Allis & Co.," faced a basic managerial problem. He was better at finances than he was at coordinating production. Hinkley was interested only in sawmilling. Gray, on the other hand, had a good head for both flour milling and finances, but not for over-all supervision and coordination. In the spring of 1877 Allis placed an advertisement in the *American Machinist* for a superintendent for the Reliance Works. Edwin Reynolds's application was accepted, and he assumed the position on July 1, 1877, at a salary of $3,500. This man willingly left a position with the Corliss Works at $5,000 per year to come to Reliance for less money, but was promised more freedom and financial support to build anything he wished. Allis willingly offered conditions he knew would be favorable to the developing and exploiting of genius. Edwin Reynolds was able not only to supervise and coordinate production but to become one of the best mechanical engineers and designers in the country.[1]

Edwin Reynolds was born into an old Rhode Island family on March 23, 1831, at Mansfield, Connecticut. Following the usual public school education, he was apprenticed to a small local machinist. After learning his trade, he worked in machine shops in Connecticut, Massachusetts, and Ohio before becoming superintendent of Stedman and Company, Aurora, Indiana. There he built engines, sawmills and drainage pumps for Mississippi plantations. During the Civil War he worked in Connecticut, Boston and New York, at one time on the machinery for Ericsson's *Monitor*. Following the war he entered the shops of the Corliss Steam Engine Company, Providence, Rhode Island, assuming in 1871 the post of general superintendent.[2]

Although Reynolds was a man of unquestionable ability, he always maintained that his success was in large part due to his early training. He had learned his trade in the days when the apprentice lived as one of the family in the house of the master. The lathe he learned on had an iron head and tail stock supported on wooden ways, most inadequate for accurate patternmaking. The skill of the mechanic, not the refinement of the tools, produced accuracy. Reynolds held that the resource and judgment of the young mechanic were taxed to contrive expedients to compensate for the inadequate facilities of the shop and still do the job. Under these circumstances the young mechanic acquired skill with his hands but at the same time he developed equal or greater skill in the analysis of the problem and in the adaptation of the means at his command to meet the necessities of his work. This was the training and experience that Reynolds brought to the Reliance Works.[3]

Reynolds's decade with the Corliss Works had been invaluable, because George H. Corliss (1817–1888) was the great engine-builder of the third quarter of the century. Corliss, like Reynolds after him, revolutionized steam engines. His first "light duty type" engine, built in 1848, had 260 horsepower, and new techniques for closing the valves and mounting the engine. With Edwin Reynolds as his general superintendent, Corliss built in 1876 the famous "Centennial Engine," a condensing beam engine with a pair of forty-inch cylinders, a stroke of ten feet, and a flywheel thirty feet in diameter. This 1400 horsepower engine, enormous for its day, stood in the center of Machinery Hall at the Philadelphia Exposition and drove all the exhibits in that building.[4]

The celebrated "Centennial Engine" was enormous, powerful, and—unfortunately—unpredictable. Like all the best engines of the time, it operated so unevenly that an electric generator driven by it produced a constantly varying current. An engine that did not provide a reasonably steady driving force was practically useless for lighting, because the light fluctuated erratically between full

Edwin Reynolds at his desk in the Reliance Works. The desk, of Honduran mahogany, was given to him by the American Society of Mechanical Engineers when he was president of that organization.

candle power and perhaps half. In the textile industry irregular power caused thread breakage. What was needed was an engine that would provide an even flow of power and greater economy. Edwin Reynolds felt that he could find the solutions to these problems if he had the right place to work.[5]

Since the shops were designed primarily for sawmilling and flour milling machinery, and were poorly equipped for work on large engines, Allis immediately launched an expansion to meet these needs. But lack of space and equipment did not deter Edwin Reynolds. He was free to improve upon the basic Corliss principles because the Corliss patent had expired in 1870. Moreover, about 1875 he had devised a new "radial" valve gear which was not only superior mechanically but operated more quietly and at higher speeds than the old releasing gears. Why this gear had been applied to only one or two engines in the Corliss shops cannot now be determined, but an opportunity to use it may have influenced Reynolds's decision to come to Milwaukee.[6]

The first announcement of "E. P. Allis & Co.'s Improved Corliss Engine" is found in the October, 1877, *United States Milling and Manufacturing Journal*. The article described this engine as *"heavier and stronger* in its essential parts than any engine of this class . . . and reliable . . . at much *higher speeds* than has ordinarily been obtained with this class of machines." Every part of the engine was held to be of the best material and workmanship. Moreover, Reynolds's new girder frame insured rigidity and smoothness of action, an improvement on the prevailing Corliss "Bayonet Type" frame which was rather light and suitable only for low steam pressures. The first Reynolds frame was essentially an "H" horizontal section with a separately attached slide member

A single stage Reynolds Corliss engine with wrought iron frame.

interchangeable for right or left-hand engines without pattern change, making for considerable savings in time and expense. The advertisement, after stressing these advantages, announced that the Allis Company was prepared to fill orders for the engines, either single or compound, of any size from 25 to 1,000 horsepower, at short notice.[7]

The engineering of the "improved Corliss" engine was not very remarkable. But in retrospect it takes on greater importance, for Reynolds's first engine had been built with untrained personnel in Milwaukee shops that were still inadequate. Some contemporary engineers held that it ranked as one of Reynolds's greatest achievements. It was designed to suit the capacity of the shops and the available men and money. This engine, then, exemplified one of Reynolds's basic maxims: "That a machine was well designed which was simple, and so designed that it could be built with the equipment available and sold for more than it cost." This engine was a great success from the very beginning.[8]

Since the Allis Company wanted more engine business and their new engines had not yet gained a reputation, they offered to fit existing engines with new cylinders and new valve gears to save twenty-five to forty percent on fuel. In fact, Allis and Reynolds were willing to guarantee "the above saving under forfeiture of any engine regulated by throttle valve." On April 25, 1878, they sent out their first widespread publicity release on the "Improved Corliss Engine Manufactured by E. P. Allis & Co., Milwaukee, Wis." By that time Allis and Reynolds apparently had such faith in the engine that they did not mention refitting old engines. The new assertion seemed almost unbelievable: "We are prepared to replace any ordinary Throttle Valve Engine and accept as our pay such an amount of money as we can demonstrate will be saved in fuel, in a stated length of time, varying from one to two years according to the grade of the Engine replaced." Many years later, John Schuette of Manitowoc, Wisconsin, reported to the company his experience with Reynolds and his guarantee in 1879.

An early advertisement of the Reynolds engine.

A cross-compound Reynolds Corliss engine installed in a factory.

Mr. Reynolds asked me what kind of an engine I had. I told him a common slide valve—high pressure, 16x36, 75 revolutions. He then took his pencil and after a little figuring said:

"I will build you a 150 horsepower Corliss, condensing engine for $3,000 and guarantee a saving of fuel from the one you have now of 50 per cent—or one-half." I was thunderstruck, but suspicious of all such claims. . . . I hesitated. Observing this he said: "What do you care? You have nothing to risk. If I don't save you 50 per cent you need not pay me a cent, and the engine will cost you nothing." "Yes," I said, "but such an intricate, complicated fine piece of machinery I fear my engineer will not be able to run." To which he replied. "I will further guarantee that your engineer after he has run the engine a day will admit that he can run it easier than the old one, and the fireman will think firing is a picnic."

Now I was being driven into a corner. But believing he had made a rash claim, which he could not carry out and not wanting to take advantage of this and act fair, I said: "Well then, we will make it $3,000 if you save 50 per cent—for every 10 per cent you save less $500 to be

Compound engines were also built in tandem. Early engines were all arranged for belt drive, with the large pulley on the engines serving as a flywheel to promote smoother operation.

deducted therefrom." He said, "Pshaw, I know what I am about," and insisted on making the contract $3,000—all or nothing.

After the engine was installed and tested it saved more than one-half as guaranteed. The first ten years after it was put in I figured a saving of over $20,000 in fuel of which the old one would have used.

This was one of the first engines Reynolds built in Milwaukee, now 30 years ago, and is yet doing service every day as when first put in.

Irving Reynolds, Edwin's nephew, said of the guarantee, "Yes, we did make that offer, but everybody paid cash rather than pay for the fuel saved because the bill would be too high."[9]

By April of 1878, when Allis sent out his letter on the new engines, the works were already running at capacity with orders ahead for the next six months. It is impossible to say whether this demand was the result of Allis's deliberate planning or if it was

Steel mills of the nineteenth century used gigantic blowing engines like this to provide large volumes of air to the blast furnaces. Reynolds greatly improved the efficiency of these engines by the use of metallic instead of leather valves.

merely chance. To satisfy the demand he was adding 16,000 square feet to his works. Now, in 1878, with power for their sawmills and flour mills, for the first time the Reliance Works could produce complete mills. The ten carloads of sawmilling equipment sent to Texas on March 19, 1878, included two mills, engines, boilers, "and everything necessary for a complete outfit." There was great advantage to both parties. The purchaser benefited from a mill and engine made by one company, and received a blanket guarantee. The Allis Company benefited from more business and increased profits.[10]

As the new Corliss engine proved itself, the company also tapped a reservoir of previous purchasers of mills. But more important than blanket contracts and old customers was the success in competition. A dozen companies lost to Allis in January, 1879, on a contract for two 400-horsepower engines for a St. Louis mill.[11]

By 1880 the Allis Company, growing more rapidly than ever before, employed over 600 men. Edwin Reynolds was preparing to move into a new field. In February, E. P. Allis contracted to build two large blowing engines for the Joliet Steel Works. Conventional designs of ponderous, inefficient blowing engines with leather valves had been submitted by leading engineers in the United States and Europe. But Reynolds had devised a direct-acting metallic valve blowing engine which proved superior to old blowing engines and revolutionized their construction. These engines became the most economical and efficient that had ever been applied to steel work. For the first time, the waste gases from the blast furnaces were more than sufficient to operate all the machinery, eliminating the need for coal and steam. More than a decade later the *Transactions* of the American Society of Mechanical Engineers cited Reynolds's blowing engines as such a radical departure from accepted practice that "the essential features of this design have not been improved upon."[12]

These mammoth blowing engines stood forty feet high, and their seventy-six inch air cylinders produced a blast of such force as "to blow a man as though he were chaff." Important as the prestige for Allis and Reynolds was the fact that these engines cost more than $30,000 each, and provided a basis for continuing orders as American steel companies produced the steel needed in a rapidly growing industrial economy. These first blowing engines were so successful that in 1881 the Joliet Steel Company ordered another pair, even though Allis's price had increased and a Pittsburgh firm had offered to furnish their machines for $11,000 less. In 1886 blowing engines which cost $86,000 were made for furnaces at Birmingham, Alabama, and by the following year five such engines were in production at one time for the blast furnaces of "Carnegie Brothers and Jones and Laughlin of Pittsburgh." The orders for blowing engines continued to increase until by 1900 Carnegie alone operated thirty-one. In that year the Allis Company had thirty-four orders for blowing engines on their books. Weighing from 400 to 650 tons each, they were sent to England, France, Russia and Nova Scotia as well as to domestic markets.[13]

The prestige of E. P. Allis and the reputation of his products were both increased by Edwin Reynolds. In 1880, when Milwaukee needed another pump for its North Point Waterworks, it looked to Allis and Reynolds. Because Corliss had never been interested in pumping engines, Reynolds apparently wanted to try his hand, and here was an excellent opportunity. It had been only six years since the two Hamilton pumping engines had been installed. Although they were considered large at the time, engine development, under Reynolds's guiding genius, had been proceeding rapidly. The single pumping engine Reynolds designed to flank the two Hamilton engines pumped as much water—twelve million gallons per day—as they did in combination. This engine, vertical beam and flywheel in a wrought iron frame, proved economical and operated almost continuously for twenty years until replaced by a larger unit.[14]

As orders for mills and engines came in to the Allis Company the plant constantly expanded. By October, 1881, over 700 men were on the payroll, and this increased to more than 800 by early 1883. The pressure for space was continuous. Unable to build rapidly enough at the Reliance Works to meet demand, and apparently unable to lease proper works in Milwaukee, E. P. Allis turned to Fond du Lac and leased the Meyer shop on November 1, 1881. This shop, until mid-1883, continued to send its finished iron work for engines to Milwaukee for assembly. By that time a major expansion of the Reliance Works had been completed, and the work and personnel of Fond du Lac was integrated into the Reliance operation.[15]

In twenty years the Allis business had grown steadily from about $31,000 in annual orders to about $2,000,000 at the close of 1882. The works by that time covered more than nine acres. There were four principle buildings, "devoted respectively to patent roller machinery for milling purposes, engines, the foundry and pattern making," and "an almost endless number of small shops, made necessary by the many minor calls associated with all manufacturing." Six large engines provided power for the works. The superintendent "under whose charge all this great business [was] conducted," was, of course, Edwin Reynolds. James F. Church acted as his assistant, with each department under the supervision of a skilled foreman. One hundred men worked on engines and steam machinery, 200 in the mill works, 300 in the foundry, 75 in the erecting room; there were also 15 blacksmiths, 15 patternmakers, 20 draftsmen, 30 carpenters, and 6 clerks in the shipping department. This number did not include the office, where "the extent of the correspondence [could] best be judged by the fact that three stenographers and three typewriters [were] constantly engaged." This work force produced about five carloads of finished products per day.[16]

The lack of precise figures makes it impossible to determine production of different lines, annual sales or other important economic factors during this period. However, working from known facts one can make estimates for the years 1878 to 1882. About twenty engines were manufactured in 1878, and in 1882, approximately eighty-five were made; even more significantly, horsepower increased from about 150 to 650. The price rose correspondingly from $2,700 per engine in 1878 to a little more than $6,000 five years later. On the other hand the cost of the engine per horsepower declined markedly with greater production and efficiency from $18.00 to about $9.40. The percentage of engine production in the total output rose sharply from six percent in 1878 to about forty percent at the close of 1882. The engine was

This vertical-compound engine was installed in Milwaukee's North Point pumping station in 1880 and operated for twenty years. It was rated at 12 million gallons per day.

making a substantial impact on the production and, we assume, on the profits of the Reliance Works.[17]

Edwin Reynolds had done very well for the company and for himself, but he was not satisfied. In 1883 he built two pumping engines for Allegheny, Pennsylvania. These large engines, mounted on a base area of thirty-six by twenty-nine feet, were of a remarkably simple design which permitted a reasonably uniform flow of water. These engines marked the first departure from the conventional type of large pumping engines, and became the forerunner of the great triple-expansion engines.[18]

Milwaukee in the eighties was confronted by a growing problem of sewage disposal. In 1884 Reynolds devised a centrifugal sewage pump, largest in the United States at that time with a capacity of seventy million gallons per day. Its twelve-foot impeller was driven by a tandem-compound Corliss engine directly connected

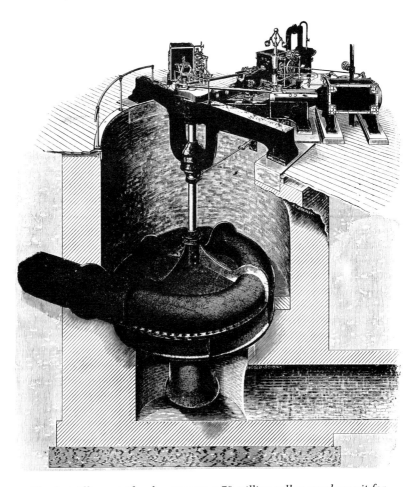

The first Allis centrifugal pump was a 70-million gallon per day unit for pumping sewage in the City of Milwaukee. It was driven by a tandem-compound engine and installed in 1884. The unit shown here was built a few years later for the City of Boston; it and succeeding units were driven by triple-expansion engines.

to the vertical pump shaft. Built at the northeastern edge of Jones Island, this successful unit was in continuous use for thirty years and was removed only when the intercepting sewage system made it unnecessary.[19]

The centrifugal sewage pump dealt with only part of Milwaukee's sewage problem. Toward the end of that decade Milwaukee's population was rapidly approaching 200,000, and the city had nearly 165 miles of sewers. The main problem was the sluggishness of the Milwaukee, Menomonee, and Kinnickinnic Rivers; virtually stagnant at some times of the year, they produced unpleasant odors referred to delicately as the "river nuisance." Resentment increased as more people recognized the public health danger of sewage flowing through the most populous sec-

Pollution plagued the Milwaukee River even in the 1880s, particularly in warm weather. This Reynolds screw pump, designed to flush the river, astounded skeptics when it delivered its rated capacity of 40,000 cubic feet of water per minute from Lake Michigan to the river.

tion of a great city. To remedy this problem, Reynolds proposed a "screw" pump to flush the Milwaukee River. Although some of the greatest engineers in the country condemned his plans, most Milwaukeeans trusted Reynolds, and he was commissioned to build a screw impeller fourteen feet in diameter with a hub of six feet. It was driven by a vertical compound engine at fifty-five revolutions per minute, with a capacity at that speed of over 40,000 cubic feet of water per minute. One skeptical alderman who had said, "I'll drink all the water that engine will throw" had to swallow his words when the screw pump astounded the engineering world and successfully accelerated the sluggish natural current, removing large amounts of waste. The new pump eased the situation while Milwaukee built an elaborate intercepting system to pump the sewage one thousand feet into the lake.

The idea of the screw pump had originally been developed by Reynolds many years earlier when he built a very small pump for raising a wrecked steamer on the Ohio River. The Milwaukee screw pump approached very nearly to plunger or piston pumps in efficiency, performing about double the amount of work with the same fuel and costing half as much as centrifugal pumps adapted for the same service. Running at fifty-five rpm, it exceeded the contract capacity of 500,000,000 gallons in twenty-four hours, pumping a greater quantity of water than any machine in the world. In 1907 Reliance installed a similar 323-million gallon per day Reynolds pump with a twelve and a half foot runner on Milwaukee's south side to flush the Kinnickinnic River into Lake Michigan. This type of pump proved very useful for large drainage projects as well as for urban sewage problems.[20]

The rapid development of an urban, industrial society in the late nineteenth century placed enormous pressures on the municipalities in their attempts to provide basic services. Among these services, the increasing demand for fresh water by industry and by private citizens posed a continuing problem. The Allis Company had provided nearly all the pumps used in the Milwaukee water system, and a great many for other cities. But while the pumps devised by Edwin Reynolds in 1880 and 1884 were a vast improvement over the Hamilton pumps of 1874, they were not efficient enough to revolutionize municipal pumping systems and thereby seize a large portion of a growing market. The man who did this was Reynolds's nephew, Irving H. Reynolds, whom the elder Reynolds brought to Milwaukee in 1884.

Born in 1862 in Brooklyn, New York, Irving Reynolds left school at sixteen to go to sea on a whaling ship. Assigned to engine maintenance, he gained enough experience to get a job following his stint at sea servicing maritime engines in New York. On the invitation of his uncle, he came to Milwaukee in 1884 at the age of twenty-two. Although without formal engineering training, he never felt handicapped for he had nothing to "unlearn," and "We didn't know that what we wanted to do couldn't be done."

Until 1884 virtually all the developments in engines and pumps by Reynolds had come in response to a particular problem or contract proposal. But Irving Reynolds was given time and support to develop a mechanical device for which there was no specific demand. Working from his experience with marine engines and the pumps Reliance had produced from 1874 to 1884, Irving Reynolds set to work to reduce pulsation in the water main, to effect fuel-saving, and to build greater durability in a pump that

The triple-expansion pumping engine, developed by Irving Reynolds, was first purchased by the City of Milwaukee in 1886. Its success led to orders from cities all over the country.

could be serviced more easily. Working almost every night and every Sunday for two years, he finally developed his own engine, which he called the triple-expansion pumping engine.

Irving Reynolds could do this within the business philosophy of E. P. Allis. "Allis, occupied with problems of management and finance, was perfectly satisfied to let the engineers and shop superintendents handle the drawing rooms and shops." Out of this freedom to experiment, Irving Reynolds came to hold six patents: three American and three foreign. The importance of this triple-expansion pump, however, made his contribution much more significant than the number of patents might indicate.[21]

Rapid development and enlargement of the city water systems in the mid-eighties made improvement in pumping necessary for the health and comfort of urban populations everywhere. Milwaukee planned a high service station and in May, 1886, advertised for a six million gallon compound engine. E. P. Allis and Company submitted three bids: one for the 1884 Reynolds model, one for a modification of it, and one for a completely new design. The third bid was the most expensive, but Milwaukee had a rule that if the low bidder placed more than one bid, the city could choose any of his bids. Since it had the privilege of choosing any of the three Allis engines, Milwaukee demonstrated its consistent faith in the engineering of the Allis Company by choosing the new design, even though it was the most expensive. The Reynolds proposal called for a vertical, triple-expansion engine guaranteeing a duty of 115 million foot-pounds of work per 100 pounds of anthracite coal, against a guarantee of 100 million for a three-cylinder compound engine. This engine had one high-pressure and two low-pressure cylinders set vertically. Below the engines, which operated on eighty pounds of saturated steam pressure, were the crank shafts and fly wheels, with the water end of the pump under the operating floor. The actual duty when the engine was installed was 118,186,312 foot pounds per 1000 pounds of steam.[22]

In 1891 Reynolds built a triple expansion pumping engine for the North Point station which had a capacity of 18 million gallons per day and developed 154,048,704 foot-pounds of work per 1000 pounds of steam. This engine proved so incredibly economical that some skeptical engineers challenged Allis's integrity, accusing him of deliberately hoodwinking the municipalities of the nation. The argument was resolved in 1890 when Professor John Carpenter of Cornell University headed a team which subjected the engine to extensive tests. The happy results of the tests appeared in a paper by Dr. Robert H. Thurston, Dean of the College of Engineering at Cornell, read at a meeting of the American Society of Mechanical Engineers. Following the vindication of the Allis position, the company rapidly gained leadership in municipal pumping which it maintained for thirty years until the triple-expansion pump gave way to turbine-powered centrifugal pumps in the 1920s.[23]

Following the Cornell test trials, the reputation of the Allis Company for quality pumping engines produced a flood of orders as growing urban populations taxed municipal facilities. Omaha, Detroit, St. Louis, Cleveland, New Orleans and Boston ordered their triple-expansion pumping engines from the Allis Company. Some municipal authorities resisted local pressures, recognized false and misleading information, and even circumvented established bidding systems to give the contracts to the Reliance Works.

Irving Reynolds's experience in dealing with Chicago in 1887 and 1888 points up some of these problems. Chicago requested bids on five pumping engines, specifying that they must be of the basic Holley design. When no bid was received from the Allis Company, Mayor John A. Roche asked why they had not responded, and learned that the company had a better engine than the specifications prescribed. The mayor then rejected all existing bids and reopened the bidding without restriction on the type of engine. The Allis bid was for a triple-expansion engine. The mayor had been told that the triple-expansion engine required nine men to start it and three men to stop it. Irving Reynolds invited the mayor to bring a delegation to the Milwaukee High Service Station to see the engine in operation. When they came, Reynolds recalled, "The engine was running like a clock." Mayor Roche said, "Now let me see you stop it." Reynolds closed a valve and it stopped. "All right!" Roche said, "Now, let's see you start it." Reynolds answered, "You do it yourself; press this lever." It started, and on April 7, 1888, the mayor of Chicago and the

Edwin Reynolds designed and Edward Allis built these powerful reversing engines for the rolls in steel mills.

commissioner of public works gave the contract for five new pumping engines to the Allis Company for $369,785. Some uncertain moments followed the placing of the contract, however, when Robert Tarrant, a well-known manufacturer of engines in Chicago, protested the decision because he had submitted a lower bid. Chicago authorities resisted these pressures, and the contract remained with the Allis Company.[24]

The pattern adopted by E. P. Allis for securing business in the 1880s was vastly different from the one he had used in the seventies. In the Milwaukee pipe contract he had underbid Eastern manufacturers on a mass production item, and had lost money. In the eighties he was manufacturing highly complicated machinery such as band saws, roller mills, engines and pumps. On these he set his own prices, based on quality and reputation, and let people come to him. When bidding was necessary, as for municipal facilities, his company frequently did not submit the lowest bid and just as frequently won the contract on the basis of quality and performance, rather than price. With imaginative engineers designing and producing quality products which could be sold at a substantial profit, E. P. Allis had found his formula for financial success.

The Joliet Steel Company was so satisfied with the blowing engines built and installed by the Allis Company in the early eighties that in 1885 they asked Reynolds to design and build engines for their rolling mills. The forty by sixty inch piston valve rolling mill engines operated at 110 rpm, driving the finishing rolls of the first continuous rail mill in the United States. These engines operated successfully for several years, and many of this basic design were built by Allis and others after the engine proved more economical and more easily regulated than other types. Each year brought new engine developments as new types were added to the list of products issued from the Reliance Works.[25]

The development of mines in Northern Michigan prompted the design and manufacture of huge hoists for lifting men and materials from deep underground.

The decade of the 1880s was exciting in Wisconsin and Upper Michigan because of the development of the Lake Superior mining area. Many Milwaukeeans were involved in this development, and in May, 1887, it was briefly rumored that E. P. Allis had purchased the Pabst iron mine in the Gogebic range for $1,000,000. This mine had shipped 18,937 tons of ore in 1886, and was expected to ship in excess of 100,000 tons that year. Although this rumor was unfounded, Allis was interested in the area and purchased some iron lands, apparently in the hope of making his operation self-sufficient. Although that hope never materialized, a great demand for hoisting machinery did come in the mid-eighties. In 1885 Reynolds designed and installed some of the largest hoisting engines in the country, all of them fitted with Corliss valve gears and large-diameter hoisting drums. His first hoisting engine was built for the Osceola copper mine. Equipped with two cylinders twenty inches by five feet, it was the first such engine ever taken into the Lake Superior country. A quarter century later the Reynolds hoisting engines were still operating successfully and efficiently.[26]

Once the iron and copper ore of the Lake Superior area had reached the surface, it had to be crushed before smelting. The

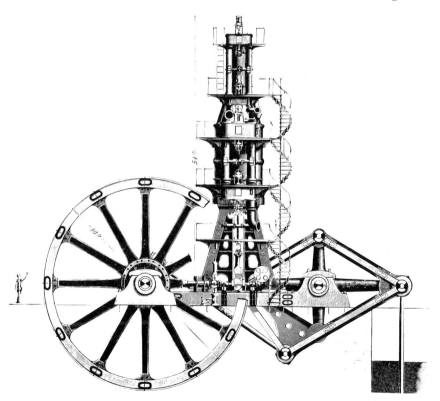

This drawing shows the largest mine pump ever built, designed for the Chapin Mining Company and installed near Iron Mountain, Michigan. The pump rod extended down the 1,659-foot shaft and lifted 3,000 gallons per minute from a depth of 1,500 feet.

existing steam stamps used heavy oak "spring timbers" to absorb the shock of the stamp blow. Edwin Reynolds saw that this practice could be improved upon to eliminate the expense and time of replacing the spring timbers. In 1887 he built the first steam stamp with a solid cast iron base for the anvil block; this stamp reduced copper rock for the Tamarack Mining Company. The success of that innovation was immediate and resulted in the prompt remodeling of all existing steam stamps.[27]

Perhaps the most famous contribution of Edwin Reynolds to the technology of the Wisconsin-Upper Michigan mine fields was the behemoth of pumps built for the Chapin Mining Company near Iron Mountain, Michigan. The operation of the "C" or Ludington shaft was made possible only by the use of a Cornish mine pump. The pump rod extended down the 1,659 foot shaft, the great depth requiring water to be elevated in stages by ten subsidiary pumping units. The capacity of this pump at a depth of 1500 feet was 3000 gallons per minute. The engine powering this great pump was of the vertical steeple compound type, with cylinders 50 to 100 inches in diameter and a 10-foot stroke. The great flywheel was 40 feet in diameter and weighed 100 tons. The working beam to which the pump rods were attached was 32 feet, center to center, and weighed 30 tons. This engine operated until the shaft was abandoned, and is still preserved as a monument, the greatest Cornish mine pump ever built.[28]

Edwin Reynolds was successful as an inventor—he owned forty-eight patents—as a builder of engines, and as superintendent of the Reliance Works largely because of the support of his employer. E. P. Allis, in turn, was successful because of his business technique and his ability. But good fortune or luck should not be overlooked as an element in his success.

Shortly after 8 P.M. on January 7, 1887, the night shift in the foundry discovered a small blaze while they were cleaning away the castings and piling up the sand for the next day's "heat." A three-alarm fire rapidly engulfed the entire north end of the foundry building, a frame structure about 420 feet long and varying from 80 to 200 feet wide. An unfortunate shift in the wind carried the fire the length of the foundry. Firemen brought the blaze under control at a fire wall which had been built only a year before to separate the foundry from the highly flammable pattern shop. Had the fire passed that point, all the interconnecting buildings, without firebreaks or firewalls, would have been destroyed. Although the works were insured for $400,000, that amount would not have begun to cover the loss, especially of patterns and complex equipment. Only two years before the end of his career, Edward P. Allis very nearly saw the work of a lifetime go up in smoke.[29]

E. P. Allis, Jr., told reporters that the fire had come at an unfortunate time because the company was particularly far behind in the engine department, which was primarily dependent on the foundry. Within two days Allis and Reynolds solved the problem. They arranged to take over the strike-bound foundry of the Pullman Palace Car Company, Pullman, Illinois, and immediately dispatched more than 200 molders along with the patterns to do the foundry work there until the new foundry was built. There was, then, no real interruption of the work, and the new foundry erected at the Reliance Works was larger and more substantial than the old, with brick walls and stone foundations placed upon piles. Much of the cost of this lasting improvement, moreover, was paid for by insurance. On April 28 the first heat was drawn from the new foundry after an interruption of only about two and a half months. Through a happy combination of foresight and luck Allis emerged from a potentially difficult situation in better condition than he had been in before.[30]

The new foundry and constant building at the Reliance Works did not keep up with demand. Since Allis and Reynolds preferred to concentrate the production of castings for engines at Reliance, they decided to build a new foundry for the special items, sawmill machinery and steam heating apparatus. The new "South Foundry" was built on made land on South Bay Street, the edge of Milwaukee's growing industrial district, a site accessible after some dredging to Great Lakes shipping and to land connections via the Chicago, Milwaukee and St. Paul Railroad. Articles of associa-

The Cornish mine pump of the Chapin Mining Company no longer pumps water but has become a tourist attraction. It is maintained by the Iron Mountain Junior Chamber of Commerce.

This composite shows typical Allis products of the late 1890s.

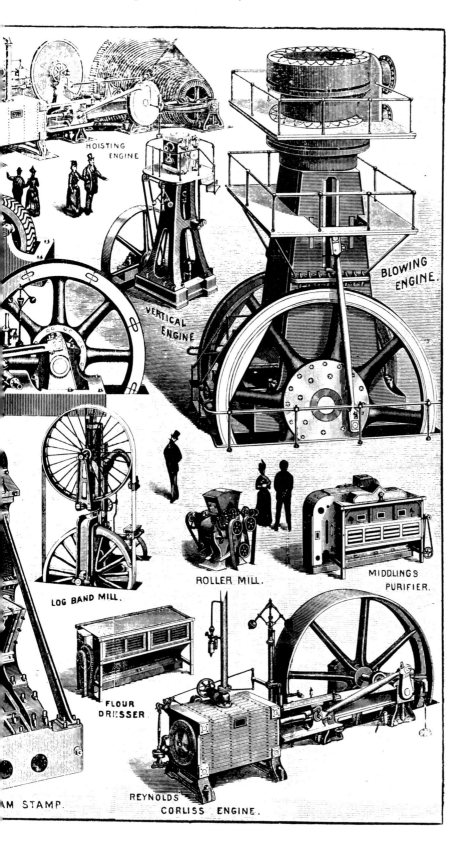

HOISTING ENGINE

BLOWING ENGINE.

VERTICAL ENGINE

LOG BAND MILL.

ROLLER MILL.

MIDDLINGS PURIFIER.

FLOUR DRESSER

AM STAMP.

REYNOLDS CORLISS ENGINE.

tion were filed in Madison on May 21, 1888, calling for a capital stock of $100,000, the incorporators being Edward P. Allis, Sr. with his three oldest sons, William W. Allis, Edward P. Allis, Jr., and Charles Allis. By mid-September 1888, the foundry was in operation, making brass castings and promising iron and babbitt metal castings in the near future.[31]

Allis and Reynolds could look forward together to 1889 as the most propitious year in their history. Sawmill equipment would carry its share of the load, accounting for about $400,000 in sales. Flour milling equipment was to remain as a principal item of production with sales at about $600,000. Steam engines and pumping engines would really constitute the bulk of the Allis Company sales with an impressive total of about $2,000,000, or two-thirds the value of all company products. In 1889 something like 250 Allis engines would be produced. By that time the cost was down to about $8 per horsepower, but the horsepower per engine was up to an average of about 1,000, making the average cost per engine about $8,000. Total horsepower production for the year would come to about 250,000. Such was the measure of success that Allis and Reynolds could anticipate.[32]

By 1889 Allis employed nearly 1500 men. About 1100 of them worked in the main shop, some 300 at Bay State and around 100 at the South Foundry. With an annual production of $3,000,000, these employees received more than $700,000 in wages. They were scattered through a complex of buildings more than a block wide and four blocks long. Approaching the Reliance Works from the Clinton Street side, one saw a solid frontage of 1100 feet. At the north or Florida Street end was the erecting shop, with its gigantic and spectacular traveling crane. South of the erecting shop was the new foundry, built after the fire of 1887, which occupied a site measuring 128 by 375 feet with a central open area 60 feet long. The open space was used for the carriage of two twenty-ton traveling cranes capable of operating separately or together. The cupolas, cove-ovens, and molding floors for lighter work were on either side of this central space. Running south from Florida along Clinton Street was the machine shop, measuring 60 by 350 feet, and filled with expensive lathes and other costly machinery. This was the heart of the Reliance Works.

South of the main machine shop was the office building and beyond it was a second machine shop of 60 by 230 feet, equipped mostly for sawmill and heater work. At the far south end of the works, fronting Clinton Street and National Avenue, was the Bay State Works, a three-story structure measuring 50 by 260 feet plus a one-story addition of 50 by 250 feet which was devoted entirely to the manufacture of Gray's patent roller mills. In the central court was a large three-story carpenter shop of 60 by 300 feet, and a mill shop of equal size. The pattern shop, blacksmith shop and boiler and engine rooms, together with a network of railroad track, completed the industrial maze.[33]

All of this was under the personal supervision of Edwin Reynolds. Tall, dignified, with a visored engineer's cap on his head and his solemn face hidden by a full beard, he moved majestically through the works with his greatcoat open to expose his gold watch chain and fob. The works and production were Edwin Reynolds's responsibility and, in a sense, his personal preserve, but owner Edward P. Allis also kept in close touch with "The Shops," as he called the Reliance Works. On Friday, March 29, 1889, the engineer and the entrepreneur walked together through the Reliance Works for the last time. By evening of Monday, April 1, E. P. Allis was dead.[34]

The relationship between owner and superintendent is recorded in an interview with Reynolds following Allis's death:

> I am proud to be classed among Mr. Allis's employes, though, like yourself, the public generally has assumed that I was his partner. Had I been financially interested in the concern as a principal, Mr. Allis could not have accorded me greater intimacy; been a kinder more warm hearted confidant. No, Mr. Allis was not a mechanical engineer; he depended wholly upon others for the upbuilding and development of his great enterprise in this regard. But he possessed no less that genius, that insight into character and grasp of capacity, which made him an unerring judge of what his mechanical lieutenants should be . . . . efficiency and faithfulness were always sure not only of recognition, but also of reward. He was the hardest working man of us all.

The death of E. P. Allis ended an era but, having anticipated the day when his guidance and counsel would no longer be available, he had made ample provision for the continuation of his work.[35]

NOTES TO CHAPTER THREE

1   *The United States Milling and Manufacturing Journal*, August, 1877, p. 1; Irving H. Reynolds to Alberta J. Price, July 31, 1945.

2   Allis-Chalmers Company *Sales Bulletin*, April, 1909, p. 42; *Milwaukee Sentinel*, April 2, 1889; *Industrial Progress*, April, 1909, p. 228.

3   Allis-Chalmers Company *Sales Bulletin*, April, 1909, p. 42.

4   Robert H. Thurston, *A History of the Growth of the Steam Engine* (Ithaca, 1939), pp. 501–503.

5   Victor S. Clark, *History of Manufactures in the United States* (New York, 1929), 2:359–360.

6   Irving H. Reynolds memo to George F. De Wein, August 28, 1940, p. 2. This was in response to an inquiry by De Wein in behalf of Professor Barnard, "as to Edwin Reynolds' work in connection with the steam engine, I give below a list of the more important engineering developments for which Mr. Reynolds was responsible, the greater part of this work occurring during the twenty years (1884–1904) of my association with him in Milwaukee."

7   *The United States Milling and Manufacturing Journal*, October, 1877, p. 1; Irving H. Reynolds to Alberta J. Price, July 31, 1945.

8   *Southern Engineer,* April, 1909, p. 42.

9   *The United States Milling and Manufacturing Journal*, October, 1877, p. 1; Letter of April 25, 1878, in Louis Allis Scrapbook, vol. 4. This handsome letter was engraved apparently to appear handwritten. Allis-Chalmers Company *Sales Bulletin*, April 1909, p. 44; Irving H. Reynolds to Alberta J. Price, July 31, 1945.

10   *Milwaukee Sentinel*, May 15, March 19, 1878. As the business in engines developed, the Allis Company was also called on to produce the boilers for the complete system. During this period reliable boilers for producing steam under increasingly higher pressures were developed, making it possible to run the engines at higher speeds. Cylindrical boilers replaced the early box type and soon the water-tube boiler was introduced. These continuing improvements in boilers made possible the great steam-driven central electrical stations built by Reynolds at the turn of the century. Although boilers were far less dramatic than great engines, they, too, provided sales and profits for E. P. Allis. As noted in the *Sentinel* on April 20, 1878, the Milwaukee Milling Company, which had been previously built by the Reliance Works, ordered an engine capable of maintaining constant production regardless of the water level of the river.

11   *Milwaukee Sentinel*, January 11, 1879.

12   *Milwaukee Sentinel*, February 4, March 20, 1880; Allis-Chalmers *Sales Bulletin*, p. 43; Irving H. Reynolds to Alberta J. Price, July 31, 1945; *Transactions*, American Society of Mechanical Engineers, 1909, 31:1052.

13   *Milwaukee Sentinel*, February 9, 1900; Louis Allis Scrapbook, vol. 1. A reprint of this article appeared in the *Sentinel*, June 28, 1900.

14   *Milwaukee Sentinel*, September 8, 9, 10, 11, 14, 15, November 5, 1880; Thurston, p. 503; Irving H. Reynolds to G. F. De Wein, p. 3.

15   *Milwaukee Sentinel*, October 5, 10, 27, 1881, August 16, 1882, February 9, June 16, 17, 1883.

16   *Milwaukee Sentinel*, February 9, 1883.

17   M. C. Maloney of the Thermal Power Division, Allis-Chalmers Manufacturing Company, has developed a tentative study of the growth of Allis's engine production from 1878 to 1899. This is based on the scattered existing figures available, projecting the number of engines per year, average horsepower per engine, total horsepower produced, cost per horsepower, cost per engine, total value of engines produced, total production by the Allis Company and the percent of engines to the total production. The cost per horsepower was computed on the basis of 5¢ per pound, which old employees held to be a fair price. M. C. Maloney, "Allis Engine Production," June 17, 1964.

18   Irving H. Reynolds to G. F. De Wein, p. 4.

19   *Ibid.*, p. 3.

20   Irving H. Reynolds to G. F. De Wein, p. 5; Allis-Chalmers *Sales Bulletin*, April, 1909, p. 43; Jerome A. Watrous, *Memoirs of Milwaukee County* (Madison, 1909), 1:32–303; Bayrd Still, *Milwaukee: The History of A City* (Madison, 1948), pp. 363–364. The test run for this engine indicated an efficiency of 87.6%. The

following is the description of the engine from Irving Reynolds's *Test Book:* "Vertical Compound Tandem Engine. High Pressure cyl. 19" diameter. Low pressure cylinder 38" diameter. Stroke 4 ft. cylinders jacketed. Air pump 17" diameter x 36" stroke. Volume of receiver 4 X high pressure cylinder not jacketed. [sic] Cylinder clearance 2.2% high pressure cylinder; 2.4% for low pressure cylinder. Propeller wheel 14 ft. diameter, 4 blades; 8 ft. pitch; hub 6 ft. diameter. Area of blade circle 113.66 square ft. Guaranteed delivery 32,000 cubic ft. per minute under 3½ ft. tread. Guaranteed duty 70,000,000 ft. pounds with 100 lbs. of coal. No deductions of any kind."

21 Irving H. Reynolds to Alberta J. Price, July 31, 1945; information on patents from Allis-Chalmers patent department. Irving Reynolds's philosophy of his relation to society is evident in these words: "Fortunately or unfortunately, I live in a society in which a man is privileged to retire and to pursue his avocational interests without doing any work for money. Retirement, however, is possible only because a man produces a sufficient surplus during his work years to justify an income for life. In my case the surplus I produced was the fuel which my engines saved for the country."

22 Irving H. Reynolds to Alberta J. Price, July 3, 1945; *Transactions*, American Society of Mechanical Engineers, 1893–1894, pp. 318, 436, 1909, p. 1052; Herbert H. Brown, *Historical Development of Pumping Machinery, Milwaukee Water Works*, notes in Allis-Chalmers files; Allis-Chalmers Manufacturing Company *Sales Bulletin*, December, 1925, p. 59, November, 1928, p. 85. The following data is from Irving Reynolds's *Test Book* in Allis-Chalmers files: "This engine is a triple expansion engine with cylinders 21" x 36" and 51" x 36". Steam jacketed, three single acting outside plunger pumps 23-½" diameter and 36" stroke. Guaranteed capacity 8 million gallons at 27.33 revolutions per minute against a head of 55 pounds on discharge and from 10 to 15 pounds on suction surface-condenser placed in suction chamber. Air pump 14" x 36" guaranteed duty 120 million foot pounds for every 100 pounds of best anthracite coal. Steam pressure 80 pounds." When the pumping station was shut down in 1924 this engine was still in good running order.

23 *Transactions*, American Society of Mechanical Engineers, 1890, pp. 654–667; Irving Reynolds to *We of Allis-Chalmers*, July 1947, p. 8.

24 *Milwaukee Sentinel*, April 6, 7, 8, 1888; Irving H. Reynolds to Alberta J. Price, July 31, 1945. The *Sentinel* for January 2, 1889, indicates that Milwaukee was enormously proud of the Allis Company when it won contracts such as this in the face of vigorous competition. "The Chicago contract was secured in active competition with all the largest manufacturers in the country, and its acquisition by Mr. Allis was a great triumph. The plant is designed to be the largest and best pumping plant in the country."

25 Irving H. Reynolds to G. F. De Wein, p. 4.

26 *Milwaukee Sentinel*, May 17, 18, 1887; Irving H. Reynolds to G. F. De Wein, p. 5; Allis-Chalmers Company *Sales Bulletin*, April, 1909, p. 43.

27 Irving H. Reynolds to G. F. De Wein, p. 5; Allis-Chalmers Company *Sales Bulletin*, April, 1909, p. 43.

28 Irving H. Reynolds to G. F. De Wein, p. 6; Paul C. Ziemke, "Preserved for Posterity," *Compressed Air Magazine*, November, 1947, pp. 276–277. When the mine's iron ore reserves were exhausted, the Oliver Mining Company (United States Steel Corporation) deeded the surface plant to the trustees of Dickinson County, Michigan, with the thought of preserving it as a memorial to the early mining days on the Menominee Iron Range. The steel headframe remained in place until World War II, when it was sacrificed to a scrap drive. Now only the engine remains as a tourist attraction.

29 Information on patents from Allis-Chalmers patent department; *Milwaukee Sentinel*, January 7, 8, 1887.

30 *Milwaukee Sentinel*, January 7, 9, 22, February 17, 23, March 16, April 28, 1887.

31 *Milwaukee Sentinel*, May 14, 22, August 9, September 12, 1888.

32 Maloney, June 17, 1964.

33 *Milwaukee Sentinel*, January 2, April 2, 1889.

34 *Milwaukee Sentinel*, April 2, 1889. E. P. Allis died of " 'gastrodynia' neuralgia of the stomach."

35 *Milwaukee Sentinel*, April 2, 1889.

Edward P. Allis

CHAPTER FOUR

EVALUATION AND STAGNATION

THE WILL OF EDWARD P. ALLIS, dated March 3, 1888, and written in his hand, was filed for probate on April 10, 1889. Keeping his wife and eleven children in mind, he concluded, "In this will I have made no outside bequests, deeming the size of my family and the central idea of a great and publicly beneficial Milwaukee industrial institution with ample means and the family name as precluding and taking the place of any other bequest."

Allis was a builder. From the Empire Leather Store of Allis and Allen through the building of the Reliance Works, Edward Phelps Allis, undeterred by fire and financial disaster, had worked constantly to create a great industrial institution. His goal achieved, he had no intention of allowing his estate to be divided or dissipated. His intention was that it should "remain in my family and that an enduring and perpetual industrial institution of great magnitude and importance may be built thereupon." To accomplish this end, Allis appointed his wife, Margaret W. Allis, his three eldest sons, William W. Allis, Edward P. Allis, Jr., and Charles Allis, and his superintendent, Edwin Reynolds, to be executors, administrators, and trustees of his estate. He directed that the trustees or their successors should be released from the necessity of bonds, taking only an oath or pledge to administer the trust faithfully and honestly. Allis appointed his friend Frederick C. Winkler as legal advisor.[1]

Andrew Carnegie's definition of the businessman as one who "plunges into and tosses upon the waves of human affairs without a life-preserver in the shape of salary; he risks all," certainly fit Edward P. Allis. He had built his Reliance Works and E. P. Allis and Company on a philosophy of constant expansion financed by borrowed money. He anticipated being in debt at the time of his death. Consequently he resolved the problem of probable debt with his desire to perpetuate his name through his company by insuring himself heavily. Although Allis had taken out some insurance in the 1840s, he took out little more until after his financial difficulties in 1876. Until shortly before his death in 1889, he added to his insurance until he was perhaps the most heavily insured man in Milwaukee, with a coverage of nearly half a million dollars. Even Allis's close friends did not know the extent of his insurance until after his death. He had policies with ten or more companies, partly to spread the risk and perhaps partly so that no one person would know of more than about $100,000 in coverage. At his death he was paying annual premiums of about $35,000.[2]

As the executors began to untangle Allis's financial obligations, they realized that the annual insurance premiums had been a very sound investment. In 1889 the real estate, tools and machinery were carried on Allis's books at about one million dollars. The business then had a net working capital of about $550,000, and an outstanding indebtedness of about $415,000 for net working capital of about $135,000. At his death Allis owed $243,000, secured by a first mortgage on the plant, to the Northwestern Mutual Life Insurance Company, and $160,000 to his sister, Mrs. Joseph T. (Lucy) Gilbert. These two debts consumed practically all the life

Expansion continued at the Reliance Works; this was the appearance at the time of the 1901 merger.

insurance. He also owed $100,000 on certain lands in Michigan and, as a leading Milwaukee art collector, between $20,000 and $30,000 in unpaid bills for paintings.

Without the insurance, the Allis family would have been in trouble. Even so, they were faced with real financial pressures. The National Exchange Bank of Milwaukee, with which Allis had done his business, had notified Allis shortly before his death that it was not satisfied with the financial condition of the company, and that unless matters were cleared up and the bank satisfied that his financial condition was sound, it would lend no more money, nor continue the current loans. Following Allis's death the Milwaukee banks refused to lend any money, and the estate had no credit. The claimants against the estate were calling for payment and, in order to secure money, the family had to make special discounts on production and get advances on contracts so that the men in the shops could be paid and business continued.

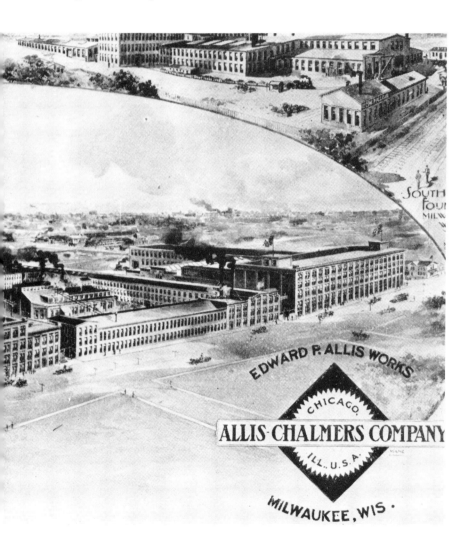

The principal value of the company was that of a going concern with three outstanding lines of manufacture which were being distributed to all parts of the world. It was the intent of Allis and was also in the interest of the family to keep the company operating, for if it went into liquidation, comparatively little would have been realized by those interested in the estate. The executors conferred with their legal advisor, General Winkler, to study how best to fulfill the intentions of the will, and how to save the estate for the heirs. They agreed that there was not time to await a court determination, and the General believed the will to be of doubtful validity. Under the circumstances, Mrs. Allis, the children and Edwin Reynolds, with the aid of General Winkler, devised a plan which would preserve the property and substantially fulfill the terms of the will.[3]

After the business was operated for some months on an interim basis, during which Charles Allis was compelled to raise money on his own life insurance to sustain it, the estate filed articles of association for the Edward P. Allis Company on April 1, 1890 thus implementing the plan suggested by General Winkler. The capital stock of the corporation was $1,500,000, divided into fifteen thousand shares of one hundred dollars par value. A board of five directors, chosen annually by the stockholders, was placed in charge of the business. According to the wishes of E. P. Allis, the shares of stock were allocated on the basis of age, with 3,061 allotted to William W. Allis, 2,755 to Edward P. Allis, Jr., and 2,449 shares to Charles Allis. These three older sons had been active in the business for some years. Ernest Allis was allotted 1,911 shares, Frank, 1,530, Louis 1,761, and Gilbert 1,530. The shares allotted to the four younger sons were held in trust, since they were "not yet settled in life." Margaret W. Allis and Edwin Reynolds were each allocated one share of stock to insure their voting rights. During the first seven and one-half years of the Edward P. Allis Company, dividends were not to exceed four percent. The articles specified that the excess of net profits over and above the four percent was to be divided among William, Edward and Charles "as an extra compensation for their past and present services in the conduct and management of the business." Upon the adjournment of the meeting of stockholders on April 1, the board of directors, including Margaret W. Allis, met and elected William W. Allis, president; Edward P. Allis, Jr., first vice president; and Charles Allis, secretary and treasurer.[4]

The abilities of the board of directors and the officers of the Edward P. Allis Company were in inverse ratio to the shares of stock held by them. In 1909 Edwin Reynolds was referred to as the "practical head of the Edward P. Allis Company," a judgment that was probably correct. This was recognized by the other officers for, during the nineties, Edwin Reynolds, as superintendent of the Reliance Works and second vice president of the Edward P. Allis Company, received the highest salary in the company, $15,000 per year. Moreover, as an indication of his position in company affairs, his contract of April 1, 1890, provided that three percent of the net earnings of the company should also go to him.[5]

The other driving force in the company was the founder's widow. In compliance with her husband's will, Margaret W. Allis received an annuity of $20,000. A woman of forceful personality and great intelligence, Margaret Allis was a stronger person than her husband in the view of some contemporaries. During his lifetime she largely confined her activities to good works, appearing at the plant only to assist in the sewing of bolting cloths when that department was pressed by rush orders. But after his death she worked in the business office nearly every day during the 1890s.[6]

No one of the sons inherited their father's breadth of interest and ability. Two of the eight, however, came to be distinguished in their own fields. Edward P. Allis, Jr., born in 1851, inherited his father's interest in science. While his father lived he worked actively in the company as an able liaison man between the management and the shops. He continued his service as an officer of the Edward P. Allis Company until he left both the company and the country in 1892 after a personal disagreement with his brothers Will and Charles. From his early years he had taken an active interest in scientific matters, and had a fully equipped laboratory in his home. Before leaving the United States he established and became editor of the *Journal of Morphology*. He moved to Montone, France, and devoted the remainder of his life to the study of fish. In 1907, France bestowed upon him the cross of the Legion of Honor.[7]

The offices in the Reliance Works in the late nineteenth century.

Louis, the seventh son, born in 1866, was the only one who distinguished himself for his business ability. Much of his childhood was spent in the shops, where he was a great favorite with the men. After earning his civil engineering degree at Pennsylvania Military College in 1888, he entered the business, rising rapidly from storeroom clerk to purchasing agent for The Edward P. Allis Company. He left the company in 1901, and two years later was elected president of the Mechanical Appliance Company, where he was better able to exercise his inventive, manufacturing and executive talents. This company soon became the Louis Allis Company, specializing in electrical equipment. As president Louis Allis also came to occupy something of the position of leadership in the civic and business community once held by his father. During the nineties his notable contribution to the Edward P. Allis Company, over and beyond his business ability, was the Allis Mutual Aid Society, which he helped develop to a position of national prominence.[8]

Charles Allis, born in 1853, spent almost his entire life in the business, first working with the finances of E. P. Allis and Company for his father, then serving as secretary and treasurer of the Edward P. Allis Company, and for four years acting as president of the Allis-Chalmers Company. The family business was his livelihood but not his primary interest, for he had not inherited his father's love for and aptitude in business. His father's avocation of art became his preoccupation. His handsome home in Milwaukee, built to house his priceless art collection, was given with the collection to the City of Milwaukee after his death.[9]

But the business ability and the interest of Charles Allis exceeded that of his older brother William, who had been born in 1849. Will, eldest of the Allis children, was "distinctly a home man and found his greatest enjoyment at his own fireside and with his family." Although he inherited none of his father's gifts, as the eldest son he was constantly thrust into business affairs. As president of the Edward P. Allis Company and later chairman of the board of directors of the Allis-Chalmers Company, he filled the offices with neither outstanding ability nor interest.[10]

Except for Louis Allis, the younger Allis sons played little part in the affairs of the company. Jere spent some time on a stock farm that his father had purchased in Minnesota. Ernest, Gilbert and Frank had desks at the Reliance Works, but Irving Reynolds and others recalled that they did little work and took very little responsibility.[11] The elder Allis sons, Margaret W. Allis, and Edwin Reynolds constituted the management of the Edward P. Allis Company. This was the group that inherited not just the Reliance Works but also the precedents that E. P. Allis had set, and they were the people who felt called upon to perpetuate the image that he had established.

One facet of the personality of Edward P. Allis that the family had to live up to, or live down, was that of Allis the radical businessman. In 1877 the generally retiring and slightly deaf Milwaukee industrialist left his normal business activities for the hectic life of politics as the Greenback candidate for governor of Wisconsin. After stumping the state in a strenuous campaign, Allis received 23,216 votes. William E. Smith, the Republican candidate, won the election with 78,759 votes. Again in 1881 Allis accepted the Greenback Party's nomination, but by that time the movement had passed its peak and was in a steady decline. A less than enthusiastic campaign brought him only 7,002 votes.

At that time and since, many have raised the question of why Edward P. Allis, a staunch Republican and Milwaukee's leading industrialist, publicly espoused views that were considered radical. In his speeches he contended that the two major parties were equally wrong on the money question, and only the Greenback Party could act as a pressure group to force the proper views on both parties. He hoped that the Greenback movement could include both Democrats and Republicans. Once its goal had been achieved and its principles incorporated into both party platforms, then the Greenback Party could disband and its members return to their original affiliations. He believed that this approach was absolutely essential, for if only one major party came to see the light, it would probably be the Democratic Party.

> If it comes to be a direct party issue, there is no escape from the fact that there will be arraigned on one side the wealth and accumulated capital and on the other the labor of the country, and it is not difficult to foresee what might be the possible result of a sharp and point-blank contest between these two forces. It would be too apt to be disastrous whichever won, for to the victor belong the spoils. It might result in absolutism and serfdom on the one side and/or communism and license on the other. No chance for the choice should ever be given. These forces should be harmonized and not placed in antagonism.

As a man of integrity, Allis saw himself as a leader striving for the truth and the preservation of harmony on a national scale.[12]

The Greenback issue harked back to the Civil War, when neither loans nor taxes kept the government supplied with ready cash, and Congress had to authorize several issues of paper money. At the close of the conflict, about $431,000,000 in fiat money was still outstanding. These dollars, worth only sixty-seven cents in 1865, had appreciated to eighty-nine cents by 1875, then ninety-six cents in 1877, and before January, 1879, to one hundred cents. This steadily appreciating value of the dollar insured comfortable profits to money lenders, especially those dealing in long-term loans. For the borrower, however, the situation was different. The farmer or the manufacturer who had mortgaged his property as security for a five-year loan found, to his sorrow, that when the time for payment came the dollars he had borrowed were worth less than the dollars that he had to use to repay his loan. Protests against dearer dollars began to quicken as the depression deepened after 1873. At that critical time both major parties were controlled by hard-money men who represented the point of view of the creditor rather than that of the debtor.

E. P. Allis described his position as one of principle. In point of economic fact, however, his support for the Greenback movement sprang from his constant debtor position. Over his entire lifetime he built his business on borrowed money. As the panic of 1873 had broadened into general depression and money became steadily dearer, the consequent pressures were great enough to drive the politically conservative but economically radical Allis into the Greenback movement. As early as September, 1875, Allis was defending the Greenback position, and defense gave way to forthright advocacy after his financial embarrassment a year later. Allis thought that the panic of '73 and the subsequent depression could have been avoided had there been sufficient currency in circulation. He had an "abiding faith in the greenback as the best adequate currency the world ever had, and its being not only an essential, but an indispensable element in our future prosperity, and necessarily in the perpetuation of our republic." He held that the manufacturer in the entire economic scheme of things occupied a middle ground between capital and labor. "It is necessary for him to extend one hand to capital and the other to labor and draw them together, and if either is absent or insufficient his occupation is gone. Backed only by faith in the government, fiat money was the basis for prosperity and the manufacturer was the catalyst who could join capital and labor to make possible unending prosperity and expansion. In that cause Allis could unleash all his eloquence, as he did in his peroration to the Milwaukee Greenback Club in 1878.

> What is it, gentlemen, that waves over our heads as a nation, flapping lazily in peace, and snapping furiously in the tornado of War? It is nothing, gentlemen, but a rag—a cotton rag—but those painted stars and stripes change it from matter to spirit, raise it from earth to Heaven. It is no longer form and weakness, but sentiment and power; 45,000,000 of people bare their heads in love for it and all humanity raise their hats in respect for it. What that rag is to us politically, the greenback ought, and will, be commercially. Long live the twin rags, and while one ever waves over us, may the other ever float among us.[13]

In 1878 Congress decided that $346,681,016 in greenbacks should remain a permanent part of the national currency, but the existence of the gold reserve made every greenback dollar as good as gold. The successful resumption without the dire calamity predicted by some, along with the return of prosperity, tended to discredit Greenbackism. After the lackluster gubernatorial campaign of 1881, E. P. Allis took no active part in the declining movement, and quietly returned to the fold of the Republican Party. When seventy-five employees of the Bay State Works organized a Blaine and Logan club in October, 1884, they received at least tacit support from the management, which held that the election of the Republican candidate James G. Blaine was necessary for prosperity. In fact, when Grover Cleveland was elected, the first Democratic president since the Civil War, Allis ordered an "equalization of salary." Allis maintained that the company had been paying higher wages than any other establishment and reduced the wages of some men, particularly skilled laborers, by five percent. E. P. Allis, Jr., said: "The results of the election had something to do with the reduction. We would like to keep salaries at that old figure, and we think we could have done so if Blaine had been elected, because I have no doubt that there would have been a considerable improvement in business generally. But as the actual results of the election gave us nothing to build a hope of a business boom on, we were compelled to bring the wages of a number of our men where we could afford to pay them." Four years later E. P. Allis felt so strongly about the issues, especially the tariff which he thought should be sustained or perhaps increased, that he distributed to his workers a printed statement over his signature discussing the merits of the Republican Party.[14]

E. P. Allis, the radical businessman, had returned to the Republican Party by 1884 and issued public statements in its behalf by 1888, but he had not deserted his Greenback principles. Even with the prosperity of the eighties, he was still a debtor and still held that the inflationists were right. Indeed, he used nearly a quarter of his 1888 statement on the merits of the Republicans to reiterate the Greenback virtues. Only because the election of the Greenback Party or the adoption of its platform by the two major parties was an impossibility had he addressed himself to the other major issues of the day, particularly the tariff.[15]

Main foundry of the Reliance Works.

Unlike their father, the Allis sons never ventured into radical politics but always remained respectable Republicans. Their father's radicalism simply did not serve their purposes. Whereas E. P. Allis had built his business career on a philosophy of expansion based on debt, the family was able to pay off the Allis debts by 1890. The family inherited a complete manufacturing unit in the Reliance Works, the Bay State Works, and the South Foundry. The family, therefore, inherited the Republicanism of E. P. Allis rather than his radicalism. They were interested largely in conserving the institution that he had left them.[16]

E. P. Allis always saw himself as a manufacturer, never as a capitalist and, perhaps because of his debtor-consciousness, a friend of labor. This sense of identification led to the founding of the Allis Mutual Aid Society. In 1883 a "Grand Ball" celebrating the opening of a new carpenter shop realized a net profit of $65.10 to the committee in charge.

After pondering what to do with the money, the committee finally proposed that it be used for an aid society for employees in case of sickness, accident, or death. With Allis's approval the committee canvassed the shops and found about 300 men willing to join such a society financed by a monthly assessment of twenty-five cents a member. Allis announced that he would immediately have a room in the works partitioned off for the society and equipped as a reading room if the members chose. More importantly, on June 15, 1883, he was prepared to match the monthly contributions of the members of the society, "which will make it in

The south erecting floor of the Reliance Works, showing two pumping engines being assembled.

fact as in name a mutual aid society, and will give you a fund always double that which comes from your own wages."[17]

That arrangement solved a problem which had concerned Allis for some time. In a day before protective devices were customary, often older employees took great pride in an accident-free career of many years in the Reliance Works. But not all were so fortunate. Men who worked with molton iron, heavy castings, and unprotected belts and wheels were bound to incur accidents, and with some frequency. Although the worker was legally responsible for his own safety, Allis recognized that accident insurance was almost prohibitively expensive for the individual. The Allis Mutual Aid Society provided a solution which Allis hoped every one of his employees would use. The initiation fee was fifty cents, with monthly dues of twenty-five cents. For absence because of sickness or disability the member was entitled to seventy-five cents a day for sixty days in any one year, for a total of forty-five dollars. The society retained a physician, selected by vote of the members from among applicants, to prescribe medicines and perform operations. An immediate death benefit to the family was one hundred dollars.[18]

Unhappily, employee response was neither immediate nor general. Although about 300 had indicated an initial interest, only 85 became charter members. By 1888 the membership had increased to 485, still less than half the workers. Following a fatal accident, a sudden influx of members usually occurred, succeeded by a decline. Of the 485 members in March of 1888, 55 had joined only recently because of two fatal accidents in February. During the first five years the treasury was never exhausted, and by 1888 there was a reserve of $1,200. On the basis of the membership, E. P. Allis was paying in about $1,400 yearly by that date, and the members were proud that "Ours is the only aid society among employees that we know of where the employer contributes."[19]

E. P. Allis did not live to see the completion in 1889 of a brick office building which offered excellent quarters for the society including a dining room, reading room and a hall for social meetings. Officers of the society continued virtually unchanged through the remainder of the century, Edwin Reynolds as president and Charles Allis as treasurer, with the remainder of the officers and directors usually shop foremen. This arrangement provided for the continuing interest of management after the death of Allis.

Louis Allis took a more active interest in the society during the nineties than did any other member of the family. He found that workmen, in efforts at emergency first aid, frequently applied tobacco, cobwebs, shavings or shellac to stop bleeding. Realizing that such emergency treatment complicated the work of the physicians and possibly endangered the recovery of the victim, Louis Allis organized an ambulance corps among the workers. By 1900 the ambulance corps had thirty-one members who received biweekly training from a physician on the treatment and care of the injured and in the use of first aid kits conveniently placed around the plant. The company also purchased an ambulance to replace the local police wagon. Preventive medicine programs made the society's physician available for consultation by any member during the noon hour of each working day. By 1900 the society numbered 1,233. It had paid out in benefits for sickness and accidents $40,778.82, and in death benefits to widows and families, $8,096.50—a total of $48,875.32. Physicians' fees alone amounted to $11,926.14. The Allis Mutual Aid Society disbursed $68,884.50 during its first seventeen years, including administrative expenses of $3,083.24.[20]

It is possible that the membership in the Allis Aid Society (1233 members out of 1862 employees in 1900) represented most of the men who were steadily employed, for there was a transience among the workers that a later age would find appalling. Payroll records for 1890 indicate that the daily average number of employees was 1,342, but the total of men hired that year was 11,101. Obviously a great many men had been hired by the day. Hiring practices were so informal that the foreman frequently hired foundry and shop men from the front porch of his home, or on the street with the simple instruction, "Come to work in the morning." Many workers of the late nineteenth century floated from job to job and from city to city, never establishing any sense of employment identity.[21]

Permanent workers in the Allis Company, and in all probability members of the Aid Society, were predominantly of northern European stock. The management was not only northern European but British in background, and proud of it. The names of the foremen indicate that they were primarily British, with an admixture of other northern Europeans such as Germans, Scandinavians and French. Until the company began to pay its men by check in the late eighties, workers from southern and central Europe did not even have a sense of individuality within the company. The foundry helpers, drawn largely from the growing Polish community on Milwaukee's south side, posed a problem to their foremen and timekeepers because of the complexity of their names. The alien name Joe Bidriski was entered on the payroll as Joe Brown. Names such as Czarsolewski, Jagodzinski, Wichlacz and Przykucki were simply listed as Smith 1, Smith 2, and Smith 3, or Schultz 1 and Schultz 2.[22]

In 1883 Allis had used the occasion of his dues-matching offer to the aid society to thank his employees for their loyalty to the company, adding, "It is to this faithful service and the absence of all contentions between us that our mutual prosperity and the magnitude of our work is due." Although Allis felt particularly and peculiarly close to his workers, he simply did not fathom all aspects of the labor situation. He did not realize, for instance, that the temporary workers never felt any identity of interest with the Allis family or with the Reliance Works. He also failed to recognize the emergence of a need on the part of labor to organize. Skilled labor, those regular workers who recognized Allis as the "Old Man," was in the strongest position to exercise pressure on management through organization. Because virtually all Allis products depended on castings, the molders were a key group—the most easily organized—and in a position to give Allis the most difficulty.[23]

Union members had worked in Allis's shops from his earliest days, but they had been a minority, applying no pressure and creating no tensions. But the organization of unions had gathered momentum in the early eighties in Milwaukee as well as elsewhere in the United States. Unions were active in all the skilled trades in Milwaukee by the mid-eighties, and as their numbers increased, there was a growing sense of injustice among some and power among others. Most were members of the Knights of Labor which, excluding only bankers, lawyers, stockbrokers, gamblers and liquor dealers, had mushroomed until it had about 700,000 members nationally by 1886. By that time there were about 25,000 members of the Knights in Wisconsin, 12,481 of them in Milwaukee. The E. P. Allis Company branch of the union, known as the Reliance Assembly, had nearly 600 members among the 1000 workers, with membership running at 100 percent among the molders. For the first time union members constituted a majority of the Allis employees.[24]

The cooperative spirit of 1883 which had led to the Allis Mutual Aid Society had given E. P. Allis the feeling that other firms might have difficulty with their workers, but not the Reliance Works. The issue that precipitated at least forty disputes in Milwaukee and involved thousands of workers was the question of the eight-hour day. Terrence V. Powderly, Grand Master Workman of the Knights of Labor, called a nationwide strike for the eight-hour day on May 1, 1886. This call affected labor across the nation, and in Milwaukee, it loomed as an ominous threat for virtually every industry in the city. The eight-hour day and the impending general strike were the principal topics of conversation in Milwaukee during the spring of 1886. The call for labor action at the Reliance Works slowly gained momentum so that by late March a petition was circulated among the Allis workers demanding the eight-hour day effective May 1, the employees to receive ten hours' pay for eight hours' work.[25]

E. P. Allis replied to the petition on April 3 in a long, thoughtful and temperate statement:

> It is not unknown to you that my sympathies are with the laboring classes, and I have for many years been conscientiously identified with what seems to me, and still seems, the true movement for their permanent benefit. Labor has many wrongs to be righted, but the existing state of things is the growth of centuries, and its wrongs cannot be righted in a

day or a year, and if no more rapid progress could be legitimately and peaceably made we might be satisfied if each generation would be an advance over the preceding one.

Allis said that he was willing to adopt the eight-hour rule, but because of sharp competition and what he felt to be a present depression of business, he could not give ten hours' pay for eight hours' work. As he saw it, the demand required a twenty-five percent increase in wages which would cause him to operate at a loss. He argued that if the employees insisted on their proposal he would have no choice but to close the Reliance Works. He concluded with the statement that:

> It would be a proud crowning of my life as an employer of labor if my works could continue to the end with the success, the harmony, and the mutual respect of the past, and be among the leading ones in the final and just solution of the vexed question of labor and capital. To attain this much desired end you may depend upon my earnest efforts and best advice, and while insisting upon safe business principles for my own protection, you may rely upon the most liberal judgment and treatment consistent with such safe business conduct of the Works.[26]

The Allis workers discussed their employer's statement and seemed to agree that they might accept the proposition of eight hours' work with eight hours' pay, which would amount to an immediate reduction for them, but with the hope that an increase would be soon forthcoming. In an atmosphere of amity Allis offered an unusual and interesting proposition to his employees in the form of a joint investment plan. He proposed that upon receipt of wages for May, the workers would have the option of signing over about five percent of their wages to an investment fund; he would then invest the same amount in the fund, the aggregate to belong equally to employer and employee. The fund would be managed by a board of directors chosen by the employees and by Allis, who would be the president and a member of the board. The fund, which he thought in time would grow large, could be used "in providing homes for members, in real estate, investment, in interest-bearing securities, or in industrial work to be operated in common." It would be governed by the practical cooperation of employers and employees. Following the announcement of the proposal, Allis told the press that he did not anticipate any trouble with his workers. For a brief time it appeared that his assumption was correct. Over 600 of his employees met and appointed a committee to draw up articles of agreement which would have continued the ten hour work day and the existing wages. The press greeted the agreement as a triumph for moderation and as evidence of Allis's progressive views toward labor.[27]

One week later on April 18, 1886, the Allis employees met again and repudiated the articles of agreement. Allis attributed the bitter change of tone to outside influence. He was probably partly correct, but time was also moving rapidly toward May 1 and a general confrontation, with passions becoming more inflamed. The press was shocked at this reversal, and Allis was both surprised and deeply hurt. He maintained that the agreement of April 12 was a contractual obligation binding both himself and his men, and made his course quite clear. He would close his works rather than capitulate to pressure and violate the articles of agreement. For this reason he refused to meet one week later with the Executive Board of the Knights of Labor.[28]

By April 30 about 5,000 men were on strike in Milwaukee, and the number swelled to nearly 12,000 on May 1. Leaders of the strike at the Reliance Works were the predominantly Polish unskilled laborers and the skilled molders who were thoroughly unionized. In the face of a growing general strike, Allis kept his works in partial operation. A general attack on the works on May 3, however, repulsed by fire hoses and police, convinced him that he should follow the advice of Mayor Emil Wallber and for his own protection close the works. But under the protection of the Watertown Rifles, the Beloit City Guards and the Darlington Rifles, provided by order of Governor Jeremiah Rusk, the Reliance Works reopened on May 5. Allis asked his workers to return. "I appeal to you all to remember that this is not a question of wages or of hours of labor, but of human rights and of manhood—of my right to run my works and your right to sell to me your time and labor." About 200 men responded to the call. Some remained away out of conviction, others out of fear. That same day, May 5, the militia fired on the strikers in the Bay View area of Milwaukee. That action, with

The Reliance Works was involved in the considerable labor unrest of spring, 1886. The National Guard was called in by Governor Jeremiah Rusk to maintain order.

the Haymarket bombing in Chicago on May 4, brought the unfortunate events to a violent climax. For all practical purposes the strike was over, but the men returned to work at Reliance slowly, reluctantly, and on Allis's terms. By May 8, 300 men had returned, and this number increased to between 500 and 600 on the 11th. Except for a few molders and carpenters, all the Allis men were back working again by May 15.[29]

E. P. Allis, the workers, and the community had suddenly and brutally been thrust into one of the great problems of the rapidly emerging urban industrial society that was the new United States. Allis's view of himself as the spokesman not just of his own workers but of workers generally was shattered, leaving him confused and perplexed. Although Bay View had been the scene of major violence in Milwaukee, Allis had been a general spokesman for management. His position after the strike was to remain the position of his company and its successors for many years. He would not discriminate between union and non-union men in hiring, but there was no mistaking the *locus* of power. "In deciding who will and who will not be employed in our works we mean injustice to no one, but we are and ever shall be the sole judges, and no man will ever be employed under duress of any kind. . . ."[30]

E. P. Allis—the radical businessman, the Greenback candidate for governor, the friend of labor had been caught in a web which he had helped create. In 1861 his employees in the old Reliance Works numbered in the twenties. Twenty-eight years later he was the employer of around 1,500 men. This was the period of the emergence of big business, and even the most considerate and humane employers were caught in the tensions of a new industrial society. There was a note of tragedy in Allis's words of October, 1886, when he said: "So far as I know my feelings and methods, they have always been for the interest of the working men, and I understand them just as far as I know how. Up to the riots of last spring I thought I was always considered their friend." E. P. Allis had lived to see the first major contest between big business and the growing labor movement.[31]

The drama of 1886 did not recur. The Knights of Labor, which had numbered over 700,000 in 1886, had dwindled to about 100,000 in 1890, and by the panic of 1893 had declined still further to about 75,000. At the height of its power it already suffered from a variety of internal weaknesses. Its diffuse membership, which included farmers and shopkeepers as well as industrial workers, provided little solidarity, Powderly proved to be a dramatic but weak leader, and the organization dissipated its energies in political efforts. The disastrous strikes of 1886 were the supreme effort of the Knights, and also their death rattle. The American Federation of Labor, which was founded in 1886, superseded the Knights of Labor; it numbered only about 225,000 in 1890 and by the turn of the century had fewer than 550,000 members. Skilled workers organized into local branches of international craft unions made up the A.F. of L. The volatile unskilled and semiskilled workers, who had accounted for a great share of the violence in 1886, were left without leadership. Until the turn of the century organized labor played little part in the affairs of the Allis Company.[32]

E. P. Allis had concerned himself with hours of labor and rates of pay but not, apparently, with the number of employees as long as the total sales continued to increase. The management of the Edward P. Allis Company in the few years before the depression of 1893 reduced the number of workers at the same time that they increased the total amount of production. In 1889 the production accounted for sales of about $3,000,000, and that total rose to $3,280,000 in 1892, the highest point reached until prosperity began to return in 1897. At the same time, the work force was reduced from 1,500 in 1889 to 1,342 in 1890 and 1,200 in 1892. Because the company was producing large equipment such as engines, blowers and pumps, the labor force did not bear a direct relation to sales since production was normally scheduled over a period of months. In 1893 the number of employees was reduced to the lowest point during the decade, 931 men, while sales declined to $2,700,000. By 1894 employment had struggled barely above the thousand mark to 1,040, while sales declined further to $2,440,000. At no point, however, did the bottom drop out of the market for the giant engines, for even in the poorest year, 1894, Reliance produced 130 engines, averaging 1,000 horsepower each.[33]

The interior of Milwaukee's North Point pumping station in 1909 shows several Allis units. A compound pumping engine is visible in the background with three triple-expansion units. The work of these pumps was later done by smaller and more efficient centrifugal pumps.

Industrialism had irrevocably changed the face of the United States and also the role of the Allis Company in society. Transportation had been revolutionized so that a company founded to service the fields and forests of Wisconsin had rapidly come to be national and international in its outlook. The production of pipe and pumping engines for the Milwaukee water system was the most spectacular local contribution by the Allis Company, and it also marked the final point at which the local market had any profound effect on the company. While other Milwaukee companies developed a national market much more slowly, from the mid-seventies the Allis Company had operated in a different context. It was the leading Milwaukee industry, E. P. Allis until his death was the greatest local industrialist, and the company was the largest employer of Milwaukee labor. But for all of that, in an industrial sense it was of Milwaukee but not dependent on Milwaukee as a market. The relation of the company to the community after the death of E. P. Allis became, in a sense, rather tenuous. The management of the Edward P. Allis Company lacked the color and public appeal of old Edward Phelps Allis. During the nineties newspapers devoted remarkably little space to the company. Except as a source of employment, the company seemed to have little connection with the city, and Milwaukee in turn seemed to have had remarkably little interest in the company. Nothing dramatic or traumatic occurred. The nineties can perhaps best be described as a caretaker period.

The one disturbing force that the Allis family had to contend with during the decade was the depression of 1893. As orders declined, the management reduced the number of workers and cut the wages of those remaining by ten percent. They took pride, however, in the fact that their wage scale remained relatively high while other employers put into effect reductions of twenty to thirty percent. Although it was not known to the employees at the time, the management maintained their salaries at pre-depression levels throughout that difficult period.

The financial basis of the Edward P. Allis Company was not as solid as it might have been from the very outset, and the depression aggravated the situation. The Allis family finally secured resources necessary to survive from the Northwestern Mutual Life Insurance Company of Milwaukee, with which they had maintained close connections for many years. E. P. Allis had been a member of its board of directors, and Charles Allis was to be on the board after the turn of the century. E. P. Allis had secured loans from Northwestern Mutual in the past but during the depression, the insurance company delayed from December 5, 1893 until June, 1894 before granting a loan for $340,000. The minutes of the board record that the money was to be used for expenses incurred "in increasing the capacity of the works, and the severe depression of business interests in the country." Except for construction of a machine shop, little building had taken place following the death of E. P. Allis, and one must assume that the loan was largely used "in reducing liabilities currant" (*sic.*). The company did survive the depression, and by July 1, 1895, was able to restore the ten percent wage reduction imposed in 1893. Charles Allis announced that the action was taken because of a great improvement in almost all the lines of work produced by the company, and the press considered it an encouraging sign of the return of "good times."[34]

Businessmen operating in the last quarter of the nineteenth century held that fourteen of the twenty years between 1873 and 1893 were years of "recession" or "depression." If those feelings record the facts with any degree of accuracy, we must conclude that the Allis Company did rather well to survive at all. That it did survive can be credited to the talents of E. P. Allis, who provided the initial leadership and also plotted the direction in which the company would move within the specialties of Hinkley, Gray and Reynolds. The five most important industries in the United States in 1890, according to the value of their products, were slaughtering and meat packing, flour and gristmill products, lumber and timber products, iron and steel, and foundry and machine-shop products. Allis had chosen well, for the three major lines of production at the Reliance Works either fell into one of the five most important products, as with engines, or serviced one of the categories, as with

In the 1890s, the most important development in the sawmill industry was the invention of the Trout power setworks. This made possible control, within very narrow limits, of the thickness of boards being sawed.

sawmilling and flour milling. Perhaps he had chosen these even more astutely than he first assumed. In the nineties the five major categories remained the same but the order of importance shifted, and the shift was indicative of the direction in which industry was going. By 1900 the products stood in the following order: iron and steel, slaughtering and meat packing, foundry and machine-shop products, lumber and timber products, and flour and gristmill products. The Allis Company still produced for three of the five major categories. The machine age, however, had arrived: the principal work of the firm, engine production, had increased to seventy percent of the company product by the turn of the century at the same time that foundry and machine-shop products had moved from fifth to third place. In 1899 engines constituted about $3,400,000 of the total company product of $4,800,000. That was not done at the expense of the other two major lines, for all other products of the Allis Company in both 1890 and 1900 constituted a total of $1,200,000. The vigor of industrial capitalism during the nineties is illustrated in the increase of total wealth of the nation from $65 billion in 1890 to $88.5 billion in 1900, the national income from $12 billion to $18 billion during the same years. When we consider the four years of prolonged depression in that period, the figures constitute a remarkable growth.[35]

In a sense the Edward P. Allis Company did not need remarkably able and perceptive leadership to survive in the 1890s. By following the path designed for it earlier and by improving on its three major product lines, the company could at least hold its own,

The sawmill industry in Wisconsin peaked in the late 1890s. It took an enormous number of logs to keep the larger sawmills busy.

and late in the decade develop in a steady if not spectacular fashion.

George Madison Hinkley had done a remarkable piece of work with the Allis sawmill department. Starting from next to nothing, sales were close to $400,000 per year about 1890. But in 1890 lumbering approached its peak in Wisconsin, with some 18,000 men engaged in logging and another 54,000 employed by the wood-processing industries. In 1890 alone the Weyerhaeuser companies cut over a billion feet of timber in Wisconsin. That this enormous amount of timber could be processed was, in part, a tribute to the perfected band saw. But no other revolution in sawmilling was forthcoming; only refinements of existing equipment were made, and the sales of the department had hit a ceiling above which they did not rise, and below which they sometimes fell.

In the early nineties G. M. Hinkley patented the Hinkley Automatic Power Swage and the Hinkley Saw Guide, both refinements of existing equipment. Perhaps the outstanding development of the decade was the Trout Power Set Works which Hinkley and William H. Trout, a draftsman for the company, devised and sponsored. This device controlled the thickness of lumber to be cut with far greater precision than any means up to that time. In fact, though it has been subjected to further improvement, the basic principle as conceived by Hinkley and Trout remains unchanged and is still in use today.[36]

One of the great improvements in flour milling in the late nineteenth century was the purifier, which increased the yield of high-grade flour.

Hinkley and the Allis Company were also prepared to initiate change in powering sawmills. Despite their commitment at the time to steam engines, they installed the first successful electrically driven sawmill in 1896 for the American River Land and Lumber Company of Folsom, California. The nine-foot band mill, running a fourteen-inch saw, was connected directly to a one hundred horsepower motor and driven through spur gears. The edger was a six-saw, sixty-inch machine driven by a fifty horsepower motor. The smaller equipment used the power of two thirty horsepower motors. This was considered the most advanced installation of the day. The big timber of the far West was being opened up, and the call went out for new and heavier sawmill equipment. Hinkley devised a whole new line of equipment to cope with the size and weight of the west coast logs.[37]

Throughout the decade the sawmill department provided a steady income for the company, but it remained the smallest of the major lines. While G. M. Hinkley loved his work, he was rough, crude, and outspoken. Perhaps that was the reason why the records indicate that his salary from 1890 to 1892 was no more than $300 per month. For all his patents and his innovations in sawmill equipment, his royalties seem to have been of little consequence.[38]

W. D. Gray in perfecting the roller mill during the late seventies had so revolutionized the milling process as to thrust the Allis Company into a primary national position. Once the great rush to convert to roller mills had taken place in the eighties, Gray was responsible only for the further perfection of equipment and the

Another important contribution of Edwin Reynolds to the production of power was the shaft mounting of generator rotors to large engines, eliminating the need for long belts.

maintenance of a position of leadership. Throughout the nineties the flour milling department could normally be depended upon for about double the sales of the sawmilling department, up to $800,000 in a good year. Next to the introduction of rolls, the greatest improvement in milling was the purifier, a machine previously unknown to the milling process. It displaced no machine but was capable of converting the "middlings" which formerly went into poorer grade flour, into a high grade flour. In combination with the rolls it permitted an efficient milling process.

Besides the introduction of new machines such as rollers and purifiers, the perfection of existing machines continued through the nineties as milling engineers worked to produce the highest possible grade of flour at the lowest possible cost and with the least waste. The old hexagonal reel gave way to the centrifugal reel and the flour dresser. The greatest improvement was in their smaller size, which made it possible to place them to far better advantage. The milling engineer was then in a position to plan a mill that was mechanically far more perfect.[39]

Gray rightly took great pride in his role: "We who have been closely identified with these great improvements may well be excused if we feel a little proud of the progress made, and the results obtained." The tall, handsome and canny Scot was worth more than Hinkley to the Allis Company. But it is also possible that he was the kind of man that Will and Charles Allis found more to their liking. His salary of $833.33 per month for the period 1891–1896 was nearly three times that of Hinkley. Gray also seemed to drive a hard bargain in patents, because during some months he earned up to $200 more than his monthly salary from his royalties.[40]

But neither sawmilling nor flour milling could compare in either mechanical drama or financial worth to Edwin Reynolds's engines. The role of engine production in the economy of the company can best be seen in the following figures.

Year	No. of Engines	Total Engine Prod. $/Yr.	Total Allis Prod. $/Yr.
1890	260	2,080,000	3,300,000
1891	300	2,400,000	3,100,000
1892	244	1,952,000	3,280,000
1893	150	1,200,000	2,700,000
1894	130	1,040,000	2,440,000
1895	200	1,600,000	3,200,000
1896	170	1,260,000	3,180,000
1897	250	2,000,000	3,560,000
1898	500	3,600,000	4,250,000
1899	500	3,400,000	4,800,000

As business began to revive in 1895, an appropriation of $35,000 was made for new tools to facilitate further production. By the fall of 1897 engine work increased to the point that a new machine shop on National Avenue became necessary. At the same time the Allis family authorized an addition to the South Foundry measuring 186 by 274 feet. Two years later they built an additional machine shop to house the manufacture of sawmill equipment and mining machinery. The Allis Company had ventured into mine machinery in 1894 with Henry Holthoff in charge, thus adding a fourth line to their products. In 1900 Edwin Reynolds said that even with the additions to the plant, "We have turned down more orders this year than we have filled on account of our inability to deliver satisfactorily."[41]

During the nineties, Edwin Reynolds continued to make significant improvements in engines. He followed his heavy duty engine frame of 1890 with the vertical engine frame of 1894, which could support the largest engines. The following year he secured a patent on a high steaming capacity water tube boiler of the four-drum type which was sold to one of the large boiler manufacturers. In 1898 he devised self-adjusting bearings which were immediately successful and came to be used almost universally on all large engines. He designed and built the first horizontal-vertical engine in 1899. But perhaps his most significant innovation of the decade was the engine-type or flywheel generator. No one in the United States had successfully mounted generator spiders directly on engine shafts until, in March of 1892, Edwin Reynolds contracted to do so. He agreed to build two horizontal 500-horsepower cross-compound engines for the Narragansett Electric Light Company of Providence, Rhode Island. Slow-speed generators were not considered practical at that time in view of the experience Ferranti brothers in London had with this type for high voltage. As a consequence, neither General Electric (then the Thompson-Houston Company) nor Westinghouse was willing to undertake the generators for the Narragansett engines.

In 1892 Reynolds announced that he would have all the work done in his shops, which would have plunged the Allis Company into the generator business. To prevent this, General Electric and Westinghouse agreed to each build one generator if Reynolds would design and build the rotors in Milwaukee, which he did. These units were so successful that belt driving was almost immediately abandoned, and direct mounting of the generators on shafts of large engines became standard practice.[42]

Reynolds also gained national and worldwide recognition for outstanding achievements in engine building. People from all over the globe attended the World's Columbian Exposition in Chicago from May to October, 1893. Among the attractions remembered by most was the "Pride of Machinery Hall," a 3,000 horsepower, horizontal quadruple-expansion Reynolds-Corliss engine, the Allis Company's contribution to the fair. 1893 was a depression year but it was a dramatic moment when President Grover Cleveland pressed the button which started the "monster" engine and the "breath of life" throbbed "in the mighty cylinders of the Giant that is the Pride of the World's Fair Machinery Hall." The crowd was "galvanized" by the sight. The flywheel was thirty feet in diameter and the entire engine weighed 325 tons. Driving two Westinghouse 750 kilowatt alternators, it supplied the current for 20,000 of the "16-candlepower incandescent lamps." Fairgoers "took home memories of electricity used on a lavish scale for the first time," and of the engine that produced the power. Powering the Columbian Exposition's famed Intra-Mural elevated electric railway was an Allis/G.E. 2000-combined engine-generator direct current unit, prototype of those soon to go to Brooklyn Rapid Transit.[43]

The Edward P. Allis Company concluded the nineteenth century dramatically by building the mammoth engines for the New York rapid transit system. This feat brought to its apex the construction of steam engines in the United States. As Bayrd Still suggests in his study of Milwaukee, this marked the coming of age of Milwaukee's industrial economy. Over several years the Allis Company supplied the engines for electrifying New York's rapid transit system. The New York City Railway Company, operating every surface car in New York City, had eleven Allis engines, aggregating about 61,000 horsepower at normal rating, installed in the Ninety-sixth Street power house; the Brooklyn Rapid Transit

The horizontal-vertical engine was developed by Reynolds to save space as engines became more powerful. In this design the high pressure cylinders were horizontal and the low pressure cylinders were vertical.

One of the attractions at the 1893 World's Columbian Exposition in Chicago was the 3,000 horsepower quadruple-expansion Reynolds-Corliss engine known as the "Pride of Machinery Hall." The engine drove two generators which supplied power for 20,000 lamps.

Company installed nineteen Allis engines aggregating 60,000 horsepower; and the Interborough Rapid Transit Company installed seventeen Allis engines aggregating more than 200,000 horsepower. Nine of the latter were in its Fifty-ninth Street power house for operating the subway and eight at the Manhattan (74th St.) station for operating the elevated roads.[44]

The "Manhattan Engines," all seventeen of them, were the largest stationary engines ever made and were definitely dramatic. That they were designed by Edwin Reynolds on a train while enroute from Chicago to New York was considered in itself spectacular. As the proportions of the monster engines took shape, however, they staggered men's imaginations. They were of the combined horizontal-vertical type with the horizontal high-pressure cylinders forty-four inches in diameter and the two vertical low-pressure cylinders eighty-eight inches in diameter, working two cranks with a stroke of sixty inches. Operating at seventy-five rpm, they were directly connected to electric generators, the

The apex of steam engine development came with the forty-seven horizontal-vertical engines built for the New York City transit companies. With the largest engine rated at 12,000 horsepower, these were the most powerful ever built. The Smithsonian Institution has on display a complete model and actual parts from these engines.

thirty-two foot rotating fields serving as flywheels. The nine engines at the Fifty-ninth Street station constituted the most powerful steam plant of that day. The battleship *Connecticut*, just launched, was regarded as America's most powerful warship, but the subway power plant was equal to a half-dozen *Connecticuts* with 4,500 horsepower to spare. At every full inspiration the nine monsters required about 150 pounds of water supplied as steam at 175 pounds per square inch. They required 1,000 tons of coal per day and more than 1,800,000 gallons of water, enough to supply the needs of a city of 30,000 to 40,000. The daily average output of the nine engines was 800,000 kilowatt hours. The entire powerhouse was nearly twice the size of the old Madison Square Garden. These Manhattan engines reached the practical limit of size for reciprocating engines, and marked the high point of Edwin Reynolds's steam engine development.[45]

Whether it was the Chicago World's Fair or the engines for the New York transit system, power for Glasgow's first electric trolley system, or blowing engines for Russia's Crimea the products of the Allis Company were serving the world at the turn of the century. In industry Allis was the name that made Milwaukee famous. But fame could not overcome certain very basic problems. One of the problems was that the company lacked capital, and the management had difficulty securing money for an expanding business. Another problem was change. When the steam engine had reached the peak of its development, the impact of new sources of power such as the steam turbine-generator units, hydro-electric units and large gas engines were beginning to make themselves felt. And all this was compounded by the competition that the Allis Company faced in great electrical age giants like General Electric and Westinghouse, which had the necessary capital and, each in its own way, was getting into the new sources of power. Management of the Edward P. Allis Company was in a somewhat limited position to face the twentieth century with confidence or equanimity.[46]

NOTES TO CHAPTER FOUR

1 A copy of the complete will of E. P. Allis is found in the hand of Charles Allis as a preface to the *Minute Book, E. P. Allis Co.*, April 1, 1890–May 25, 1938, in the First Wisconsin Trust Company, Milwaukee. The important provisions of the will along with a layman's interpretation of them is found in the *Milwaukee Sentinel*, April 11, 1889.

2 The Carnegie statement is quoted by Edward Chase Kirkland in *Dream and Thought in the Business Community, 1860–1900* (Ithaca, 1956), p. 163; *Evening Wisconsin*, April 2, 1889; *Milwaukee Sentinel*, April 3, 1889; *Weekly Statement Leaflets*, May, 1889, in Louis Allis Scrapbook, vol. 1. At the time of his death Allis held the following policies:

Mutual Life of New York	$121,836
New York Life	100,000
Equitable	100,000
Northwestern Mutual	50,000
Massachusetts Mutual	30,000
Mutual Benefit of N. J.	20,000
Connecticut Mutual	20,000
Home Life of New York	10,000
National of Vermont	5,000
Penn Mutual	5,000
Total in ten companies	$461,836

Apparently there were also one or two small policies not included in the above list. William H. Metcalf, an intimate friend of Allis, said of him, "He failed in business some years ago, and since that time had but slight confidence in the perfect solidity of any business venture. In recent years he became very cautious, and was extremely solicitous about his family. This led him to place a heavy insurance on his life, and he probably carried more insurance than any man in Milwaukee. I remember his having told me more than a year ago that he carried policies aggregating $100,000, and was going to increase that amount." *Milwaukee Sentinel*, April 2, 1889.

3 The most complete report on the financial situation of E. P. Allis at the time of his death and the subsequent arrangements by the family regarding the business is found in the Appellant's Brief, State of Wisconsin in Supreme Court, January Term, 1916, No. 56, "In the Matter of the Last Will and Testament of Edward P. Allis, Deceased."

4 *Ibid.; Minute Book, E. P. Allis Co.*, April 1, 1890. Since this is a study of the Allis Company, no attempt has been made to give a complete study of the Allis family. However, it might be helpful to list the Allis children and when they were born.

William Watson	November 14, 1849
Edward Phelps	September 14, 1851
Charles	May 4, 1853
Jere	March 4, 1855
Maud	July 3, 1856
Ernest	July 11, 1858
Mary White	June 26, 1860
Frank Watson	July 10, 1865
Louis	December 30, 1866
Margy	August 23, 1868 (died in infancy)
Margaret Watson	October 26, 1869
Gilbert	January 4, 1871

Charles Allis, *Allis Genealogy* (Milwaukee, 1893), Louis Allis Scrapbook, vol. 2.

5 Allis Chalmers Company *Sales Bulletin*, April, 1909, p. 42; *Wages Ledger*, The Edw. P. Allis Co., 1890–1897; contract signed by Edwin Reynolds and William Allis, April 1, 1890, photostat in Allis-Chalmers files.

6 *Evening Wisconsin*, December 20, 21, 1909; *Milwaukee Journal*, December 20, 1909.

7 *Milwaukee Sentinel*, April 2, November 3, 1889, January 22, 1899, February 9, 1907; *Marquette Medical Review*, "The Allis Lake Laboratory," March 1956, pp. 141–144; J. M. J. Keogh to Alberta J. Price, August 1, 1945; Irving Reynolds to Alberta J. Price, July 31, 1945. Although it is difficult, if not impossible, to interpret the Wages Ledger of the Edw. P. Allis Co., 1890–1897, it is clear that

during that entire period William and Charles Allis each received a salary of $1,000 per month with an additional undetermined amount in dividends, and, for some unexplained reason, E. P. Allis, Jr., was paid only $83.33 per month, which meant that he had to live almost wholly off his dividends. This could well have been part of his reason for leaving the company.

8   John G. Gregory, *History of Milwaukee, Wisconsin* (Chicago, 1931), 4:146–151; William George Bruce, ed. *History of Milwaukee City and County* (Chicago, 1922), 2:11–12; Clippings from *Electrical World*, December 20, 1924, and *Business Advertising Digest*, May, 1943, in Louis Allis Scrapbook, vol 6.

9   Bruce, 2:15–16; *Milwaukee Journal*, July 26, 1918; Irving H. Reynolds to Alberta J. Price, July 31, 1945.

10 Bruce, 2:12–15; Irving H. Reynolds to Alberta J. Price, July 31, 1945.

11 Irving H. Reynolds to Alberta J. Price, July 31, 1945; J. M. J. Keogh to Alberta J. Price, August 1, 1945.

12 The role of E. P. Allis in the Wisconsin Greenback movement is not explored in detail. Only his philosophy as it related to his business is included. *The Blue Book of the State of Wisconsin* (Milwaukee, 1895), p. 347; the *Milwaukee Sentinel*, August 18, 20, 22, 25, 28, October 6, 26, 1877, records his campaign in the state; *Speech of Mr. E. P. Allis Before the Milwaukee Greenback Club*, 1878, photostat of original in Allis-Chalmers archives.

13 *E. P. Allis Before the Milwaukee Greenback Club*, 1878; The *Address of Edward P. Allis, of Milwaukee, to the Fourth Dist. Congressional Convention of Wis.*, Sept. 30, 1878, was largely a restatement of the above principles. E. P. Allis was invited to present his views to the National Bankers' Association in New York City during their meeting September 12, 13, and 14, 1877. After he arrived, however, they denied him the floor and his prepared address was read into their minutes. *National Finances: Speech of Edward P. Allis*, was then published and distributed at 50 cents per 100 or $4.50 per 1,000. In October, 1878, he issued a statement *To the People of Wisconsin, and More Particularly, My Old Political Associates, the Republicans*, in an attempt to convert them to his position. Photostatic copies of all four pamphlets are found in the Allis-Chalmers archives. The originals are in the archives of the State Historical Society of Wisconsin.

14 *Milwaukee Sentinel*, October 11, December 1, 1884.

15 *To My Employees*, August 17, 1888, Louis Allis Scrapbook, vol. 1; this statement was reprinted in the *Sentinel*, August 18, 1888.

16 *Minute Book, E. P. Allis Co.*, October 9, 1897. On November 21, 1899, an addition to the machine shop on National Avenue was approved.

17 *E. P. Allis Before the Milwaukee Greenback Club*, 1878; *Milwaukee Sentinel*, February 9, 1883, March 13, 1888.

18 *Milwaukee Sentinel*, July 21, 22, 23, 1883. In an editorial, the *Sentinel* praised Allis for his "recognition of the principle that an employer's concern as to his employees should not end with their simple employment at the lowest price to which the market can be reduced; that the advantage which the employer holds over the employee is not his to be used entirely in his own interests, but that there is a sense of justice which demands of the employer the distribution among his workmen in some form or other of the profits which are considered more than reasonable."

19 *Milwaukee Sentinel*, March 13, 1888.

20 *Milwaukee Sentinel*, July 21, 1883, April 2, 1887, April 10, 1888, April 14, 1889; Gregory, 4:146–149.

21 *Milwaukee Journal*, March 31, 1900; Data from payroll books, 1876 to 1902, extracted by Alberta J. Price. Unfortunately, the original payroll books have been lost and only the notes taken by Mrs. Price remain. Corroboration for this data is found in the unpublished Ph.D. thesis, "A Social History of Industrial Growth and Immigrants: A Study with Particular Reference to Milwaukee, 1880–1920," by Adolf Gerd Korman, for the History Department, University of Wisconsin, 1959, p. 127. Korman had an opportunity to see the payroll books before they were lost.

22 J. M. J. Keogh, Factory Notes, 1902; Data from payroll books, 1876 to 1902, extracted by Alberta J. Price.

23 Data from payroll books, 1876 to 1902, extracted by Alberta J. Price; *Milwaukee Sentinel*, July 22, 1883.

24 *Milwaukee Sentinel*, May 2, 15, 1886; Bayrd Still, *Milwaukee: The History of A City* (Madison, 1948), pp. 291–296. Still offers an excellent summary of the political struggle at this time between the militantly aggressive Central Labor

Union, led by Paul Grottkau, editor of the *Arbeiter Zeitung,* and the more moderate approach of the Knights of Labor under the leadership of Robert Schilling, editor of the *Volksblatt.* J. M. J. Keogh to Alberta J. Price, July 27, 1945. No attempt is made here to provide a complete study of Allis in relation to the strike of 1886.

25 *Milwaukee Journal,* March 26, 1886; *Milwaukee Sentinel,* March 27, 1886.

26 *Milwaukee Sentinel,* April 4, 1886.

27 The *Sentinel,* April 8, 1886, carried the following statement from a "prominent member of the Knights of Labor." "I believe that the men will accept Mr. Allis' conditions, as they are reasonable." *Milwaukee Sentinel,* April 11, 14, 17, 1886; *Milwaukee Journal,* April 14, 1886.

28 *Milwaukee Sentinel,* April 19, 1886. The articles of agreement were repudiated on April 18 on the grounds that the "Special Shop committee" had met with Allis one day earlier than had been originally agreed upon and the articles were consequently null and void. *Milwaukee Journal,* April 19, 27, 1886; *Milwaukee Sentinel,* April 20, 27, 1886.

29 *Milwaukee Sentinel,* April 30, May 1, 2, 4, 5, 7, 8, 11, 13, 15, 1886. The *Sentinel* of May 3, 1886, offers a fine description of the parade and demonstration by 3,000 workmen in Milwaukee in behalf of the eight-hour day. An estimated 25,000 people watched the parade sponsored by the Central Union. Only one assembly of the Knights of Labor, about 300 men, participated. *Milwaukee Journal,* May 3, 4, 5, 6, 7, 11, 1886; Irving H. Reynolds to Alberta J. Price, August 31, 1945, gives a graphic description of how Reynolds, E. P. Allis, Jr., and the foremen hosed down the strikers. The *Sentinel* of May 16, 1886, reported that "The backbone of the great strike is broken, and nearly two-thirds of the men who left their work on or about May 1, have returned to their respective employment. Allis' moulders on Friday (May 14) concluded to return to work under the old conditions of 10 hours' work for 10 hours' pay. In nearly every instance employers have refused to discharge the men who took the places of strikers, and the result has been to throw a number of ring leaders permanently out of good employment." Allis was among the employers who refused to discharge men who took the place of strikers.

30 *Milwaukee Sentinel,* May 21, 1886.

31 *Milwaukee Sentinel,* October 16, 1886.

32 Selig Perlman, *A History of Trade Unionism in the United States* (New York, 1922), pp. 180–186.

33 Notes taken by Alberta J. Price from payroll books of the period; M. C. Maloney, "Allis Engine Production," June 17, 1964.

34 *Evening Wisconsin,* June 27, 1895; *Milwaukee Sentinel,* June 28, 1895; *Wages Ledger,* The Edw. P. Allis Co., 1890–1897; *Minute Book,* E. P. Allis Co., December 5, 1893; June, 1894. As secretary of the Company, Charles Allis provided some interesting variations on normal spelling.

35 Kirkland, p. 7, quoting Willard L. Thorp in *Business Annals* (New York; National Bureau of Economic Research, 1926), pp. 131–137.

36 Robert F. Fries, *Empire in Pine* (Madison, 1951), pp. 84–99; *Allis-Chalmers Saw Mill Catalogue,* 1902, pp. 38, 180; E. C. Shaw to Alberta J. Price, September 14, 1954.

37 *Pioneer Power* (Milwaukee, 1942), p. 29; E. C. Shaw to Alberta J. Price, September 14, 1954. On a tour of the Pacific Coast in 1946, Shaw found a considerable amount of sawmill equipment produced by the Edward P. Allis Company still in use.

38 *Wages Ledger,* The Edw. P. Allis Co., 1890–1897.

39 W. D. Gray, "A Quarter Century of Milling," Part XX (conclusion), *The Weekly Northwestern Miller,* March 21, 1900, p. 564.

40 *Ibid.;* John Storck and Walter Dorwin Teague, *Flour for Man's Bread: A History of Milling* (Minneapolis, 1952), pp. 261–262; M. C. Maloney, June 17, 1964; *Wages Ledger,* The Edw. P. Allis Co., 1890–1897.

41 M. C. Maloney, June 17, 1964; *Minute Book,* E. P. Allis Co., September 5, 1895, October 9, 1897, November 21, 1899; M. C. Maloney, 98 *Years of Thermal Power History,* 1963, MS., p. 35.

42 Irving H. Reynolds to G. F. De Wein, August 28, 1940, pp. 6–7.

43 *Ibid.,* p. 6; *The Chicago Sun,* April 23, 1893. The "Reynolds gear" received the following citation [of award] at the Chicago Exposition of 1893: "For excellence of a type of engine releasing gear which has shown its exceptional merit; not only by the success in its use on engines of the inventor's own design, but by being copied in all essential features by nearly all of the prominent builders of

Corliss engines in this country." Robert H. Thurston, *A History of the Growth of the Steam Engine* (Ithaca, 1939), p. 504.

44 Still, p. 339; *London Times*, June 4, 1900; *The Four Powers*, Allis-Chalmers Company, 1904.

45 Thurston, p. 504; Irving H. Reynolds to G. F. De Wein, pp. 7–8.

46 *London Times*, June 4, 1900; *Milwaukee Sentinel*, August 13, 1900; Appellant's Brief, State of Wisconsin in Supreme Court, January Term, 1916, No. 56.

Charles Allis
President 1901–1904

CHAPTER FIVE

PROBLEMS OF MERGER AND EXPANSION

IN 1900 THE EDWARD P. ALLIS COMPANY entered a new century. Seventy-six million Americans were anxious to forget the troubles of the nineties. The depression of 1893, which had sharply curtailed the progress of industry and had resulted in thousands of bankruptcies and millions of people hungry and unemployed, had passed into history. The country had been victorious in the war with Spain, and through it had acquired overseas possessions stretching halfway around the world. The concept and psychology of "empire" were still in the air. Free silver, the focal point of the bitter campaign of 1896, was no longer an issue, and business had been reassured by the formal adoption of the gold standard in 1900. The discovery of a new supply of gold in the Yukon had both stimulated the spirit of adventure and added more gold to the currency, causing both wages and prices to rise.

The United States had also become the greatest manufacturing nation in the world. More than seven million people were engaged in manufacturing and mechanical pursuits, and the value of manufactured products in 1900 exceeded $13,000,000,000, far surpassing the $4,717,000,000 estimated for agriculture. The impact of this trend was seen in Milwaukee. In the nineties the product of the city's metal trades alone increased from $5,568,445 to $14,495,362, making it the city's leading industry. In forty years the annual production of the Allis Company had increased from about $30,000 to a figure in excess of $5,000,000. The Allis Company had succeeded handsomely because nowhere else in the world were the same vital ingredients for industrial success found in such force and profusion. Inexpensive foodstuffs, ample raw materials, and cheap immigrant labor were combined with a rapidly expanding market and a high protective tariff. Moreover, no other culture held the same high respect for business values. The young American man at the turn of the century was convinced that the greatest rewards in life were to be found in the business world.[1]

By the turn of the century, the telephone and the electric rapid transit system had made possible the decentralization of urban society. In the Milwaukee area a number of manufacturers had purchased land on the edge of the city for expansion. The Allis family and Edwin Reynolds had discussed such a move because the Reliance Works had utilized all immediately available land for factory space and was still unable to satisfy the demand for its products. They took no action until November, 1900, because they felt that one dark cloud from the nineteenth century remained as a threat to the business prosperity. That threat was the possibility of the election of William Jennings Bryan to the presidency. On the day before the election Edwin Reynolds, representing the business community, stated the case for Republicanism. He said that there were millions of dollars worth of business suspended until the election "with the understanding that the business goes forward if Mr. McKinley is elected." The election of William McKinley promised continued business prosperity and therefore the Allis Company was prepared to expand.[2]

Anticipating possible expansion of the Allis plant, a number of promoters had offered inducements on behalf of various locations. On November 25, 1900, Charles Allis looked at property in the North Greenfield area west of Milwaukee, and the next day showed it to Edwin Reynolds. With authorization from the other directors, they purchased 100 acres of land for $25,000, with the assurance of railroad connections with the Chicago & North Western and the Chicago, Milwaukee, and St. Paul roads which ran within a mile of each other through that area. Their decision led to the establishment of the community of West Allis, which rapidly attracted other industries as well, since it was now assured adequate transportation.[3]

The business community, in an atmosphere of optimism and expansion, found sheer "bigness" more attractive than ever. An increasing number of businessmen believed that a consolidation of similar firms would control production and, therefore, prices and profits. They held that such consolidation would permit economies in the purchase of machinery and raw materials, transportation services and commercial credit. Bigness would also increase the power of management in dealing with organized labor. As a result, the consolidation of American industry, which proceeded at an increasing rate, was the outstanding feature of economic life at the beginning of the new century. In 1901, forty-six industrial combinations with a capital of $1,000,000 or more were formed; sixty-three were formed in 1902, and eighteen in 1903. Clearly, the Allis-Chalmers merger of April and May, 1901, came at about the peak of this activity.[4]

As the largest single manufacturer of steam engines in the United States the Edward P. Allis Company was caught up in the tide of national economic developments. Milwaukee papers reported rumors of combinations involving the Allis Company for nearly two years prior to the actual merger. The general feeling seems to have been that a combination involving the Allis Com-

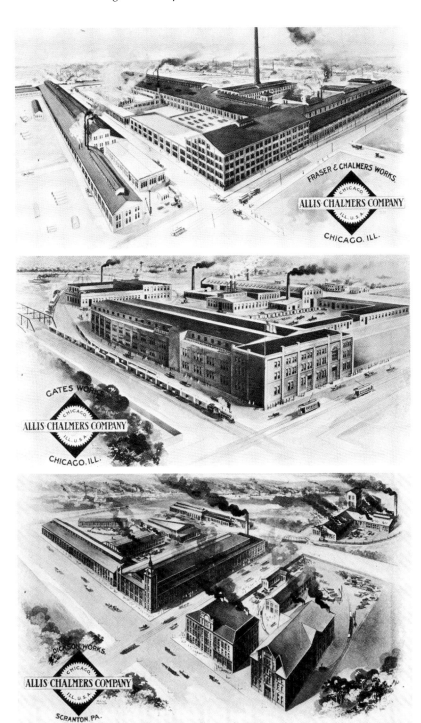

*Four companies combined to form the Allis-Chalmers Company in
1901. In addition to the Reliance Works of the Edward P. Allis
Company and its Bay State plant, these included the Fraser &
Chalmers plant of Chicago, the Gates Iron Works of Chicago, and the
Dickson plant of Scranton, Pennsylvania. The combination resulted
in total manufacturing space of over three million square feet.*

pany was inevitable. The only questions were when and how it would occur. Edwin Reynolds broke the suspense when he announced on April 30, 1901, "Yes, the Edward P. Allis Company has sold out to the machinery combine."[5]

Edwin Reynolds was the catalyst in the development of the merger. During the last months of 1900 and in early 1901, he had represented his company in a series of meetings of the National Metal Trades Association, held to formulate a uniform management policy for dealing with the machinists' union. After one meeting Reynolds fell into conversation with William J. Chalmers, president of Fraser & Chalmers Company of Chicago. Sitting in a hotel lobby, Reynolds outlined his plan for building bigger and better steam engines in the proposed West Allis plant. In fact, Reynolds felt that the market warranted three such factories: one in Milwaukee and one on each coast to satisfy the worldwide demand for engines created by the tremendous development in electric transit lines. But he recognized that such expansion demanded access to capital far beyond the resources of the Allis Company. Chalmers told him that he would see who might be interested in backing such a venture. To the amazement of Edwin Reynolds, he received a telegram from Chalmers only a few days later, asking him to come to New York and begin discussions.[6]

The discussions with Chalmers and William Marburg, a New York promoter, resulted in the merger of four companies to form the Allis-Chalmers Company. The most important of these was the Edward P. Allis Company, with a valuation of $5,120,000 placed on its property. The Fraser & Chalmers Company of Chicago was well known for its mining machinery, including hoists, Riedler pumps, stamp mills, and smelting and concentrating machinery. Their equipment was used in all the world's largest and most important mining areas at the turn of the century. The Fraser & Chalmers plant was valued at $3,205,000. The Gates Iron Works of Chicago, the third company involved in the merger, brought to the combine its famous Gates Gyratory Rock and Ore Crusher, as well as general mining machinery and a fine line of cement-making machinery. The valuation placed on the Gates property was $410,000. The Dickson Manufacturing Company of Scranton and Wilkes-Barre, Pennsylvania, built blowing engines, compressors, sugar mill machinery, coal mining machinery, tanks and locomotives. Excluding the locomotive works, which was acquired by the American Locomotive Company, this property was valued at $1,200,000.[7]

Milwaukeeans were impressed by the size and scope of the plants in the merger. They were more impressed by the men on the Board of Directors. Reading like a Who's Who of American business at the turn of the century, the board included Elbert H. Gary, Edward D. Adams, Mark T. Cox, William L. Elkins, Jr., Henry W. Hoyt, William A. Read, James Stillman, Charles Allis, William W. Allis, Edwin Reynolds, William J. Chalmers, James H. Eckels, Max Pam, and Cornelius Vanderbilt. Milwaukeeans were proud that Charles Allis was elected president of the Allis-Chalmers Company, with Will Allis as chairman of the board, and Edwin Reynolds chief engineer of the company. Civic pride should have been tempered by recognition that the powerful Finance Committee was made up of William A. Read as chairman with Adams, Gary, Cox and Chalmers. W. J. Chalmers, who had played an important part in arranging the merger, had a prominent place in the new company as vice president and treasurer. He was also chairman of the Executive Committee, composed of Charles Allis, Hoyt, Adams, Vanderbilt, Eckels and Pam. Significantly, the offices of the company were to be in Chicago.[8]

The contracts held by William Marburg with the four companies participating in the merger authorized the Allis-Chalmers Company to issue capital stock to the extent of $50,000,000: $25,000,000 to be preferred stock with seven percent cumulative dividend, and $25,000,000 in common stock, the shares having a par value of $100 each. However, when the articles of incorporation were filed in New Jersey on May 8, 1901, the actual amount of stock issued was reduced to $16,250,000 in preferred stock and $20,000,000 in

One of the important products obtained in the merger was the Gates gyratory crusher, used in the mining, cement, and rock crushing industries for reducing rock to smaller sizes.

common stock, although the authorized capital stock still remained at $50,000,000. Of the total preferred stock, $7,850,000 was accepted by the participating companies in lieu of cash in part payment for their properties, and the remaining $8,400,000 was offered for sale. This stock was preferred in cumulative dividends to the extent of seven percent per annum, and had a further preference of one percent, non-cumulative, after the payment of seven percent on the common stock.[9]

The Marburg contracts declared that "Vermilye & Company, managers of a syndicate . . . . has been formed to finance the affairs and float the securities of said proposed corporation." In promoting the stock, Vermilye and Company of New York and Boston announced that the officials of the four companies estimated that their net profits for the year ending May 1, 1902, would be sufficient to pay dividends at the rate of seven percent on the preferred stock, and about four percent on the common stock. They also expected sufficient economies from the merger to be able to pay seven percent on both classes of stock. In their enthusiasm they also estimated that when the West Allis Works was completed, they would realize a return of eight percent on the preferred and ten percent on the common stock. These dividends were thought to be well within reach.[10]

Many great business combinations were consummated in 1901, including the formation of the United States Steel Corporation. But many Americans were becoming concerned over the increasing control of the country by big business. President Charles Allis felt compelled to issue a statement of purpose and intent. He assured the press, "We have not entered a trust." The reason for the combination of the four companies, he said, was simply to combine their capital resources so that they could handle greater work and larger contracts than they could singly. "There is no idea of grinding the workingman or forming a trust in the common acceptation of that word."[11]

The statement by Charles Allis cannot, however, be taken at face value. Something more potent than efficiency brought about the merger, and that thing was gain. The principals involved in the merger found it distinctly profitable. Edwin Reynolds hoped to profit not so much in finances as in his first love, machines. Some claimed that he was the driving force behind the merger because the Allis Company lacked both space and capital to build the giant machines that Reynolds had in mind. The vision of manufacturing mammoth engines in three great factories spanning the continent was more than enough to secure the cooperation of Edwin Reynolds.[12]

The Allis boys, unlike their father who had viewed the company as operating for the benefit of society, had regarded Reliance largely as a source of income. Under the terms of the Marburg contract the Allis family received $5,375,000 par value of the preferred stock and $1,663,631 in cash. The directors of the Edward P. Allis Company promptly declared a 25 percent dividend on June 25, 1901, a 10 percent dividend on August 13, and another 35 percent dividend on September 9. In fact, between June 25, 1901, and November 3, 1902, the family declared dividends to themselves totaling 138 percent. The Allis family had ample reason to consent to the merger.[13]

The motives of William J. Chalmers were even more transparent. Chalmers's eagerness to promote this merger, or any other merger for that matter, arose from the fact that his firm was headed for serious financial difficulties brought on by intense competition. In 1889 the firm of Fraser & Chalmers, Chicago, had been sold to Fraser and Chalmers, Ltd. of England. Because of the poor business prospects of the American branch, the English company was eager to unload its interests in the Chicago plant. It was reported that W. J. Chalmers was given a handsome present by the English owners for successfully engineering the merger. Having accepted stock in the Allis-Chalmers Company in exchange for its properties, Fraser and Chalmers, Ltd. disposed of this stock as rapidly as possible. Although no documents exist concerning the negotiations with the Gates and Dickson people, one may reasonably assume that the terms were not unprofitable to the owners.[14]

The promoters and the New York interests on the Board of Directors also profited handsomely. According to Irving Reynolds, who acted as assistant to his uncle in the negotiations, the promoters accepted the common stock at twenty-five cents on the dollar to give them a chance to unload at a profit.[15]

The valuation of $9,935,000 placed on the Allis, Fraser & Chalmers, Gates, and Dickson plants for public consumption by consulting engineer Julian Kennedy was apparently a generous figure. The appraisal of the four plants entered in the Minutes of the Allis-Chalmers Company Executive Committee reads very differently: "E. P. Allis Company $1,669,312.60; Fraser & Chalmers, Inc. $2,594,845.74; Gates Iron Works $362,474.45; Dickson Mfg. Co. $776,872.90; Total $5,403,505.69." In light of the latter total, the amount of stock issued on the Allis-Chalmers Company, to the extent of $36,250,000, seems even more generous on the part of the promoters. It is abundantly clear that for the new company to succeed when it was capitalized at nearly four times its actual worth (according to the advertisement) and nearly seven times that worth (according to the figures of the Executive Committee), optimum business conditions had to exist within the company and in the national economy. There would have to be a long, uninterrupted period of prosperity. Both labor and plant facilities would have to combine to maintain continuous production at a capacity level. The products would have to be among the best and most advanced on the market. The officers and directors would have to be farsighted, prudent, economical and have a clear and consistent business philosophy.[16]

The Allis-Chalmers Company was formed during a period of economic prosperity which had been growing since 1897. Although President Charles Allis could issue a very rosy report in April, 1902, neither he nor the other directors knew that the momentum of the wave was nearly spent. Later that year the money market began to tighten at the same time that the stock market became choked with "undigested securities." All of this led to the stock market panic of 1903. The economically unsound basis for issuing stock and the unsettled economic conditions of the nation were reflected in the dividend payments by the Allis-Chalmers Company. From its inception the company paid quarterly dividends on its preferred stock of 1.75 percent through the quarter ending January 31, 1904. No dividends were ever paid on the common stock. From January, 1904, until after the reorganization of the company in 1913, no dividends were paid on either form of stock.

A slow recovery from the depths of the panic of 1903 was noticed in 1904, but in early 1905 conditions were sufficiently bad to warrant the closing of the South Foundry. At the Reliance Works foremen were laid off for the first time in many years. A Wall Street summary of the net assets of the company, 1902–1905, indicates something of the nature of the decline.

Year	Current Assets	Current Liabilities	Net Current Assets
1902	10,626,561	1,423,876	9,202,676
1903	9,235,388	1,364,064	7,781,342
1904	7,875,511	1,048,396	6,827,116
1905	7,623,986	1,080,950	6,643,086

But in the fall of 1905, the South Foundry was reopened with 500 men employed, while iron castings were produced at a record rate in the West Allis foundry. By mid-1906, the president was able to report that some business was being turned away "due to inability to make shipments required by the trade."[17]

Unhappily, the prosperity of 1906 was short-lived, for an acute banking panic occurred in 1907 and was followed by a general business depression. At the height of the panic, when the Westinghouse Company failed and the Pittsburgh Stock Exchange suspended operations, the Allis-Chalmers Company very nearly went under. A headline in a Milwaukee paper on August 16, 1907, read, "Dastardly Attack on Milwaukee Institution: Rumor of Receivership Circulated on New York Stock Exchange Meets with Prompt Response from Company." The Milwaukee Journal went on to record how Allis-Chalmers common stock dropped as low as 4 and the preferred to 15½. On the corresponding day a year earlier this stock stood at 17⅞ and 48. Walter H. Whiteside, then president, was perfectly correct in stating that the company had not asked for a receiver, but the public never knew that this had been a very close call.[18]

The company floated a $12,000,000 first mortgage bond issue in July of 1906, which was supposed to be a prosperous year. In November, E. D. Adams arranged a $5,000,000 short term loan through the Deutsche Bank of Berlin, and this loan was extended on March 7, 1907. By May 3 the Executive Committee strongly urged on the Finance Committee the "immediate necessity of arrangements to procure money to liquidate obligations of the Company now past due." They accomplished this through a $1,000,000 loan for six months from the Illinois Steel Company. In short, Allis-Chalmers was not in the best shape to weather the panic. "A prominent banker in Chicago" estimated that Allis-Chalmers credit was about forty-nine percent.[19]

The company faced its real test in December, when a number of its outstanding notes fell due, and others could not be renewed. Although Allis-Chalmers had weathered the financial difficulties of August, it had little credit with the banks and had to meet notes to the extent of nearly $2.5 million in December. Although the Eastern interests involved in the merger were responsible for much of the overcapitalization and subsequent mismanagement of the company, still they were the men who pulled the company through its December crisis. At a meeting of the directors on December 6, Edward D. Adams, Mark T. Cox, E. H. Gary, William A. Read and Cornelius Vanderbilt each delivered $200,000 face

Edwin Reynolds directed the construction of the new Allis-Chalmers West Allis plant in 1901.

value of the Allis-Chalmers ten to thirty year Sinking Fund Gold Bonds, issued in 1906, as collateral security for loans to be arranged principally with the First National Bank of Chicago. Adams, Gary, Read and Vanderbilt also each advanced the company $250,000 in cash, taking unsecured notes of the Allis-Chalmers Company in return. It was the Eastern interests under the leadership of Judge Gary, then chairman of the Executive Committee, rather than the Western interests who helped to save the company at that critical time.[20]

There was a gradual improvement the following year. Because the foundry at the new West Allis Works was then in full production, the old South Foundry built by E. P. Allis was no longer needed; it was sold in February to the Pfister & Vogel Leather Company for $67,380. In March there was an increase in sales, and for most of the remainder of the year the plants were reported running at 85 percent of capacity. Unfortunately this upswing was brief, for by December production was down to about 50 percent of capacity. In February of 1910, Adams, Gary and Vanderbilt were requested to extend their $250,000 loans, and generally depressed conditions continued through 1911.[21]

It is clear that the Allis-Chalmers Company had been launched in a period of great economic uncertainty. When economic stability and continuing economic growth were absolute requisites for success, these had not been forthcoming.

The new company gave up Edwin Reynolds's plan for three great new factories, one on each coast and one in Milwaukee, and concentrated all building efforts on developing the West Allis site. Reynolds was determined to avoid the inefficiency in most ex-

Construction of the pattern storage, pattern shop and foundry in 1901, viewed from Greenfield Avenue.

isting plants of the day (including the four plants in the merger) which resulted from the haphazard erection of buildings to meet current needs. He developed a basic plan for a scientifically designed factory. The buildings, from the office and engineering division to the shipping platform, were arranged so that the work, from blue print to finished product, moved in one direction—eastward. Irving Reynolds and C. Edwin Search went to Europe in 1901 to observe factory planning and production techniques, and their findings were incorporated into the final construction. The work on the West Allis site was pressed with vigor, and by August, 1901, men were working around the clock with the help of electric lights strung over the site. The builders, using the most modern equipment such as a "gigantic" two hundred fifty-one horsepower, cubic yard steam shovel "which tosses the earth about like some great giant," worked so rapidly that the foundry was able to pour the first heat in September, 1902. The total cost of initial construction amounted to about $3,000,000. It is doubtful whether any other plant in the United States was better designed, tooled and equipped for building heavy machinery.[22]

The economic upswing in 1905 so encouraged the directors that they authorized extensions at West Allis which more than doubled the floor space and cost an additional $3,500,000. By 1907 the West Allis Works provided 1,416,000 square feet of floor space on 113.22 acres. This modern plant produced Corliss, hoisting, and pumping engines as well as steam turbines, gas engines and electric generators. The new West Allis Works was in addition to a total of 1,456,500 square feet of factory floor space brought to Allis-Chalmers through the merger of the four original plants. The Gates

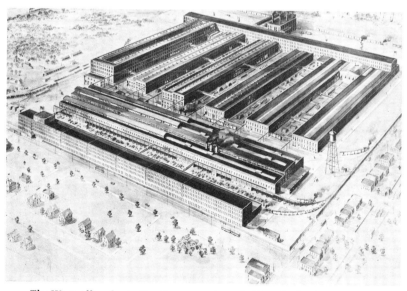

The West Allis plant after the expansion of 1906. The building at left, designed for pattern storage, was later remodeled for office space.

plant had 121,500 square feet of floor space set on nearly 5 acres, and was used in 1906 for the production of crushing and cement machinery. Fraser & Chalmers brought to the new company 375,000 square feet on 11.6 acres and produced the mining machinery. The Dickson Company had 240,000 square feet on 7 acres and produced tanks and sugar machinery. The Reliance Works, including the South Foundry, had 720,000 square feet on 18.21 acres, which were engaged in the production of saw and flour milling machinery and hydraulic turbines. The Bullock Works in Cincinnati, which had been acquired in 1904 so the Allis-Chalmers Company would have the facilities and experienced personnel to manufacture electric motors, generators and transformers, brought 327,000 additional square feet to the company and covered 18.5 acres. The total Allis-Chalmers factory space in 1907, then, was 3,119,500 square feet, covering a total of 173.45 acres. These facilities in the first decade of the twentieth century were considered enormous.[23]

The company, however, was robbed of the advantages it had hoped to obtain from the efficiency of the West Allis Works and the domination of the market seemingly promised by the total productive capacity. A *Sales Bulletin* of September, 1905, noted that "buildings have been completed for some time, but failure on the part of the manufacturers to deliver machine tools within specified dates has seriously delayed starting up the new shops." The situation was largely unchanged by June, 1907.[24]

If, during the most prosperous period of the decade, production was often held up because of lack of equipment, the fact remains

The erecting floor of the Bullock plant in Norwood, Ohio, showing motors and generators in process of manufacture.

that for the remainder of the period the Allis-Chalmers plants had been overexpanded. For the year ending June 30, 1909, the works were operating at about 50 percent of capacity. In the following year conditions were little better. The *Annual Report* of 1910 estimated that the works had been operating at about 60 percent of capacity. The plain fact was that with one of the finest and most efficient works in the land, the Allis-Chalmers Company had been unable to use it to advantage.[25]

The principal consideration of the Allis family and Edwin Reynolds in locating the new plant on the outskirts of Milwaukee had been to find a sparsely settled area where sizable profits could be made from the sale of land for homes and factories. In this they were most successful. From its incorporation as a village in 1902, when the census recorded a population of 1,080, to 1912, West Allis grew at a rate of almost 900 persons per year. The Central Land Improvement Company, which represented the Allis interests, had annual land sales ranging from $50,000 to $100,000 during the first decade of the century.[26]

Although the West Allis Works proved to be a model of industrial efficiency, the location plunged the company into a new set of problems that it had not been required to face in its established plants. Benjamin Warren, who succeeded Charles Allis as president of Allis-Chalmers in 1904, faced critical public reaction in both Milwaukee and West Allis. Hard feelings had developed between Allis-Chalmers and a number of Milwaukee and West Allis companies over discriminatory rate charges for the use of

Even in the early days of the electrical industry many women were employed in various manufacturing processes. These women are winding coils in the Norwood plant.

(Top) In 1894 Bullock was a pioneer in the application of motors to machine tools. (Center) This Card motor was built about 1890. (Bottom) In 1904 Bullock built this 8,000 hp frequency changer for Calumet Hecla Mining Co.

Benjamin Warren,
President 1904–1905.

Allis-Chalmers connections to the railroads. Tensions were eased only by the establishment of uniform charges. Other hard feelings stemmed from the involvement of some Allis-Chalmers men, including purchasing agent Edgar Dickson, in the organization of a lighting company which was accused of collecting "exorbitant charges for the public lighting of the village streets, etc." The Allis-Chalmers personnel involved were forced to publicly resign their connections with the lighting company. A "serious condition" had developed because of the financial practices of the Central Land Improvement Company and its known connection with the Allis family. Warren proposed to alleviate the situation by having Allis-Chalmers buy land for development from other companies, and thus publicly dissociate itself from the private interests of the Allis family. He also proposed that Arthur Warren, his brother and director of publicity for Allis-Chalmers, cultivate the Milwaukee and West Allis newspapers to attract as much favorable publicity as possible. The policy of rapprochement toward West Allis continued under the presidency of W. H. Whiteside when, in 1906, he was responsible for a $1,500 donation to the city to assist in the building of a public library and city hall. In the long run the company was able to rebuild its image in the area and play a leading part in further development.[27]

Building the new works west of Milwaukee caused numerous problems for the Allis-Chalmers Company, but the construction proved a stimulus for growth by bringing municipal services to the area far sooner than would otherwise have occurred. During the last months of 1901 the company worked with Milwaukee authorities to plan the extension of water mains to the West Allis Works. The officers of the company also worked with their counterparts from The Milwaukee Electric Railway and Light Company to bring the rapid transit system to West Allis. Until fares were reduced to five cents, the company, save during the recession of 1903, subsidized its employees by buying car tickets at seven and one-half cents each and selling them to the men at four cents. This was not simple philanthropy but necessity. The Allis family may have profited from speculation in land, but the entire area profited through the dispersion of industry and the extension of municipal facilities.[28]

On October 1, 1900, the four companies that were soon to become Allis-Chalmers employed 4,720 men. With the West Allis Works in production and the peak of prosperity for the company achieved during the latter part of 1906, the total number of employees increased to 8,493. Of the men employed in December, 1906, 450 were earning $100 per month or more, with the largest number of this group, 197 men, in sales. Those classed as "general" numbered 41, while 59 were in engineering and 153 in manufacturing. The old E. P. Allis Company was the only one of the four that had a mutual aid society. This had proven so successful over the years that in 1902 the Allis-Chalmers Company instituted a "Relief & Aid Association" on the old Allis principles in all the plants for the benefit of all personnel.[29]

Employer-employee groups such as the aid society had proven insufficient in 1886 to prevent workers from organizing and striking; they were even less effective after the turn of the century. The rapid industrialization of the United States in the latter half of the nineteenth century multiplied the number of factory workers more than ten times, a rate far too rapid for adequate social adjustment. Just as capital was increasingly combined at the turn of the century, a strong, nationally-organized labor movement arose swiftly. Between 1897 and 1904 the American Federation of Labor grew from 264,000 to 1,676,200 members, and the total trade-union membership from 447,000 to 2,072,700. The craft unions, where this growth was largely concentrated, were taking advantage of the wave of prosperity following the Spanish-American War. A study by the Commissioner of Labor revealed that the total number of

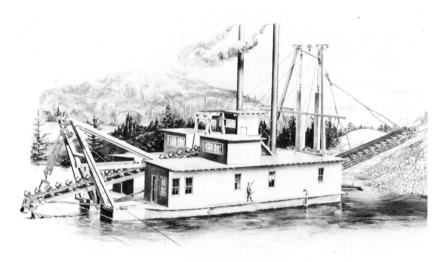

Extensive gold mining in Alaska in the early 1900s was aided by Allis-Chalmers dredges like this which scooped gold-bearing sand from river bottoms, processed it, and discharged the tailings.

strikes in the six years from 1893 to 1899 was 7,129, and that this more than doubled to 15,463 in the next six years, 1899–1904.

The organization of the Allis-Chalmers Company occurred in a period of increasing union activity. In fact, the first strike occurred almost immediately after the merger in 1901. On Saturday, May 18, the local union of the International Machinists' Association served very short notice on the Allis-Chalmers Company that, unless their demands for a 12.5 percent increase in wages and a reduction of working hours were granted, they would stop work the next

(Top) A cement mill of the early 1900s, showing in the foreground Gates type peripheral discharge ball mills and in the background tube mills for fine grinding. (Bottom) A group of vertical blowing engines providing air to blast furnaces in a steel mill.

Monday. The reaction of Edwin Reynolds was simple: "The men have taken the bit into their mouths, and I guess we shall have to let them chew it awhile." On Monday around 800 men, about 90 percent of the Allis-Chalmers machinists, walked out of the Reliance Works. "Passersby are now gazing in through the screened windows at the monster machines—all silent, looking hard and cold and there seems to be no activity around the shop inside or out." In a surprise move, fifty patternmakers walked out on May 25. At that point the company published a notice that all workers who failed to appear for work on Monday, June 10, would be "considered no longer in our employ." The central office of the International Machinists' Association declared that "We cheerfully pick up the gauntlet and hurl back the defiance."[30]

The nature and power of the machinists cast this strike in quite a different light from most previous actions. In 1898 the International Association of Machinists numbered 15,000 members, and this number shot up to 60,000 by 1900. The rapid growth of this craft union gave it an aggressiveness which led to its choice by the American Federation of Labor to head the drive for the nine-hour day. As a counterpoise to the new force of the machinists, their major employers had formed the National Metal Trades Association in 1899. Discussions between the employers' organization and the union had begun in 1900. Virtually all questions between the two groups were settled, save for the crucial matter of whether hourly wages would remain the same or be increased to compensate for the proposed nine-hour day. When both union and management stood firm, the strike idled some 50,000 machinists

The process of extracting gold from ore in the early 1900s required the use of vanners, as shown in this giant plant.

throughout the nation. Allis-Chalmers, along with other Milwaukee companies such as Bucyrus-Erie, was struck in 1901.[31]

Although the union swore that "we never will accept any modification of our demands or resume labor until the cause for which we struggle is triumphant and a shorter working day is an accomplished fact," the strike in Milwaukee ended in late July 1901 when union strike funds gave out. An editorial said that "Probably no strike of any account has ever been managed with less resort to intimidation or violence." But the situation was far different in the Allis-Chalmers plants in Chicago where the strike continued for fourteen months and was frequently marked by violence.[32]

To have the Allis-Chalmers Works in Chicago and Scranton strikebound and their production seriously impeded immediately following the merger was most serious. When there was no prospect of settlement by August 5, 1901, the Executive Committee of Allis-Chalmers decided to bring in strikebreakers to do the work of the striking machinists. Arrangements were made with the Pinkerton National Detective Agency to secure qualified machinists and bring them to Chicago at company expense. The company had offered thirty cents per hour in August and raised this to thirty-two and one-half cents in January, 1902. The arrangements with the Pinkerton Agency were unsatisfactory. On May 16, 1902, the manager of the Fraser & Chalmers plant in Chicago reported that of 176 machinists provided by the Pinkerton Agency from February 1 to May 6, 19 failed to report after their arrival in Chicago, 17 were

Copper converters like these refined copper by blowing large quantities of air through it.

discharged for being "utterly incompetent," 3 for continually absenting themselves from work, and 5 for devoting their time to attempts to unionize the shop. While one-quarter, 44 men, were discharged, 45 men quit work as soon as their contracts expired. In short, this source did not provide a competent and continuing work force. Production under these circumstances proved to be not only difficult but inordinately expensive.

The four companies involved in the Allis-Chalmers merger had been members of the National Metal Trades Association, which was supposed to assure solidarity during labor disputes. Unanimity of action did not take place and "after a full and complete discussion" the Executive Committee on January 31, 1902, "decided that the National Metal Trades' Association had not been of any benefit to the Allis-Chalmers Company in their labor difficulties." In disgust the Committee authorized the resignations of Edwin Reynolds as president of the Association, Irving Reynolds as chairman of the sixth district, and W. J. Chalmers as chairman of the fifth district.[33]

All Allis-Chalmers plants narrowly averted strikes in 1903. A strike by the molders and coremakers in 1906, however, did assume major proportions and was marked by considerable violence. The molders regarded themselves as powerful because they required a four-year apprenticeship to qualify for all types of work. The company countered with a radical new approach which ultimately helped break the strike. Allis-Chalmers discovered that it could take unskilled men from the streets and boys from high schools and with good foremen who thoroughly understood the business, teach the new men to make excellent castings of repeti-

The transporting of heavy machinery required the use of many horses at the West Allis Works.

tive types in no more than four months. This training for molders attracted wide attention and altered industrial approaches to training.[34]

The other important outcome from this strike rested in the legal fight between the striking molders and the Allis-Chalmers Company. After legal maneuvers by both sides which began on June 16, 1906, Judge Joseph V. Quarles granted an injunction on September 24 restraining the strikers from violence. The judge based his decision on the intensity of feeling which had sprung up between the parties, the extent of violence, and the system of picketing responsible for it. In October the company argued that the union should be punished for contempt in violating the terms of the injunction. On December 10, 1906, Judge Arthur L. Sanborn of the western district of Wisconsin held the officers of the union and the strike committee guilty as charged and sentenced six men to the county jail.

The legal battle in Milwaukee was closely watched by both manufacturers and union men across the land. Judge Sanborn finally issued a sweeping order in May 1907, prohibiting the strikers from compelling or inducing any of the Allis-Chalmers employees to refuse to work for the company and from preventing others from entering its employ. It also prohibited union men from congregating about the plant or upon its approaches. When the case finally reached the United States Circuit Court of Appeals in Chicago in the fall of 1908, some of the ablest lawyers in the country appeared: "Trust Buster" James M. Beck of New York and Max W. Babb for the company, attorneys Frederick N. Judson of St. Louis and William B. Rubin of Milwaukee for the union. The company was well satisfied with the decision upholding the injunction issued by Judge Quarles and sustained by Judge Sanborn. Babb claimed that the company had fought for two years to validate what it felt was a fundamental policy: the right of the company to manage its affairs without outside interference.[35]

Allis-Chalmers employees did make gains, however, during this period. In February, 1902, the company increased wages for molders to $2.50 for a ten-hour day, with time and a half for overtime. In March all workers received Saturday afternoons off with no reduction in pay, and by September the company was prepared to grant a nine-hour day. The comfort and convenience of the workers also received increasing attention. In 1905 the company built a suitable clubhouse at West Allis for the workers.[36]

As prosperity declined after the panic of 1907, union membership and the frequency of strikes throughout the country also dropped. Although there was little labor difficulty in the last years of the decade, major strikes had often held up production, decreased efficiency and increased costs during the period to 1907.[37]

At the turn of the century Americans were expansive and seemingly obsessed by the dimensions of their country, its enormous production, the size of its corporations, the extent and efficiency of such factories as the West Allis plant, and its great products. Under the guidance of Edwin Reynolds, Milwaukee had become the world center of giant engine production. By 1903 the Allis-Chalmers Company asserted that it had 6,000 Reynolds-Corliss engines at work in almost every part of the world. Their prestige was enormous.[38]

It is perhaps not surprising that Edwin Reynolds dreamed of building bigger and better engines in the great, new West Allis plant. As late as 1906 an advertisement said that the "specialty" of the company was "the celebrated Reynolds-Corliss engines, known all over the world for their durability, technical finish, economy and efficiency." But Edwin Reynolds represented past greatness, not the pattern of the future. Charles Parsons of England and others had sounded the death knell of the large reciprocating engine by inventing practical steam turbines in the early 1880s. By 1896, George Westinghouse had purchased the American rights from Parsons and adapted the steam turbine to the production of low-cost electricity.[39]

Edwin Reynolds, chief engineer of the Allis-Chalmers Company, was an engine builder and continued to be even after the steam turbine had cut into engine sales. A basic difference between the Eastern and Milwaukee interests in the company became apparent on this issue. On April 12, 1902, the Board of Directors passed a resolution affirming that, "it appears from many reports that have come to the knowledge of the members of this Board that gas engines and steam turbines have reached such a commercial stage of development in this and foreign countries, as

The company band played concerts in front of the clubhouse in 1909.

to justify this Company in preparing to manufacture the most efficient types of such machines." The board exercised leadership and power at this juncture through the New York banker, Edward Dean Adams, who was aware that the company had lagged behind others in technical developments. W. J. Chalmers supported Adams; by late 1902, he noted, General Electric had orders for Curtiss Turbines totaling 250,000 horsepower and Westinghouse had contracted for 175,000 horsepower of steam turbines, making the value of orders in hand of these two companies in steam turbines and generators $10,625,000. When the elder Reynolds did

(Top) In May of 1909 the West Allis plant was visited by German Count Von Bernsdorf and his party. (Bottom) In September of the same year the Honorary Commercial Commissioners of Japan also visited the plant.

not move rapidly enough to suit the board, it appointed him, early in 1903, to the new post of consulting engineer. Irving Reynolds, who succeeded his uncle as chief engineer, was apparently no better liked by the board as it cleaned house, and he left later the same year in April.[40]

The board signified the new order by naming E. D. Adams chairman of the Executive Committee on February 15, 1904. The board declared: "The great and growing changes in conditions applicable to the business of the Allis-Chalmers Company seem to make it absolutely necessary for the protection and promotion of its interests that the Company aggressively and promptly enter into various new and additional lines of business and also change some of the methods which have heretofore obtained." The election of Benjamin H. Warren as president of the Allis-Chalmers Company in 1904 was part of Adams's program to bring the company in line with the times. Charles Allis, who was replaced as president, represented the point of view of Edwin Reynolds. President Warren had worked with steam turbines at Westinghouse.[41]

As part of the new program, Allis-Chalmers joined the Steam Turbine Advisory Syndicate of England. This group was composed of the Yarrow Shipbuilding Company, Tweedie (Vulcan) Shipbuilding Company, Willans and Robinson, engineers and well-known engine builders, and former Parsons chief engineer, H. F. Fullagar. The license from the syndicate conceded to Allis-Chalmers all rights to manufacture steam turbines in the United States, Canada and Mexico, with equal rights and privileges in South America. During the summer of 1905 it was arranged to get turbine data from C. A. Parsons & Company directly and to acquire American rights under the later Parsons' patents.[42]

To obtain experienced personnel in the steam turbine field Allis-Chalmers hired Robert A. McKee away from the Westinghouse Company to head the steam turbine department. He assembled a well-trained group of steam turbine engineers, with the intent of placing an efficient steam turbine on the market as soon as possible. Hans Dahlstrand came to Milwaukee in 1904 from postgraduate work in Sweden, Arthur C. Flory came from Lehigh University in 1905, and James Wilson arrived in 1906 from London. Once these men began to operate as a team, employing the most advanced ideas on steam turbine design, this department developed amazingly quickly.[43]

The first Allis-Chalmers steam turbine built at West Allis was installed in the Williamsburg Power Plant of the Brooklyn Rapid Transit Company in 1905. It was of the Parsons horizontal type rated 5,500 kilowatts, 750-rpm. By early 1906, the company also had orders for steam turbines from The Milwaukee Electric Railway and Light Company, the Utica (New York) Gas & Electric Company, the New York Edison Company, the Memphis Consolidated Gas & Electric Company, and the Western Canada Cement and Coal Company of Calgary, along with half a dozen others.[44]

As early as June, 1905, the *Annual Report* noted that there had been some decline in the business of the steam engine department, "due largely to the introduction of steam-turbines, especially those of large capacity for which certain manufacturers had been preparing for several years." Steam turbines rapidly cut into Allis-Chalmers steam engine sales. Steam engines comprised 23 percent of the total orders on hand on June 30, 1906 and steam turbines only 8.5 percent. During April of 1907, however, the company booked orders of $273,352.00 in steam turbines compared to $265,154.11 in steam engines. Although Allis-Chalmers had "decided to recognize and meet the preference shown for the rotary over the reciprocating engine, and to bring the products of the Company fully up to the highest state of engineering development by preparing for the manufacture of steam turbines," this had entailed a large expenditure for expanding facilities, securing patent rights and obtaining personnel. In 1906 Edward Adams estimated that these expenditures were largely responsible for the $300,000 loss the company sustained that year.[45]

The resolution of the Board of Directors on April 23, 1902, which launched Allis-Chalmers into the steam turbine business, also directed the officers and engineers to look into gas engines. The gas engine had been operated in Europe since the 1880s, but engineers in the U.S. had paid little attention because the price of coal was so cheap. In the early twentieth century, however, as management became more conscious of profit through economy, the American manufacturer became increasingly receptive to devices eliminating or reducing waste. This shift in attitude coincided with a search for new product lines and led to a careful

The first of many Allis-Chalmers built steam turbine generator units. This 5,500 kilowatt unit was installed in the Williamsburg plant of the Brooklyn Rapid Transit Company in 1905.

investigation of different types of gas engines in Europe. Early in 1903 Edwin Reynolds announced that Allis-Chalmers had bought the patents of the Nuremberg Machine Company, Nuremberg, Germany, and had become the sole licensee of their gas engines in the United States, Canada and Mexico.[46]

At a time when production of giant steam engines was diminishing, the new West Allis works was especially designed for the manufacture of heavy machinery and was ideally suited to the production of the gas engine. The average gas engine frame weighed in excess of 90 tons, and the giant crankshaft, weighing more than 100 tons, dwarfed a man. The engineers and workmen, experienced in the building of the largest pumps and steam engines, swiftly redesigned the European plans to meet American needs. The Allis-Chalmers Company became by far the largest builder of gas engines in the United States.

By 1904 Allis-Chalmers was prepared to produce the horizontal type, double-acting, four-cycle gas engine as a tandem in units of 250 to 3,000 horsepower, and as a twin-tandem in units of 520 to 6,000 horsepower. Although the market for these engines was largely limited to industries wishing to make use of waste gas from blast furnaces, eighty-one units were produced between 1905 and 1930. Of these, fifty-one were produced between 1905 and 1912. The importance of the gas engines to Allis-Chalmers at this point is obvious because in June, 1906, they constituted 21 percent of the company's unfilled orders.[47]

The large volume of gas produced by the blast furnaces in steel mills can be used to power engines. These twin-tandem gas engines in the Illinois Steel Company are driving blowers which supply air to the blast furnaces.

The first Allis-Chalmers gas engine to be placed in service was a 1,000 horsepower unit installed in the Milwaukee works of the Illinois Steel Company. The last and also the largest of these engines (10,000 horsepower) was produced in 1930. The most impressive gas engine installation, however, was the result of a contract with the Indiana Steel Company for its immense Gary, Indiana, plant in 1906. The seventeen gas engines, rated at 4,000 horsepower each and installed in a single powerhouse 966 feet long and 105 feet wide, were double the size of any previous gas engines installed either in the United States or abroad. This first installation proved so successful and economical that another seventeen gas engines were purchased by Indiana Steel between 1910 and 1918. These engines, equipped with Allis-Chalmers generators, used waste gas to generate all the electrical power for the entire steel plant. In fact, until the heavy manufacturing demands of World War II, the plant also furnished power for the city of Gary.[48]

Expanding their new product lines in still another direction, Allis-Chalmers announced in February, 1904, that they were prepared to manufacture hydraulic turbines. For over a year the company had been working with Clemens Herschel, the hydraulic engineer who had developed the famous Holyoke testing flume for the Holyoke Water Power Company about 1880; he agreed to become the manager of the new department. Securing Herschel was considered something of a *coup*, and establishing a connection with the Swiss firm of Escher-Wyss still another, for Escher-Wyss

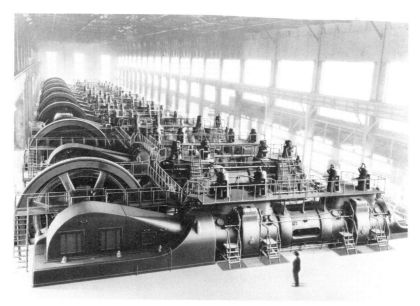

Gas engines utilizing blast furnace gas were also used to produce power for running the plants. These units are in the Illinois Steel Company plant in South Chicago. The largest units, in the foreground, are rated at 10,000 horsepower and have cylinders of 60 by 64 inches.

was one of the best-known turbine manufacturers in the world and had only recently installed equipment at Niagara Falls.[49]

The hydraulic turbine department was originally established in New York City, since Herschel lived there. Herschel made the first proposal drawings with a square ruler for lack of a T-square. The first hydraulic turbines were manufactured in the old Dickson plant at Scranton, Pennsylvania, with the first one going to the Rochester (New York) Railway and Light Company. This turbine, in continuous use until 1952, was a single horizontal plate steel, cylindrical casing, Francis type, rated at 1,250 horsepower under an 88 foot head at 360 rpm. Allis-Chalmers found it quite inefficient to have the drafting done in New York and the manufacturing at Scranton. In 1905 the entire hydraulic turbine department was moved to Milwaukee and installed in the old Reliance Works. Clemens Herschel became consulting engineer.[50]

Arnold Pfau, an Escher-Wyss engineer, was assigned the task of converting Swiss designs into American practice. He soon found himself "between the devil and the deep sea. My Swiss boss accused me of deviating from their original designs, and my American superiors expressed their disappointment at not being able to use their many Swiss drawings for turning out water wheels like making doughnuts." But within a year the engineers overcame the problems, and Allis-Chalmers adopted a policy of building each hydraulic turbine to suit its own application. The prestige of

The first Allis-Chalmers hydraulic turbine was installed in a plant of the Rochester Gas & Electric Company in 1904. It was rated at 1,250 horsepower with an 88-foot operating head.

the Allis name and word of high turbine performance brought an increasing number of orders to the company. By July 1906, no less than forty-three hydraulic turbines, aggregating 166,000 horsepower, were in various stages of construction.[51]

Since 1882 when Thomas Edison built the Pearl Street Station in New York City, electrical engineering had been expanding rapidly. The few tiny central power stations of the 1880s had grown to more than two thousand in 1898, and this number increased to more than five thousand by 1914. By 1905 nearly all new factories and shops of any size in the United States used electric drives extensively. Traditionally Allis-Chalmers had launched into new fields through the purchase of patents and the securing of licenses; electricity, however, presented a somewhat different problem from that of gas engines, steam turbines, and hydraulic turbines.

Allis-Chalmers was forced into the electrical field to maintain its existence. General Electric Company sold its dynamos and steam turbine units as complete generating plants which excluded the reciprocating engine to which Allis-Chalmers had hitched its star in the past. The *Electrical World and Engineer* reported that General Electric had rejected an Allis-Chalmers proposal to manufacture steam turbines and leave the electrical work to General Electric. Thus, Allis-Chalmers found itself cut off from a large portion of its work by General Electric as well as by Westinghouse, which was also manufacturing complete units.[52]

Allis-Chalmers therefore had no alternative but to enter electrical manufacturing. The new department secured the services of three key men from the Stanley Electric Company: John F. Kelly as head of the department, William Stanley, Jr. as consulting engineer, and John H. Kelman as department superintendent. Electrical experts could be brought together readily enough, but company leaders quickly realized that it might take up to two years to get into production.[53]

In searching for some way to more speedily enter the field, Allis-Chalmers turned to the Bullock Electric and Manufacturing Company of Cincinnati, Ohio, and made arrangements to lease it as an operating unit in 1904. This well-established firm had been founded in 1887 as the George F. Card Manufacturing Company of Cincinnati for the production of small electric motors. In 1890 the name was changed to the Card Electric Motor & Dynamo Company. A reorganization in 1897 created the Bullock Electric Manufacturing Company. In 1899 this growing electrical firm moved to the Cincinnati suburb of Norwood, Ohio. In 1898 Bullock built its first alternating current generators and in 1900 strengthened its position by securing from the Oerlikon Company of Switzerland patent rights to alternating current equipment.[54]

The E. P. Allis Company had worked with the Bullock Electric in the production of steam engine-driven generator units. This relationship was continued after the formation of the Allis-

SWEPT THE FIELD!
Three Grand Prizes
Highest Awards
at
World's Fair, St. Louis

ONE—for 5000 horse-power Allis-Chalmers Engine
ONE—for 5000 horse-power Bullock Generator
ONE—for Allis-Chalmers Mining Exhibit including Breakers and Hoist

The
Big
Reliable

The
Big
Reliable

Gold Medal
for
Bullock Multiple Voltage Balancing Sets
Silver Medal
for
Bullock Street Railway Apparatus
The Engineering Triumph of the World's Fair

Allis=Chalmers Co., Milwaukee, Wis.
The Bullock Electric Mfg. Co., Cincinnati

All the electricity for decorative use at the St. Louis World's Fair in 1904 was supplied by a Reynolds horizontal-vertical engine driving a Bullock generator. This advertisement was produced to impress the engineering world.

Chalmers Company. Bullock supplied the generator which was directly connected to the 5,000 horsepower "Big Reliable" at the St. Louis fair. When the new West Allis plant was being built, it was equipped throughout with Bullock electrical products. A union of these two well-established and reputable companies which had worked together so closely in the past was quite natural.

The Bullock Electric Manufacturing Company had a new plant for building the largest sizes of generators. Allis-Chalmers took over the entire business on March 1, 1904, as a "going concern," to become the electrical department of the Allis-Chalmers Company. By the terms of the twenty-five year lease, the Allis-Chalmers Company guaranteed to pay the 6 percent dividends on the $1,000,000 of preferred stock of the Bullock Company, and 6 percent on the $1,000,000 of common stock for all assets of the old Bullock Electric Company over and above the valuation of the preferred stock. Allis-Chalmers also took over Bullock, Ltd., a small plant in Montreal, Canada.[55]

The importance of this venture in the collective mind of the Board of Directors cannot be overestimated. Indeed, on June 1, 1904, the board voted to amend the Certificate of Incorporation to read "the name of the corporation is Allis-Chalmers Bullock Company." But by September, Edward Adams had cooled to this idea and requested that the change be postponed, and he eventually asked that it be dropped. The board had recognized that electricity was rapidly emerging as a prime source of power for the twentieth century and had incorporated electrical generators and motors into its growing list of products. They had saved the expense of a proposed $500,000 addition to the West Allis Works for the manufacture of electrical apparatus, but the company had incurred fixed costs in payment on the Bullock stocks which had to be deducted from the merger and from the sometimes non-existent profits during the coming years. These fixed costs were increased when in September, 1905, Allis-Chalmers took over complete control of the Bullock Company by buying out the original owners for $3,000,000 in capital stock. Another persistent liability throughout the decade was Allis-Chalmers-Bullock, Ltd. of Montreal, Canada. In 1904 this unit employed 175 men, producing principally Ingersoll Sergeant drills; it was in constant financial difficulty.[56]

The three years following the merger were devoted to an agonizing reappraisal of product lines and an attempt to provide a secure foundation for the future based on stable, profitable products. In 1904 Allis-Chalmers was finally in a position to issue its *Book of the Four Powers*, which introduced its new trademark, "Ours the Four Powers: Steam, Gas, Water, Electricity." In 1904 its manufacturing was conducted in six large plants in Milwaukee, Chicago, Scranton and Cincinnati, employing up to 10,000 men. No other company in the world was then in a position to design and build, under a single management and in its own shops, prime movers actuated by any and all of the four great powers for any condition of locality or circumstance.[57]

The electrical field seemed to be one of the most promising in the history of American industry. During the fifteen years after 1902, electrical energy for industrial use alone multiplied nearly sixteen times. Despite this opportunity, Allis-Chalmers faced most formidable competition. The two principal firms manufacturing electrical equipment in the late nineteenth century were General Electric, formed in 1892 by the merger of Edison General Electric Company and the Thompson-Houston Company, and the older Westinghouse Electric Company. To minimize the intense competition between them, the two firms entered into an agreement in 1897 which recognized the patents of each company and "the right, subject to certain exclusions, to a joint use thereof." This combination largely succeeded in preventing the rise of a general competition in electrical equipment.

While local newspapers viewed the entrance of Allis-Chalmers into the electrical field as part of a struggle between giants, this was not actually true. An unmarked company memorandum of about 1908 says that "Under the conditions extant before the recent business depression, General Electric, Westinghouse and Allis-Chalmers Companies together did electrical business to the extent of approximately $130,000,000 per annum, shared respectively in 54%, 44% and 2%."[58]

To compound the problem, a General Electric representative estimated that his company was in a position to secure about 50 percent of its business without competition, and without the expense of solicitation. In large part this was done through its holding companies. The Electric Bond and Share Company, the United Electric Securities Company, and the Electrical Securities Corporation had acquired, by 1913, a dominating interest in the local electric light plants in seventy-eight cities and the local gas companies in nineteen cities to the benefit of the sale of General Electric equipment. Since Westinghouse also had its affiliated financial associations, an unidentified Allis-Chalmers officer proposed that his company organize an engineering association which would select public service corporations for acquisition and reorganization, and would provide a non-competitive outlet for Allis-Chalmers electrical equipment and a financial organization to supply the necessary funds through the sale of bonds and securities. This proposal was never put into effect and, considering the financial condition of the company after 1907, there was no real possibility of it, leaving Allis-Chalmers competing with two giants on very uneven terms.[59]

Entering into competition with established companies which had great resources available sometimes posed legal problems even when the Allis-Chalmers claims were clearly valid. On July 28, 1910, Judge Ernest W. Bradford of the United States Circuit

Court for Eastern Pennsylvania signed a decree sustaining Patent Number 546,059 owned by the Allis-Chalmers Company, and holding that electric railways installed by the General Electric Company infringed on it. The decree directed the issuance of an injunction against further infringement and an accounting for profits and damages for past infringement. This patent had come to Allis-Chalmers with the purchase of the Bullock Electric Manufacturing Company, and the Company had been fighting this case through the courts since 1904. The major companies not only infringed on Allis-Chalmers patents but also brought numerous

Allis-Chalmers Co
MILWAUKEE WIS USA

Competitors Scoffed

When the greatest Engine Building and Mining Machinery makers in the country entered the Electrical Field

After that there were rumors of buying up, "amalgamating," and the rest of it,—rumors without a fraction of foundation in fact.

Then, when we won, at St. Louis Exposition, Electrical, Engine, and Mining Machinery Grand Prizes, and other high awards, the country took thought, for

Now these facts are clear: there is

Real Competition at Last

in the Electrical Industry

The market can no longer be controlled by one or two companies

This Company is equipped to Build

THE LARGEST GENERATORS
THE MOST POWERFUL MOTORS
THE LARGEST POWER UNITS

We have the Works   The Machinery—The Men
We have offices in most large cities   Six large works
Our Electrical Headquarters are at Cincinnati

ELECTRICAL DEPARTMENT
The Bullock Electric Mfg Co
CINCINNATI OHIO

Canadian representatives Allis-Chalmers-Bullock, Ltd., Montreal.

This advertisement showed the world that there were now three companies, not two, that could produce large electrical equipment.

suits against the company. In 1944 W. W. Nichols wrote, "Allis-Chalmers was a great sufferer . . . . for a long time after organizing its Electrical Department by the attacks of General Electric and Westinghouse Companies." It is not surprising that Nichols called these attacks a "conspiracy," for as early as 1905 there were pending fifty-one patent litigations involving the newly acquired electrical department. In 1910 General Electric alone had filed or was about to file some thirty claims against Allis-Chalmers. Obviously, Allis-Chalmers had to fight very hard to maintain its relatively small position in the market for electrical products.[60]

If it ever appeared that a conspiracy of forces operated to prevent the success of the Allis-Chalmers Company, the venture into the manufacture of air-brakes would seem to substantiate this appearance. The electrification of street railways and the development of the interurban system not only produced a revolution in American city life by relieving housing congestion and making suburban life possible, but also created an almost unprecedented source of business for some industries. The annual increase in track averaged well over two thousand miles from 1902 to 1907. By World War I there were 1,280 companies operating 40,000 miles of track and capitalized at $5,000,000,000. The electric cars, built high off the ground to allow room for the bulky motor control units underneath, served people in all climates and in all weather throughout the year. The Allis-Chalmers Company had been profiting from the business by supplying their great Reynolds-Corliss engines to many of the electric railway power stations. With the acquisition in 1904 of the Bullock Electric Manufacturing Company, Allis-Chalmers also began supplying electric motors for the trolley cars. To supply even more equipment for the thousands of cars should have been a most profitable venture.[61]

Niels A. Christensen, inventor of the air-brake for electric cars and founder of the Christensen Engineering Company, had sold his patents and assets to the National Electric Company. When this

An interurban car of the early 1900s. Note the high step for entering.

company went into bankruptcy through mismanagement, they forfeited the licenses for the Christensen air-brake patents. Although the records do not indicate the price paid by the Allis-Chalmers Company for the exclusive rights to the manufacture and sale of the Christensen air-brake, the *Milwaukee Journal* reported the amount as $1,500,000. On May 24, 1906, the company said that the significance of the acquisition could "hardly be overestimated," for "Christensen air-brakes are used on over 80% of the electric railways of this continent. Henceforth these systems can purchase Christensen brakes and repair parts, either for old or new installations, only from the Allis-Chalmers Company." Allis-Chalmers prepared extensive quarters for manufacturing the air-brakes in the Reliance Works and hired inventor Christensen as engineer in charge of the new department. The board of Allis-Chalmers was convinced that the possession of a superior product which was in great demand would insure success.[62]

Yet shortly after the railway compressor department had become a beehive of industry, sending compressors all over the United States, this line suddenly collapsed because of the introduction of the "hobble-skirt" in September, 1909. This new fashion, introduced in Paris, was completely incompatible with the old high street cars and interurban cars. As the style spread, women found it impossible to get in and out of the cars safely. The number of serious accidents rose, and public utility commissioners across the land demanded that the cars be lowered. The Allis-Chalmers boxed-in air compressors, however, hung too close to the ties already. Engineers made every effort to redesign the compressor to bring a smaller, lighter model into production. But street railways and interurban lines could usually buy new cars for less than the cost involved in lowering the old ones. General Electric and

A typical interurban railway company electric substation in 1907 showing a steam engine driven generator, a rotary converter to change alternating to direct current, two exciter sets to provide current to the coils of the generator, transformers and switchboard.

Westinghouse, the two principal competitors, saw their chance to squeeze Allis-Chalmers out of the electric railway business. They were willing to sell a complete motor control unit, including a new lightweight compressor, for little more than the cost of the motor control unit alone.[63]

The new Allis-Chalmers department became obsolete and its expert personnel virtually useless almost overnight. The fickle and irrational hand of fate had added a touch of wry humor to company affairs, turning a sizable investment into a liability.

The Allis-Chalmers Company had been organized on an unsound financial basis in a decade lacking in economic stability and plagued by work stoppages from union activity. Yet it had spent millions of dollars building a great new plant and enlarging old works. Technological changes compelled the company to revolutionize its product lines. It had been forced to spend millions of dollars for the licenses, franchises and plants needed to produce the new products. In fact, by 1908 the new products constituted about 65 percent of company sales. The many inevitable and pressing problems were made more acute through lack of sound management.[64]

The officers of the new Allis-Chalmers Company had been chosen because they held large blocks of stock in the company rather than for their managerial ability. The Allis interest was one of the largest in the new company, and the name represented continuity; Charles Allis was elected president and Will Allis chairman of the board. The former maintained consistently good relations with the workers—as a young man he used to play checkers in the foundry during the noon hour—but as noted earlier he did not inherit his father's business acumen. "Charlie" Allis was regarded by many as something of a playboy. The weaknesses of Charlie and Will Allis were nowhere more apparent than in the early board meetings in New York when Charlie had real difficulty arriving at decisions and Will consistently fell asleep.[65]

William J. Chalmers almost immediately incurred a virtually unparalleled degree of dislike and distrust in the new organization. His recommendations to the Executive Committee in 1901 indicate awareness of the inefficient operation of the plants involved in the merger and the desperate need to integrate the company into a working unit. Those affected, however, particularly the Milwaukee interests, felt that he was "out to get them." This feeling apparently filtered down to the shops where machinery sometimes came out stamped *All is Chalmers*. An even more intense rivalry and distrust existed between Chalmers and the management of the old Gates Iron Works. The growing success of the Gates concern had placed the Fraser & Chalmers Company on the brink of bankruptcy in 1901. One employee put the Chalmers and Gates relationship in an interesting light when he remarked, "When you put the two together it was like a parrot in a monkey cage." Some men left the company in protest. Resentment and distrust at higher levels stemmed in part from the fact that Chalmers was in sole charge of paying the executive salaries. So that no record of the fairly large salaries could be kept, he paid them by personal check and then drew the total amount to be deposited to his own account. His fall from power was anticipated as early as March, 1904, when the board cancelled his $10,000 annual "entertainment fund." On December 1, 1905, the board accepted Chalmers' resignation. Rumor had it that he was dismissed.[66]

During the years immediately following the merger, there was a notable lack of loyalty, considerable personal rivalry, and a general looseness of organization. As evidence of the first, on September 10, 1901, Max Pam, legal counsel for Allis-Chalmers, reported to the Executive Committee that a corporation was being formed by Benjamin T. Leuzarder, comptroller, William D. Gray, manager of the flour mill department, and Henry C. Holthoff, manager of the mining machinery department, and that $300,000 of the proposed $600,000 capital had been subscribed. The Executive Committee called the men in separately, and they admitted that Pam's statement was true, that they intended to divert orders from Allis-Chalmers and then leave the company. Leuzarder and Holthoff resigned on the spot. When Gray refused to resign, he was dismissed. In another instance, George M. Hinkley's successor as manager of the sawmill department, a man named Skeith, took with him "all the plans he could lay his hands on" when he left to join the Sumner Iron Works. The record of the board itself was not

When William Howard Taft was campaigning for the presidency in 1908, his train was run directly into the West Allis plant. He then addressed a meeting of the workers in one of the machine shops.

unblemished. Director Frank G. Bigelow, president of the First National Bank of Milwaukee, was relieved of his position on the board when Judge Joseph V. Quarles sentenced him to ten years in federal prison for misappropriation of $1,648,909.63 in funds from his bank.[67]

During almost the entire period from 1901 to 1911 there was a constant struggle for power and control between the Allis and Milwaukee interests and the New York financiers. By 1904 the Eastern interests had clearly gained control. Charles Allis was replaced as president by Benjamin H. Warren on March 9, 1904, and the New York banker, Edward D. Adams, was elected chairman of the Executive Committee. A graduate of the United States Naval Academy, Warren came to Allis-Chalmers from the Westinghouse Company to supervise the manufacture of the new product lines. He did his best to break the corporate ties with the past. The Edward P. Allis Works was once again the Reliance Works. The name of the West Allis Works remained the same, but the Gates Works became Chicago Works No. 1, and the Fraser & Chalmers Works became Chicago Works No. 2. The name of the Dickson Works was changed to the Scranton Works. In his personal relations, however, Warren was not the man for the job. One of the department managers commented that "he had all the arrogance of a commanding officer on a ship, but he was an ignoramus when it came to machinery." The affairs of the company did not go well. By late February, 1905, Judge Elbert H. Gary, as chairman of the board, inquired into President Warren's health. On March 3, Walter H. Whiteside, general manager of sales, was appointed vice president and general manager of the company and placed in full charge of operations. Late that month Warren, at the request of the board, left for Europe and on September 1, Whiteside was elected president. After Whiteside assumed the power, although not the title of president in late March, 1905, he ordered all offices brought to Milwaukee in the hope of securing greater economy and efficiency.[68]

Walter H. Whiteside also owed his position largely to the influence of Edward Dean Adams. Whiteside was another of the men referred to in Milwaukee as the "Westinghouse crowd," since he, too, had gained much of his electrical experience with that company. Coming to Milwaukee as manager of sales in July, 1904, he was soon elected vice-president and then president, serving in that capacity until January 1, 1911. A handsome, easy-going man, Whiteside brought into the company too many of his family and friends and was quite the reverse of Warren in developing a relaxed and congenial atmosphere. Perhaps the atmosphere was too relaxed, for old-timers recall that they could requisition virtually anything they wanted and get it. While business conditions for the company worsened after 1907, the president's letters give us an indication of him as an administrator. They deal with little things while the company was rapidly approaching receivership.

Walter H. Whiteside,
President, 1905–1911.

He cautioned the employees about extravagance in the use of electric lights, personally checked expense accounts for "extraordinary items," ordered the office boys not to throw snowballs, and as president cautioned all officers and department managers not to allow employees to misuse envelopes, carbon paper, rubber bands, pencils and erasers. Small matters seem to have been his major concern.[69]

E. D. Adams, chairman of the powerful Executive Committee, in 1906 proposed a bonus system "for purposes of stimulating and encouraging the efforts of the officers to increase the business of the company." The plan proposed a distribution of from $25,000 on net earnings of $851,000, running up to $150,000 on net earnings of $3,000,000, half to be given to President Whiteside and the other half divided among W. W. Nichols, vice president and secretary, Henry Woodland, treasurer, and Lahman F. Bower, comptroller. When this plan did not produce the desired results more generous rewards were offered in 1909, and the managers and heads of the various departments were included to the extent of three-fifths of the hoped-for amount. Unhappily, Allis-Chalmers was too far gone for such incentives to produce the desired results.[70]

While President Whiteside concerned himself with matters of small economy within the several plants and the directors attempted to stimulate the management to greater efforts, the struggle between the Eastern and Western interests continued. In April, 1907, the Executive Committee was concentrated in Milwaukee with Charles Allis as chairman, and including Whiteside, Bower and Woodland. But this group, representing the Western interests, was unable to contend with the financial panic and on August 29, Judge Elbert Gary, chairman of the board, was also elected chairman of the Executive Committee and the Finance Committee. On September 12, the Finance and Executive Committees were discontinued, and virtually complete power was vested in Gary. Three years later the Finance Committee was reconstituted with specific provision for the inclusion of "Western Directors in order to secure a greater distribution of responsibility."

Hope was expressed by the Milwaukee papers that men with such important connections and great personal prestige as Gary, head of the United States Steel Corporation, Mark T. Cox, a banker connected with Robert Winthrop and Company, William A. Read, head of the banking firm of William A. Read and Company, James Stillman, of the National City Bank of New York, and Cornelius Vanderbilt, of the family of New York capitalists, would direct the company in brilliant fashion. Unfortunately, prestige did not insure profits. Judge Gary set the pattern for the Eastern directors when on October 6, 1911, he resigned as chairman of the board and from the board itself "in view of the prior claims of other large interests with which he had long been connected now requiring his undivided attention." Others followed.[71]

In 1909 a reporter noted that "the policies of the company are determined in eastern financial circles, among men in touch with affairs the country over. Their orders are therefore an index of national conditions." This was largely true. Such control, however, produced neither corporate health nor profits. There was a most unfortunate by-product of this as well. Throughout the period 1901 to 1911 there was pride in the community that Allis-Chalmers was a great industrial institution directed by famous men. But the company and its leaders did not reciprocate. Never before or since was there so little contact with or concern for the community and the state by the company.[72]

The direct effect of the division among the directors and lack of executive leadership was immediately visible at the departmental level. The hydraulic turbine department during the years 1904–1911 is a case in point. The rapid succession of department managers is quite amazing.

Clemens Herschel	January 1904 to June 1905
W. S. Doran	June 1905 to Fall of 1905
O. O. Stranahan	Fall of 1905 to February of 1906
Will J. Sando	February 1907 to May 1910
E. F. Cassel	May of 1910 to Fall of 1910
L. F. Bower	Fall of 1910 to Spring of 1911
A. H. Whiteside	Spring of 1911–1940
*W. M. White	Spring of 1911–1940
*Manager and Chief Engineer	

The hydraulic turbine department survived the succession of managers and found an able leader in Dr. William M. White. That eleven hydraulic turbines of record size were successfully installed in a period when there was no consistent leadership and when management had no sound and consistent philosophy is little less than miraculous.[73]

The extent to which the customer may have suffered during this period is difficult to determine with any precision, but some indications do exist. In 1906 President Whiteside sent out a circular letter which noted that a number of instances had come to his attention where "caustic and unbusinesslike replies, and other cases where no replies whatever, or where answers entirely inadequate, were made to communications sent us by customers and by prospective customers." A year later William S. Heger wrote first vice president Lahman F. Bower concerning a shipment of Allis-Chalmers display materials sent from the West Allis Works for the Jamestown Exhibition. The materials had been shipped part of the distance on an open flat car without protection and had encountered a rain storm. Heger concluded his second letter to Bower on a plaintive note, "The above cases naturally raise the question, how do shipments fare that we do not see and that go direct to the customer, when out of two shipments to ourselves both are ruined." The Board of Directors faced this question in June of 1911, when a list was submitted to them of 128 cases of defective machinery "installed during recent years, the correction of which it is conservatively estimated will necessitate an expenditure of probably more than $227,000."[74]

The company was placed in an increasingly difficult position by a wide variety of forces and factors. Finances, however, sent it into receivership. Despite deceptively optimistic *Annual Reports*, a special meeting of the stockholders was called for July 16, 1906, to authorize an issue of bonds not to exceed $15,000,000. Of this amount, only $12,000,000 was actually issued, the remainder being held in reserve. The proceeds of the bonds were "to be issued to reimburse the treasury of the company for expenditures made and to be made in connection with the extension and equipment of the works at West Allis, and purchase and extension of the Bullock Works and their equipment at Norwood, Ohio, and to increase the working capital of the company." The specific security of the bonds was "a First Mortgage upon the real estate and manufacturing plants owned by the company, including new productive property acquired, buildings erected and machinery installed, since its organization." These 5 percent bonds were offered first to the stockholders at 80 percent of par value, but they took only 902,000 bonds. Shearson, Hammill and Company, acting in behalf of a syndicate, underwrote 9,648,000 bonds at 75 percent of par value. Most of these were never distributed. An option for the remaining 1,450,000 bonds of the $12,000,000 issue was never exercised by the syndicate. Because temporary loans took up $6,460,000 of the proceeds and the West Allis extension most of the remainder, the purchase of the Bullock Company could not be made.[75]

In the unstable economic period following 1907 the condition of the company worsened, and on January 1, 1911, W. H. Whiteside resigned as president and was replaced by Delmar W. Call. Primarily a railroad man with experience on the Erie road, President Call asked for economy of operation within the company. But this was not enough. Anticipating the failure of the company to pay the interest due January 1, 1912, on the First Mortgage Bonds, a bondholders' committee began to draw up a plan of reorganization. The Plan and Agreement of Reorganization dated March 18, 1912, was deposited with the Central Trust Company of New York. The bondholders clearly stated the issue as they saw it: "The purpose of the bondholders Committee has been to preserve the preferential rights of the bondholders and to secure the necessary new funds and working capital by means of assessments on the stockholders." They were convinced that their interests could be best protected by a reorganization of the company rather than by foreclosure.[76]

The reorganization plan was relatively simple and distinctly favorable to the bondholders who were entitled to:

(1) the amount of the coupons due January 1, 1912, in cash, (2) the par amount of their bonds in preferred stock of the new company at par, and (3) in addition 35 per cent. of the par amount of their bonds in the common stock of the new company. This amount of the new common stock is allotted to bondholders for the release of the mortgage lien and fixed charge.

Every preferred stockholder who paid a cash assessment of 20 percent on his stock would receive $20 in preferred stock and $90 in the common stock of the new company at par. For every $100 of common stock deposited, on which a cash assessment of 10 percent was paid, the stockholder would receive $10 in the preferred stock and $35 in the common stock of the new company at par. This simply meant that a new company would be organized with an authorized capitalization of $16,500,000 par value of 7 percent cumulative stock and $26,000,000 par value common stock. The capitalization of the old company, including the outstanding First Mortgage 5 percent bonds, amounted to over $47,000,000. This was reduced by nearly $10,000,000 as $5,192,000 par value of the new preferred stock was to represent the payment of an equal amount in cash to provide the badly needed working capital. Dividends on the new preferred stock were planned more realistically than under the old plan of merger, for they were to be cumulative from January 1, 1913 at 5 percent per annum, from January 1, 1915 at 7 percent, and from January 1, 1917 at 7 percent, but entitled to 7 percent from the beginning if earned by the new company and declared by its directors.[77]

Circular Letter No. 1, dated April 8, 1912, stated that Delmar W. Call and Otto H. Falk had been appointed receivers. Call soon resigned, and it became clear that a new name—that of General Otto H. Falk—had entered the history of Allis-Chalmers. A new and better day in company history was about to begin.[78]

NOTES TO CHAPTER FIVE

1 Bayrd Still, *Milwaukee: The History of A City* (Madison, 1948), p. 337.
2 *Milwaukee Journal*, November 5, 1900. Bruno Nordberg, a graduate of the Polytechnic Institute in Helsinki, Finland, came to the United States in 1880 and was one of the few trained engineers in the Allis Company during the decade of the eighties. He left the Allis Company in 1889 to found the Nordberg Manufacturing Company. *Dictionary of Wisconsin Biography* (Madison, 1960), p. 268.
3 J. H. Burbach, *West Allis* (Milwaukee, 1912), pp. 19–23; *Minute Book*, E. P. Allis Co., April 26, November 30, 1900. Irving Reynolds and others suggest that there was a move on foot to name the new community "Reynoldsville," but this was scotched by the Allis family. The name "West Allis" resulted from the fact that the Chicago and North Western and the Chicago, Milwaukee and St. Paul roads had changed the name of the National Depot, on the east side of the city, to Allis as a compliment to the company in April, 1900. By 1912 West Allis had sixteen manufacturing companies besides the Allis-Chalmers Company.
4 John Moody, *The Truth About the Trusts* (New York, 1904), p. 488.
5 *Milwaukee Journal*, April 8, 30, 1901.
6 Irving Reynolds to Alberta J. Price, August 31, 1945.
7 *Milwaukee Journal*, May 3, 1901.
8 *Minute Book*, E. P. Allis Co., May 9, 1901; Minutes, Meetings Board of Directors, Allis-Chalmers Company, (Hereafter referred to as A-C Co. Board of Directors) May 9, 16, 1901. Actually Stuart Lyman of New York was elected the first president of the Allis-Chalmers Company, John F. Charlton, secretary, and Henry S. Wardner, treasurer, on May 8, 1901. These men were no more than corporate functionaries, however, and resigned their offices when the Board of Directors was formally organized on May 16, 1901, and duly elected the officers of the company.
9 *Minute Book*, E. P. Allis Co., May 9, 1901; W. E. Hawkinson, *Organization of Allis-Chalmers Company*, MS. dated March 17, 1939. Hawkinson for many years was treasurer of the Allis-Chalmers Manufacturing Company.
10 *Minute Book*, E. P. Allis Co., May 9, 1901; *Milwaukee Journal*, May 17, 1901.
11 *Milwaukee Journal*, May 8, 1901. Earlier, on May 3, he had said, "The new combination is not a trust. . . . but an organization the purpose of which is to benefit all concerned. . . . It means great activity in the machinery-making world and employment for thousands of men."
12 *Milwaukee Journal*, May 9, 1901.
13 *Minute Book*, E. P. Allis Co., May 9, June 25, August 12, September 9, October 10, 28, November 11, 1901, February 1, 18, March 9, August 4, November 3, 1902. The 53,750 share of preferred stock in the Allis-Chalmers Company held by the Edward P. Allis Company were divided in the following way:

William W. Allis	10,624 shares
Charles Allis	9,968
Edward P. Allis, Jr.	9,562
Estate of Ernest Allis, Dec'd	6,746
Louis Allis	5,329
Gilbert Allis	4,252
Frank W. Allis	3,457
Mrs. Margaret W. Allis	2,312
The Milwaukee Trust Company	1,500

Minute Book, E. P. Allis Co., May 16, 1903.
14 C. R. Beck to author, July 24, 1959; Charles Allendorf to author, August 10, 1959; Herman Schifflin to Alberta J. Price, August 9, 1945; *History of Fraser & Chalmers*, MS.
15 Irving Reynolds to Alberta J. Price, July 31, 1945.
16 *Milwaukee Journal*, May 17, 1901; A-C Co. Executive Committee, August 12, 1901. It is interesting to note in passing that the United States Steel Corporation, which was organized at almost the same time as the Allis-Chalmers Company, did finally succeed, but only after passing through some difficult years and interrupting payment of its dividends on common stock. Yet United States Steel was capitalized at only twice its actual worth.
17 A-C Board of Directors, April 23, 1902. The recommendations for dividends on the preferred stock had all carried unanimously until the meeting of the Board on January 15, 1904. At that time the board split 5 to 4 on the question of a dividend for the quarter ending January 31. Adams, Allis, Chalmers, Eckels, and

Gary voted for it; Gates, Hoyt, Read and Vanderbilt voted "No." *Milwaukee Journal*, August 19, November 25, 1905; *Special Meeting of Stockholders of the Allis-Chalmers Company*, July 16, 1906.

18 *Milwaukee Journal*, August 16, 1907. In retrospect, the *Journal* made an interesting and erroneous appraisal of the condition of the company on August 17, 1907. "Everyone who is informed as to the situation speaks in the most encouraging terms of the way the company has worked out under President Whiteside. It is well known that it has finally reached the turning point in its career against odds that seemed almost insurmountable. It is now in better working condition than ever before in its history, and what is more, it has plenty of work to do at prices that are worth while."

19 A-C Co. Executive Committee, November 15, 1906, May 3, 9, 1907; A-C Co. Board of Directors, December 6, 1907.

20 A-C Co. Board of Directors, Dec. 6, 1907.

21 A-C Co. Board of Directors, November 14, 1907, February 4, 6, March 5, December 3, 1908; *Milwaukee Journal*, April 10, 1909.

22 *Milwaukee Journal*, August 24, November 7, 1901; *Second Annual Report* for year ending April 30, 1902, p. 9; "The Great West-Allis Plant of the Allis-Chalmers Company, Milwaukee, Wis.," *Machinery*, February, 1903, pp. 285–6; A-C Co. Executive Committee, September 3, 1901; A-C Co. Board of Directors, April 23, 1902; A-C Co. Special Meeting of Board of Directors, June 8, 1903. The West Allis Works was for some time a model for new factory construction in the United States. In *Industrial Progress*, September, 1909, p. 530, Henry T. Noyes, Jr., Secretary, German-American Button Company, Rochester, New York, in an article, "The Planning of a New Manufacturing Plant," mentions that the plan for their new plant was based on the Reynolds plan at West Allis; *Milwaukee Leader*, August 31, 1929; *Chicago Journal of Commerce*, March 27, 1930.

23 A-C Co. Executive Committee, June 2, October 25, 1905; *Special Meeting of Stockholders*, July 16, 1906.

24 A-C Co. Board of Directors, December 1, 1905, June 27, 1910; *Sales Bulletin*, July, 1906, p. 1, September, 1906, p. 1; *Eighth Annual Report*, for year ending June 30, 1909, p. 7; *Ninth Annual Report*, for year ending June 30, 1910, p. 8.

25 *Eighth Annual Report*, for year ending June 30, 1909, p. 7; *Tenth Annual Report*, for year ending June 30, 1911, p. 8.

26 Irving H. Reynolds to Alberta J. Price, August 31, 1945; *Milwaukee Journal*, January 27, 1902, April 13, 1907; Burback, pp. 41–43.

27 A-C Co. Executive Committee, November 23, 1904, February 1, 1906.

28 A-C Co. Executive Committee, March 24, 1903, December 3, 1901; A-C Co. Board of Directors, October 15, 1903; A-C Co. Special Meeting of Board of Directors, June 8, 1903. Bayrd Still in *Milwaukee* (pp. 307–310, 373–375) details the problems of development and regulations of the Milwaukee street railway system during this period.

29 A-C Co. Board of Directors, October 15, 1901; A-C Co. Executive Committee, October 28, 1902, October 11, 1906, March 14, 1907; *Milwaukee Journal*, November 22, 1902. In the first decade of the twentieth century the attitude of management had become more impersonal. The company supported the aid society but the officers no longer attended such functions as the Thanksgiving Ball, as E. P. Allis and his family had once done.

30 *Milwaukee Journal*, May 18, 20, 25, 31, 1901. For a general discussion of the rise of labor during this period see Thomas C. Cochran and William Miller. *The Age of Enterprise* (New York, 1956), pp. 228–248; Harold U. Faulkner, *The Decline of Laissez Faire, 1897–1917* (New York, 1951), pp. 289–291; George E. Mowry, *The Era of Theodore Roosevelt, 1900–1912* (New York, 1958), pp. 6–12.

31 Selig Perlman and Philip Taft, *History of Labor in the United States, 1896–1932* (New York, 1935), 4:115. For a discussion of the strike in relation to the Bucyrus-Erie Company of South Milwaukee, see Harold F. Williamson and Kenneth H. Myers II, *Designed for Digging* (Evanston, 1955), pp. 62–63.

32 *Milwaukee Journal*, June 8, July 26, October 18, 1901, January 24, 1902, September 23, 1903.

33 A-C Co. Executive Committee, August 5, 1901, January 28, 31, May 16, 1902.

34 A-C Co. Board of Directors, August 6, 1903; A-C Co. Special Meeting of Board of Directors, June 8, 1903; A-C Co. Executive Committee, May 24, 1906; *Milwaukee Journal*, October 27, 1906.

35 *Milwaukee Journal*, January 4, May 20, 1907, October 10, 1908.

36 A-C Co. Executive Committee, February 4, March 11, September 9, 1902, September 6, 1905.

37 A-C Co. Special Meeting of Board of Directors, June 8, 1903; *Second Annual Report* for year ending April 30, 1903, p. 9.

38 M. C. Maloney in a summary of "A-C Engines by Industry and Use—1881 to 1914," suggests that the number 6,000 is misleading since it includes the units of compound engines. The number of separate engines was somewhat less.

39 *Milwaukee Journal*, February 17, 1906; *Sales Bulletin*, January, 1907, section 19, p. 1.

40 A-C Co. Executive Committee, December 16, 1902; A-C Co. Board of Directors, April 23, July 24, 1902; A-C Co. Special Meeting of Board of Directors, October 25, 1902; Irving H. Reynolds to Alberta J. Price, August 31, 1945. For development of Parsons and Parsons-Westinghouse steam turbines see Henry G. Prout, *"A Life of George Westinghouse"* (New York, 1922), pp. 184–190.

41 A-C Co. Board of Directors, February 15, 1904.

42 Adams first contacted the Escher-Wyss company in relation to control of the Zoelly patents, but decided that these were not practicable. A-C Co. Executive Committee, September 30, 1902, January 6, May 2, May 31, 1904. Chalmers announced the entrance of Allis-Chalmers into the turbine field in a statement to *Electrical World and Engineer*, (1904), Vol. XLII, No. 9, p. 410. All steam turbine patents held by Max Patitz were purchased by the company for $10,000; A-C Co. Exec. Comm., March 28, 1904. Arrangements were concluded for the use of certain Parson patents by the Executive Committee on October 15, 1903 and June 28, 1905.

43 *Sales Bulletin*, November, 1906, Section 17, p. 1, September, 1915, p. 11; "Autobiography of Hans P. Dahlstrand," Allis-Chalmers files.

44 *Sales Bulletin*, November, 1905, p. 5; *Milwaukee Journal*, March 5, 1906. Although it could not be proved conclusively, all evidence pointed to Westinghouse employees as those involved in the sabotage of the Williamsburg turbine; A-C Co. Executive Committee, May 3, 1906.

45 *Fourth Annual Report*, for year ending June 30, 1905, p. 9; *Special Meeting of Stockholders*, July 16, 1906; A-C Co. Executive Committee, August 15, 1906, April 29, 1907; A-C Co. Board of Directors, June 30, 1909.

46 A-C Co. Board of Directors, April 23, July 24, 1902; *The "Nürnberg" Gas Engine*, Allis-Chalmers Co., Chicago, n.d. pp. 7–8; *Milwaukee Journal*, May 1, 1903.

47 *Industrial Progress*, March, 1909, p. 147, June, 1909, p. 335; *The A-C News*, October, 1945, p. 2; A-C Co. Executive Committee, August 15, 1906.

48 *Industrial Progress*, January 1909, pp. 13–14, May, 1910, p. 1048; *Gas Engines*, A-C Bulletin No. 1535, September, 1915, *passim*. During a modernization and improvement program at the Gary plant following World War II these giant engines, having run nearly forty years, were completely overhauled. For a more detailed but undocumented account of the gas engine, see Walter F. Peterson, "Built to Last—The Allis-Chalmers Gas Engine," *Wisconsin Academy Review*, Spring, 1961, pp. 63–66.

49 *Electrical World*, February 17, 1904, p. 410; Edward Dean Adams, "Early Hydraulic-Turbine History," *Mechanical Engineering*, April, 1930, p. 395; A-C Co. Board of Directors, October 15, 1903.

50 *Sales Bulletin*, July, 1905, pp. 1–2, January, 1930, p. 703; *Circular Letter* No. 52, June 13, 1905; W. H. Whiteside to O. A. Stranahan, June 2, 1905.

51 *Milwaukee Journal*, July 10, 1906; *Sales Bulletin*, January, 1930, p. 703.

52 *Electrical World and Engineer*, February 13, 1904, p. 304. Allis-Chalmers had considered entering the electrical field as early as 1902 when a committee of the board composed of E. H. Gary, E. D. Adams and Edwin Reynolds was appointed to consider the matter. As a result of the work of this committee, negotiations were entered into with the Crocker-Wheeler Company to manufacture the Brown-Boveri Generator of West Allis, but no agreement was achieved. A-C Co. Board of Directors, January 23, 1902, October 15, 1903; A-C Co. Executive Committee, February 11, 1902.

53 *Electrical World and Engineer*, February 13, 1904, p. 304.

54 *Milwaukee Journal*, March 7, 1904; "Brief Description of Bullock Electric Manufacturing Company's Factory," n.d., n.p.

55 *Third Annual Report* for the Year Ending April 30, 1904, p. 9; A-C Co. Board of Directors, March 9, 1904; *Electrical World and Engineer*, April, 1904, p. 510.

56 A-C Co. Board of Directors, June 1, 1904, June 1, 1906, October 2, 1907, October 1, 1908; A-C Co. Executive Committee, July 13, August 24, 1904. In the Board of Directors meeting of March 16, 1905, "Mr. Bullock presented, for himself and associates interested in the lessor company, certain criticisms concerning the

management." The Board may have bought out Bullock to stop such criticism.

57 *The Book of the Four Powers*, Allis-Chalmers Company (Chicago, 1904), p. 5; *Milwaukee Journal*, August 5, 1905; Unmarked, undated memorandum, c. 1908, p. 3., A-C Mfg. Co.

58 Edward C. Kirkland, *A History of American Economic Life* (New York, 1951), p. 412; Unmarked, undated memorandum, c. 1908, p. 1; John T. Broderick, *Forty Years with General Electric* (Albany, 1929), pp. 14–35, *passim*.

59 Unmarked, undated memorandum, c. 1908, pp. 1–4; General Electric and Westinghouse, the giants of the industry, were the primary source of problems for Allis-Chalmers as a new and small producer of electrical equipment. By contrast, studies of the two major companies totally ignore Allis-Chalmers as a producer of electrical equipment. See John Winthrop Hammond, *Men and Volts, the Story of General Electric* (New York, 1941), and Henry G. Prout, *A Life of George Westinghouse.*

60 A-C Co. Executive Committee, December 20, 1905, February 15, 1906; *Milwaukee Sentinel*, July 29, 1910.

61 Cochran and Miller, p. 31.

62 A-C Co. Executive Committee, April 26, May 9, 1906; *Milwaukee Journal*, May 24, July 16, 1906.

63 *Sales Bulletin*, January 1934, pp. 4–5; Speech by Arch J. Cooper, A-C New York Office to Newark Kiwanis Club. For a slightly expanded but undocumented treatment of this matter, see Walter F. Peterson, "The Whim of Fashion: Allis-Chalmers and the 'Hobble Skirt,' " *Tradition*, December, 1961, pp. 13–16. Mabel E. Brooks ingeniously lampooned the "hobble skirt" in delightful fashion; see Mark Sullivan, *Our Times* (New York, 1932), 4:550.

64 *Milwaukee Journal*, October 9, 1908.

65 E. C. Shaw to Alberta J. Price, August 23, 1954; Irving H. Reynolds to Alberta J. Price, August 31, 1945; C. R. Beck to author, July 24, 1959. On January 23, 1902, the Board of Directors accepted "with regret" the resignation of Will Allis. Elbert H. Gary was thereupon elected chairman of the board.

66 A-C Co. Executive Committee, June 21, 1901; A-C Co. Board of Directors, March 9, 1904, December 1, 1905; Irving H. Reynolds to Alberta J. Price, August 31, 1945; J. M. J. Keogh to Alberta J. Price, July 27, 1945; Herman Schifflin to Alberta J. Price, August 9, 1945. Milwaukee employees asserted that Chalmers maintained a small staff of personal informers, and that he used his position on the board to reduce the influence of the Allis family and the old Allis men. Chalmers maintained that he was trying to find out what was going on in the company with the aim of bringing efficiency into the operation.

67 A-C Co. Executive Committee, September 10, 1901; *Milwaukee Journal*, May 17, September 12, 1901, May 1, 1905; Ernest Shaw to author, August 3, 1959. The Gray, Holthoff, Leuzarder company became the Power Mining and Machinery Company.

68 It is interesting to note that two years before his appointment as president, Warren had applied for the position of works manager and after an interview had been rejected by Reynolds and others as unsuitable. Irving H. Reynolds to Alberta J. Price, August 31, 1945; Herman Schifflin to Alberta J. Price, August 9, 1945; *Circular Letter*, No. 7, June 3, 1904; A-C Co. Executive Committee, March 9, May 31, September 14, 1904, February 28, March 3, 15, 17, 31, 1905; A-C Co. Board of Directors, March 9, 1904, September 1, 1905; B. H. Warren biographical file, A-C Co. files. In "Allis-Chalmers: 'America's Krupp' ", *Fortune*, May 1939, p. 54, there is the statement, "During its first twelve years the company had not much business but a lot of Presidents, one of whom had to be cashiered with a year's salary when he insisted on building up a colossal inventory of small electrical equipment that was sure to become obsolete before it could be sold." The first half of the statement is quite correct, but the author finds no support for the latter half.

69 Obituary of E. D. Adams in *Electrical World*, May 30, 1931, clipping in A-C biographical files; Obituary of W. H. Whiteside in *Sales Bulletin*, April to July, 1931, p. 1052; *Milwaukee Journal*, September 7, 1905; C. R. Beck to author, July 24, 1959; *Circular Letter* No. 68, April 23, 1906; Letters of W. H. Whiteside, August 23, 1907, December 18, 1908, July 22, 1909, March 19, April 15, 1910.

70 A-C Co. Executive Committee, March 22, 29, 1905, September 20, 1906; A-C Co. Board of Directors, November 4, December 2, 1909.

71 A-C Co. Executive Committee, April 4, August 29, 1907; A-C Co. Board of Directors, August 29, September 12, 1907, October 6, 1910, October 6, 1911;

Milwaukee Journal, April 12, October 24, 1907. After August 29, 1907, there are no minutes of meetings of the Executive Committee until April 3, 1913.

72 *Milwaukee Journal*, August 23, 1909.

73 *Sales Bulletin*, January 1930, pp. 703–704; A-C Co. Board of Directors, April 12, May 5, 1911.

74 *Circular Letter* No. 66, March 26, 1906; W. S. Heger to L. F. Bower, letter of 1907 with illegible date; Letter, July 16, 1907; A-C Co. Board of Directors, June 12, 1911.

75 A-C Co. Executive Committee, June 14, 1906; A-C Co. Board of Directors, June 15, 25, July 2, 1906, June 4, October 1, 1908, October 7, 1909, June 2, 1910; *Special Meeting of Stockholders*, July 16, 1906; *Milwaukee Journal*, August 17, 1907.

76 A-C Board of Directors, December 22, 1910, December 29, 1911; *Circular Letter* No. 155, January 9, 1911, No. 156, January 17, 1911; *To Holders of Allis-Chalmers Company First Mortgage Bonds Preferred and Common Stocks and Certificates of Deposit*, March 26, 1912.

77 *Plan of Reorganization*, March 18, 1912; *To Holders of Allis-Chalmers Company First Mortgage Bonds, Preferred and Common Stocks and Certificates of Deposit*, March 26, 1912.

78 A-C Co. Board of Directors, March 22, December 2, 1912; *Circular Letter* No. 1, April 8, 1912; Press release, April 10, 1912.

General Otto Falk
President 1913–1932

CHAPTER SIX

GENERAL FALK
TAKES COMMAND

THE YEARS FROM 1913 TO 1920 posed many problems but still offered great opportunities to American businessmen. A period of industrial consolidation during the previous decade had resulted in unassimilated combinations and organizations. Technological change had created demands for new equipment at the same time that older products became obsolete. The normal industrial pattern was soon to be upset by a great war which would create enormous demand for industrial products at the same time that it produced acute shortages of skilled labor and raw materials. Profits were available, but often at the price of a loss of industrial balance and disruption of long-range plans.

When Allis-Chalmers entered this period, its name, products and great West Allis plant were its only real assets; a decade of mismanagement in a period of relative prosperity had resulted in receivership. Otto H. Falk acted as the sole receiver of the Allis-Chalmers Company from June 30, 1912, when Delmar W. Call resigned as joint receiver, until the Allis-Chalmers Manufacturing Company was formally incorporated on April 16, 1913. To overcome its problems and to profit from its opportunities, Allis-Chalmers required leadership that was firm, balanced and far-sighted. General Falk ensured the company's success by bringing these qualities to Allis-Chalmers.[1]

Born into the Franz Falk brewing family on June 18, 1865, Otto Herbert Falk attended the German-English Academy of Milwaukee and Northwestern College in Watertown, Wisconsin. Interest in a military career led him to enroll in the Allen Military Academy of Chicago from which he graduated in 1884 with the rank of captain. He served briefly as adjutant of the 4th Infantry, Wisconsin National Guard and aide-de-camp to Governor Jeremiah Rusk before being promoted to quartermaster of the state militia. Falk earned promotion to state adjutant general at the age of twenty-eight because of his outstanding work in suppressing riots in both Milwaukee and Chicago. Having served as president of the National Guard Association in 1894, General Falk, at his own request, was retired in 1895 by Governor William Henry Upham.

During the Spanish-American War, Otto Falk was called into active service as lieutenant and special inspector of the quartermaster department. He was discharged in 1899, but was then ordered by Governor Edward Scofield to take command of the 1st Infantry at Milwaukee. In 1905 Governor Robert M. La Follette refused to retire General Falk at his own request, stating that he was too valuable to the national guard. Having weathered major changes in the political views of Wisconsin's governors from Rusk to La Follette on the basis of his personal ability, Falk finally retired from the national guard in 1911 with the rank of brigadier general. "To my mind," General Falk once said, "there's nothing like a season of military training for any young man, whether he becomes a lawyer or a manufacturer or a doctor. It teaches him the value of discipline, self-control, fairness and self-reliance."[2]

Military interests did not preclude participation in the family business. After selling the former Franz Falk Brewing Company, then the Falk, Jung, and Borchert Brewing Company, to the Pabst Brewing Company in 1892, the family developed a steel foundry company, known as the Falk Corporation, of which Otto Falk was vice president. If military service brought him a knowledge of men and organization, the family business provided invaluable business experience. "The small company is the place for the average young man to start in," Falk said. "There a chap just naturally has to do a greater variety of things, and so he discovers what it's all about in the shortest possible time."[3]

President Otto Falk at his desk.

When General Falk was awarded Milwaukee's Cosmopolitan Club gold medal for "distinguished service to the community" in 1937, his comment was, "I haven't done any more than any other man in my position should do for his community." Responsibility to society early became part of his philosophy. He made a creditable record as president of the Public Safety Commission of Milwaukee and also as a board member of the Fire and Police Commission. Otto Falk represented Wisconsin at the National Tariff Commission Convention in 1909, the National Peace Congress held the same year, the Lakes to the Gulf Deep Waterways Convention in 1909–1910, and the National Irrigation Congress in 1910. From 1909 to 1912 he was president of the Merchants and Manufacturers Association of Milwaukee. Such a variety of interests, breadth of experience and concept of responsibility had not been found in the Allis-Chalmers leadership since the days of Edward P. Allis. Falk brought stature to his position; others had derived stature from it.[4]

The Falk Company and the Allis-Chalmers Company, for some time before 1912, had a rather close operating relationship. Herman Falk, Otto's brother and president of the Falk Company, was a member of the Board of Directors of the Allis-Chalmers Company. As vice president of the Falk Company, Otto Falk handled the arrangements with Allis-Chalmers for making some of the largest steel castings produced in the United States for Allis-Chalmers turbines. The direction Falk provided the Milwaukee business community as president of the Merchants and Manufacturers Association was so excellent that he was the unanimous choice of Milwaukee financial leaders to reorganize Allis-Chalmers in 1912. Otto Falk accepted the post of receiver because

President Falk liked to get the feel of the tractors the company was producing. Here he tests one on his farm near Oconomowoc, Wisconsin.

he felt it a civic duty to help put the company back on its feet. But he regarded it as a temporary association, for he stipulated that he must have two hours a day for his duties as vice president of the Falk Company. Even after he had agreed to remain as president of Allis-Chalmers, he reserved the right to devote up to two hours a day to Falk Company business.[5]

The Allis-Chalmers Company had been controlled by the Board of Directors before 1912. A strong board had chosen weak presidents over whom it had exercised direct control. The board was made up mostly of Eastern men; except for a few months in 1907, the Executive Committee was composed almost completely of Easterners, and meetings of the Board and of the Executive Committee were generally held in New York City. The Allis-Chalmers Company had been subject to absentee control.

The reorganization of 1912 was thorough. In 1913 eight men were elected directors by the preferred stockholders. Five of these were from Milwaukee's top business leadership: Oliver C. Fuller, president of the Wisconsin Trust Company; James D. Mortimer, president of The Milwaukee Electric Railway and Light Company; Gustave Pabst, president of the Pabst Brewing Company; Fred Vogel, Jr., president of the First National Bank; and Otto Falk. Max Pam and F. O. Wetmore were Chicagoans, and John H. McClement, who had been largely responsible for the legal work involved in the reorganization, was a New Yorker. Directors representing the common stock were Arthur W. Butler, Charles W. Cox, Oscar L. Gubelman, R. G. Hutchins, Jr., Arthur Coppell and William C. Potter, all of New York City, and James P. Winchester of Wilmington, Delaware.[6]

On March 17, 1913, the Board of Directors elected an Executive Committee of five members, all Milwaukeeans: Fred Vogel, Jr., chairman, Otto Falk, Oliver Fuller, James Mortimer and Gustave Pabst. Until August, 1915, when it was decided to hold weekly meetings at the West Allis Works, the Executive Committee met at the Pfister Hotel, the First National Bank, or the West Allis Works. By contrast, the Board of Directors was less important in the operations of the company and did not meet so regularly. In 1914 there were no meetings during half the year and during 1915 no meetings during five months. Control of Allis-Chalmers had been returned to Milwaukee businessmen through the reorganization.[7]

Allis-Chalmers after 1913 had a professional executive. An increasing measure of power and control came to General Falk and, to the extent that the executive control was strengthened, that of the directors was lessened. While this was part of a developing pattern in American business organizations, it was particularly conspicuous in the Allis-Chalmers Manufacturing Company. General Falk largely operated on the theory that the directors' function was not to formulate or to veto policy but to advise. He believed, as he once stated, that the executive should "thoroughly control the business."[8]

Otto Falk had the perspective and business sense to recognize the problems which had plagued the company from 1901 to 1911 and eventually placed it in receivership.

A majority of the officials of the old companies stayed on after the merger. Plants in the various cities were continued. Perhaps a certain lack of co-ordination was inevitable, and following that, perhaps some jealousy and bickering was natural. At any rate, this poison spread into some parts of the management, sapping the vitality of the organization at the fountainhead, and creeping downward through the ranks.

By 1912 the situation had become critical. When Falk took over as receiver he found that not only were the company plants operating at least one-third below capacity, but that the shops had not been properly maintained and did not permit efficient production.[9]

Despite these shortcomings, the board of the old Allis-Chalmers Company had put the company in the mainstream of twentieth century industrial society. In 1913 the company owned or controlled 806 United States patents, 224 applications for United States patents and about 100 foreign patents. Among these were key patents on electrical equipment, which was largely manufactured at the Norwood Works, patents for steam and hydraulic turbines, as well as others to a fine line of crushing, cement, mining, flour milling, and sawmilling machinery. This diversity

Another Falk hobby was the raising of full-blooded goats on his farm.

153

supported the company through the first half of the century. Although it had never been used efficiently or effectively, the West Allis Works was potentially one of the finest plants in the country. The good name of the company had suffered somewhat, but Allis-Chalmers was still recognized in almost every civilized country as a manufacturer of quality products. The five million dollars in cash secured through reorganization would be sufficient to bring the plants to a high state of efficiency and to reestablish the company's credit. In a report to the board on March 24, 1913, Otto Falk announced that the break-even point was about $1 million per month in orders. If business to the extent of $1.5 million could be done each month, "in my judgment substantial profits can be made." To the department managers he said, "Our interests are identical, and with all working together we are bound to succeed."[10]

The careful appraisal of assets and liabilities indicated to Otto Falk that the former outweighed the latter. The realism of his evaluation would have been of little consequence, however, without a plan, a philosophy of management, which was formulated concisely by the General:

> I looked upon it as my function, the true executive function, to secure co-ordination of effort throughout the organization in accord with well-established principles of management: such as thrift in the use of all physical properties—fair and square treatment for every individual—the winning of the kind of co-operation that accompanies loyalty and interest—financial good health—excellent quality in products—and the final test, *profits*.

With confidence in his ability and with control firmly in the presidential office, General Falk set to work to put into effect his business philosophy.[11]

The first "principle of management" effected by General Falk was "thrift in the use of all physical properties," by which he meant the concentration of manufacturing operations. The Allis-Chalmers Company, as formed in 1901, was a collection of four plants in three cities. After the merger the West Allis Works had been built and the Norwood Works acquired. As matters stood in 1912, there was duplication of equipment and personnel and imbalance in shop facilities. This lack of balance stood in the way of the low manufacturing costs necessary for profit.

The No. 1 Chicago Works, the old Gates Iron Works, had been closed just before the Falk administration, and the last of the equipment was moved to Milwaukee during the Falk receivership. The company began to close the Scranton Works during the summer of 1911 and offered the property for sale in 1914. Concentration of industrial operations was part of the new pattern of American business because the consolidations around the turn of the century had left many concerns with the same plant duplications found at Allis-Chalmers. With Allis-Chalmers operating at about 60 percent of capacity from 1909 until the receivership in 1912, the necessity for concentration was conspicuously apparent.

Otto Falk speeded up the concentration already begun. A study by C. Edwin Search, commissioned by Falk, indicated that the closing of the Chicago Works No. 2, the old Fraser & Chalmers Company, would save Allis-Chalmers about $160,000 per year. It was therefore closed during 1913 and 1914. The principal manufacturing lines of the No. 2 works—mining, crushing and cement machinery—were transferred to the West Allis and Reliance Works and the property placed on the market to be sold in parcels.[12]

The Reliance Works in Milwaukee, moved to its new location by Edward P. Allis in 1867, had been enlarged over the years until it occupied four full blocks. The random placement of buildings made difficult any efficiency in manufacturing, and by 1920 the shops and much of the machinery were obsolete. Moreover, operation of both the Reliance and West Allis Works entailed a good deal of expense in traffic between the two locations. The last heat at the Reliance foundry was poured on April 29, 1921, and the decision to close the works was made in September. A large new shop was

This combined hydroelectric unit is a good example of Allis-Chalmers "undivided responsibility," with turbine, generator and governor all supplied by one company.

erected at the West Allis Works to take over the manufacture of flour and sawmill machinery. When this shop began operation early in 1923, the old plant was closed and the land was offered for sale; the company was unable to dispose of it until the summer of 1944.[13]

By 1923 General Falk could be proud of his record of consolidation. He had virtually taken over six plants and discontinued operations at four. Moreover, the capacity and efficiency of the two remaining plants, the Norwood Works and the West Allis Works, had been increased until they surpassed the previous combined capacity of the six plants.[14]

A second principle, which Falk referred to as a "guiding policy" and which he saw as integral to the prudent use of all physical properties was the use of good machinery and methods. Because of continuous financial troubles in the years before receivership, too little advantage had been taken of economies possible from the best modern machinery and industrial methods. In 1912 and 1913 many machines and tools were obsolete or badly in need of repair. To regain the high standards set in earlier years, a sizable portion of the $5 million cash fund supplied by the reorganization program was used to place the shops in first class condition and to replace old machinery with the most modern equipment.

The improvement in American industrial efficiency and business management had come with incredible speed during the first decade of the century. One study indicates that 240 volumes on business management were published during this period. Studies and research to increase productivity came with equal speed. For

General Falk enjoyed going into the shop to see new machines in operation.

example, the introduction of Taylor-White high speed steel in 1906 more than doubled the productivity of machine tools. Such developments demanded increased efficiency through modernization at Allis-Chalmers. Otto Falk had grown up with industrial society and was more an activist than a theoretician. Although there is no evidence that he took great interest in the business literature or in new production techniques, he was alert to new machines. To him the simple solution was to get a good man and to give him good modern tools. He was practical, however, and took pride in counting the cost rather than modernizing for the sake of change. As he pointed out, it was sometimes better economy to use old-fashioned methods and machines for rarely used processes. Modern machines, if idle a good share of the time would use up more overhead than old ones.[15]

By reorganizing and modernizing the company physically General Falk solved one set of problems. But perhaps a more difficult problem was how to establish a new system of personal relations to restore confidence in management. In this he was attacking another of his principles of management, "fair and square treatment for every individual." Part of this problem, as the General saw it, was to "fit myself into the organization so that my coming would tend to a better adjustment of the people already there." To become acquainted with as many people as possible and allow them to become acquainted with him, he deliberately made himself immediately accessible by telephone to everyone. "It was because I did not want to be aloof, or even to give the appearance of being aloof, that I instructed the operator to connect my telephone calls direct." He also tried to see that every caller could come into his office immediately. If he could not possibly see someone, he often went to the waiting room to explain and to arrange another appointment. The deliberate sacrifice of his own time projected his strong, quiet personality through the entire organization.[16]

The general rules adopted by President Falk in the administration of the company were those of common sense and ordinary fairness, yet they were in essence the application of the golden rule to business conduct. A personal rule he brought from his army experience was never to humiliate a man by reprimanding him in front of others. To eliminate dissension over pay rates and salaries he did his best to adjust them by a common scale throughout the organization. To set an example for the employees he himself worked full days regularly. He believed that there should be no excuse for anyone not starting work promptly when the president himself was at his desk at or before eight-fifteen. To be available to anyone almost every day, he took no regular summer or winter vacation, only a day or two at a time until he had used up a reasonable vacation allotment. His was a conscious program followed conscientiously. "Analyze these rules and you will find that they, and others like them, fall under the head of fair dealing— treating others the way you like to be treated."[17]

As receiver and later as president, General Falk had the power to make extensive changes in personnel. But he felt keenly that the company's problems stemmed from both lack of leadership and lack of confidence in that leadership. He came to his position resolved to make as few changes as possible. He was intent on winning "the kind of co-operation that accompanies loyalty and interest," and thus placing in action another of his principles of management.

Having attempted to re-establish confidence within the organization, General Falk was determined to make Allis-Chalmers as much as possible a self-contained industry by offering advancement to his men within the organization. In 1903 the company had established a Graduate Student System open to graduates of recognized schools of technology. The first class entered training in 1904. The system embraced three different courses: mechanical, electrical and mining engineering. Each course took two years, with 5,500 hours in the shops and the work arranged to give the best possible experience in both breadth and depth. For the first 1,375 hours the student was paid fifteen cents an hour, for the second 1,375 hours, eighteen cents and for the final 2,750 hours twenty cents. At these wages in 1904 Allis-Chalmers could attract graduate engineers from universities such as Harvard, Cornell, Massachusetts Institute of Technology, McGill, Illinois, Michigan and Kentucky. Each student who successfully completed the course received a bonus of $100. Whenever possible the successful students were placed in subordinate positions in the engineering and commercial departments or district offices. Although the system was excellent, the graduates faced an uncertain future both in possibilities for promotion and in the future of the company itself. During the uncertain years from 1909 to 1911, many looked

From this group of graduate and co-op students in 1925 came a vice president, a district service manager, engineers, and salesmen. Many stayed with the company until retirement.

elsewhere for security. By refining and expanding the existing training programs, Falk proposed to train enough men to fill every supervisory position in the shops and all positions in the engineering and sales departments.[18]

Having assured greater confidence and security by reinforcing the leadership training, Falk also had to integrate the separate parts to make the company "a smoothly running and profit-earning whole." He had taken over a company which manufactured a variety of products. Each of the main departments had come to operate as an autonomous unit, the departmental managers assuming the proportions of "little tin gods," as they were sometimes called. Interdepartmental competition was extremely keen and apparently sometimes cut-throat.[19]

To suppress the rivalry the General instituted weekly meetings of department managers, despite his real dislike for conferences. In the president's office the problems of the ailing business came to light. Differences were thrashed out. "A spirit of compromise and conciliation began to replace the spirit of jealousy and mutual recrimination in the company, and the new spirit was due to the simple expedient of getting the men to face one another and to tell the truth under such circumstances that the facts could not fail to get a fair hearing." Votes on the questions raised were usually unanimous, and the General believed that the men were "pulling together for the company." Evidence indicates that he was somewhat optimistic over the success of these conferences. But, although rivalries and competition did continue until about the time

The department managers in the 1920s: (l to r) Herman Schlifflin, crushing and cement; Dr. William M. White, hydraulic turbine and centrifugal pump; Allan Hall, flour mill and sawmill; Gustaf L. Kollberg, pumping engine; Irving Reynolds, engine; Arthur C. Flory, steam turbine; Henry C. Holthoff, mining; and John R. Jeffrey, electrical.

of World War II, relationships improved enough that Allis-Chalmers could begin to make the kind of progress that had been impossible before then.[20]

A similar system of conferences was used to arbitrate cases of disproportionate profits when two or more departments were engaged in a joint project. The conference was a means of alleviating the traditional hostility between the engineering departments and the shops. To the General the conference technique was a means to an end, for he held that "many minds working together towards a common objective can achieve marvelous results. The same minds working equally hard but with divergent objectives can produce confusion—and receiverships." These meetings gave him insight into the affairs of the company that he would have had difficulty obtaining in any other way.[21]

As an inducement for General Falk to accept the presidency of Allis-Chalmers, the Reorganization Committee had offered him an option on a large block of the company's stock at advantageous terms. Certificates representing 6,505 shares of preferred and 2,292 shares of common stock were held in escrow by the Central Trust Company of New York under option agreement for sale to Otto H. Falk and Associates for purchase at a cost of $406,135.50, something less than half of par value. This option Falk and his "associates," who were actually the remainder of the Executive Committee, surrendered on September 30, 1915. In declining the stock option the General proposed an alternative which would establish an "additional compensation account." The resolution establishing this account, passed December 6, 1915, stipulated that, after the company paid a stock dividend, a sum equal to 7 percent of the dividend on preferred stock and 10 percent of the common stock dividend, if paid, be deposited in the account.[22]

Two classes of persons were designated to share in the account. Class A consisted of the president of the company and the members of the Executive Committee. Thirty percent of the account was to go to this class, with one-third set aside for the president and the remainder divided equally among the other four members of the Executive Committee. The remaining 70 percent of the total was set aside for Class B, which "shall consist of such officers, heads of departments and members of the Operating and Commercial Departments of the Corporation as the Executive Committee may . . . from time to time determine." In 1917 this meant that the executive officers and the general works manager (Group 1) each received an additional compensation of $2,000; the commercial department managers (Group 2) each received $1,500; "Other Managers and Superintendents" (Group 3) received 15 percent of their salary; and district office managers and division heads each received 10 percent of their annual salary. Since the first dividend paid by the Allis-Chalmers Manufacturing Company had been declared December 6, 1915, this additional compensation plan went into effect immediately.[23]

This was not a true profit-sharing plan because it did not include the wage earner. However, that was not the intent. At this point General Falk was willing to sacrifice his personal gain to ensure the establishment of his principle of management which called for "the winning of the kind of co-operation that accompanies loyalty and interest" and he deliberately confined it to the upper echelon of the company. Perhaps no single act by Falk did so much to secure the continuing support of all personnel, even those who did not share in it. At a later date Falk's public explanation of his action on the stock option was, "No executive is worth the huge sum represented by that offer." This was for public consumption and it had a wonderful effect. But the effect he most of all desired, loyalty bordering on devotion, had come about through some personal financial sacrifices.[24]

Financial good health accomplished through a conservative fiscal policy was one of the cardinal principles of Falk's administration. From the beginning, he eliminated unnecessary frills. Office carpeting, which had been a mark of status, was removed. The Allis-Chalmers brass band, which had been giving concerts in the clubhouse and the erecting shop as well as getting time off for practice, was discontinued. District managers were directed to stop using taxi cabs and to ride the street cars instead. Mail service could easily replace most of the long-distance telephone calls. The

In the early days of the automobile Allis-Chalmers supplied starting and lighting equipment for, among others, Grant and Ford cars. This view shows the equipment on a Ford.

company's legal representative in New York, C. V. Edwards, had received $750 per month as a retainer fee. Falk sent him a letter April 11, 1912, offering him a retainer of $100 per month and a per diem of $50 for services rendered the company by Edwards, and a lesser amount for services by others in the firm. This letter, which amounted to an ultimatum, was accepted.[25]

All these actions were in line with the basic policy set forth by the General. "Fundamentally, our policy is to buy what we need and have the money to pay for. I suppose you might call that old-fashioned. Assuredly it is conservative." One fiscal officer of the company summed up the Falk policy most succinctly: "Even after Allis-Chalmers was financially in the clear, General Falk still continued to run the company as though it were in receivership." The reason for economy was clearly stated by the General in a *Circular Letter* of January 24, 1914: "The money of the company must be expended as carefully as our own and, in fact, with more care, for the interests of many are affected in company expenditures." Otto Falk properly regarded himself as trustee for the stockholders.[26]

To consolidate the financial interests of the company, effect future financial economies, and "permit direct operation as a mere department of this company," the plant and properties of the Bullock Electric Manufacturing Company were conveyed to the Allis-Chalmers Manufacturing Company on April 30, 1914. Under the agreement, the preferred stock of the Bullock Company was exchanged for first mortgage 6 percent bonds of that company at the rate of $80 in bonds for $100 of preferred stock and the payment of the accrued and past due dividends on the preferred stock of the Bullock Company to January 1, 1914, amounting to 7.5 percent. When this exchange was completed, Allis-Chalmers, which had already owned the common stock excepting the directors' qualifying shares, was sole stockholder of the Bullock Company. The Allis-Chalmers Manufacturing Company then purchased $893,700 par value of the total issue of $936,000 6 percent mortgage bonds at 95 percent and accrued interest. The importance of this complicated financial deal really lies in the fact that in 1914, when American business was sinking into a marked depression, President Falk was prepared to purchase Bullock for nearly one million dollars so that the financial interests of the company could be consolidated.[27]

Falk's final principle of management was "excellent quality in products." Falk insisted that "nothing be spared to make the product of the proper design and every machine as good in its class as it was possible to build." To ensure this, a Development Committee was formed in May of 1913. Made up of Lahman F. Bower as chairman with C. Edwin Search and J. F. Max Patitz, it met with the manager and engineer of each department for deliberations relating to their respective areas. The purpose was "to keep up the standard of our products, to decrease the cost of production and to develop new lines." Development was combined with standardization of lines and enlarged production of standard units, to reduce engineering expense and risk in contracts. Falk recognized that manufacturing profits come from having the right men operating the right machines under proper supervision. For purposes of both development and production, Search, named works manager during the period of reorganization, provided real leadership. His tremendous knowledge of machine tools and his experience in their use, combined with his ability in handling men, proved invaluable to the company until his retirement in 1924.[28]

Although Otto Falk was referred to as "General" more often than "President," he was not one to issue orders out of the isolation of his office. He was seriously interested in the plant and went into the shops to find things out for himself. The experience of Ernest Shaw in trying to perfect a sawmill machine was not unusual. One day when Shaw was down on his knees in dirt and sawdust with his head inside the machine, he found a head beside his—it was Otto Falk wanting to see what was going on. Spurred on by the president's interest and constant pressure, the manufacturing departments were reorganized, and the company was ready to profit from industrial competition.[29]

This sound and consistent philosophy of management, with executive control well established, mutual confidence increasing, and product development systematically conducted, showed almost immediately in the product lines. The hydraulic turbine department was allowed to develop without interference for the next several decades under the able leadership of William Monroe

Allis-Chalmers 84 by 66-inch jaw crusher with rope drive furnished the Phelps Dodge Corp. for crushing copper ore in Arizona about 1925.

White. Having come to Allis-Chalmers just before the Falk administration, he selected outstanding young American engineering graduates to meet the engineering and sales needs of the department. White sponsored the first Allis-Chalmers apprentice drafting school and established one of the first hydraulic testing flumes ever undertaken by a manufacturer for pump and hydraulic turbine runner testing. He inspired others in the department by becoming the holder of some fifty United States and foreign patents.

Forrest Nagler was encouraged in his development of the propeller turbine. Installations of this type were made as early as 1916, when two 600 horsepower, 200 rpm, 15 foot head propeller type turbines were designed and built for the Geddes Plant of the Detroit Edison Company. In 1917 Allis-Chalmers produced the first hydroelectric unit for the production of aluminum. The Cheoah Plant of the Tallahassee Power Company furnished the electrical energy for the aluminum company. All equipment, turbines, governors, generators and exciters were designed and built under the Allis-Chalmers roof and under one combined guarantee of performance. Allis-Chalmers was the only company in the world prepared to build all three types of hydraulic turbines: Francis, propeller and impulse.[30]

Another example of Allis-Chalmers "undivided responsibility": steam turbine generator unit, steam condenser, circulating water pump, and the motor to drive it.

The steam turbine department was made a separate department in 1913 with Robert A. McKee as its first manager. With such men as Patitz, consulting engineer until 1935, the department made rapid progress. Metallurgical work at the West Allis laboratories was stepped up to make possible cast steel for steam chests and cylinders by 1915. This permitted the use of higher steam pressures and temperatures. Power ratings also began to climb, and turbines were built that generated 10,000 kilowatts. Regenerative feedwater heating meant gains in thermal economy, and in 1917 the first Kingsbury thrust bearings were used, which improved performance. These developments made possible the installation of a 10,000-kilowatt 1,500 rpm, turbogenerator for the Calumet and Hecla Mining Company in 1918. This exceeded the 7,500 kilowatt unit installed in 1913 and remained the largest mixed pressure steam turbine for over twenty years.[31]

Electrical products, which had first been part of the steam and electrical department and later of the power and electrical department, were made separate and independent in 1913. John R. Jeffrey, who was appointed manager, was to provide continuity and direction for this department until 1929, when Robert S. Flesheim became manager. Although competition remained intense between Allis-Chalmers and the electrical giants of General Electric and Westinghouse, W. W. Nichols helped to solve the problem of constant patent disputes after 1912. He arranged an amicable settlement of policy with Edwin Wilbur Rice, Jr., president of General Electric, and Edwin M. Herr, president of Westinghouse. This action apparently marked the acceptance of Allis-

This is the last of the horizontal cross-compound pumping units supplied by the company; it went to an Hawaiian sugar plantation.

Chalmers as a permanent manufacturer of electrical equipment. In a period when the total electric power produced in the country rose from 3,343,000,000 kilowatt hours in 1902 to 74,567,000,000 kilowatt hours in 1923, a twenty-fold increase, the company was better able than ever before to produce for the electrical revolution in the United States.[32]

Although the greater portion of the country's steel mills had been erected during the first decade of the century, the production of gas engines held up remarkably well from 1912 through 1920. Though only twenty-four engines were produced, as compared with fifty-one from 1905 through 1911, all were equipped with generators of 3,000 kilowatts or more, while those of the previous period were equipped with generators of 750 to 3,200-kw. From 1911 on the production of these giant engines was to be negligible in Allis-Chalmers history, for in the decade of the twenties the company produced only six. However, gas engines were a significant and stable item during the years when Falk was placing the company on the high road to success.[33]

Otto Falk could not have taken charge of the company at a less auspicious time, for the economically unstable years from 1907 to 1912 had seen little revival of trade and industry. Rather than improving, the American economy declined into a recession in 1913. Despite this, through prudent management Allis-Chalmers showed a net profit of $755,124.73 for its first eight and one-half months of operation. General Falk and the board of the new company followed a policy on declaring dividends that was diametrically opposed to that of the earlier Allis-Chalmers management. After carefully considering the possibility of declaring a dividend on the preferred stock out of the surplus profits, in view of the depressed conditions they "concluded to adopt the conservative policy and defer taking action thereon until such time as current profits and future prospects would justify a reasonable assurance of the continuance of dividend payments."[34]

In early 1914 the recession sank into depression. Otto Falk wrote his salesmen acknowledging that there might not be any large orders, but urging them to search out any orders, however small. The immediate effect of the European war in August was a partial paralysis of existing business. But Percy A. Rockefeller of New York, in Milwaukee in late September for the annual meeting of the directors of the Chicago, Milwaukee, and St. Paul Road, had a hopeful thought. "The great cost of modern warfare will have a tendency to bring the conflict to an early close, but whether it will go on for weeks or months I cannot say." By mid-November the United States Steel Corporation was working a three-day week. Foreign trade was violently disrupted and as unemployment spread, the winter of 1914–1915 became very lean indeed. In Milwaukee, Mayor Gerhard Adolph Bading issued an appeal on December 9 for the city's poor, reporting that twice as many applications had been made to the Associated Charities than in

1913. Allis-Chalmers sustained a net loss for the year of $25,068.40, which was in keeping with the pattern of the eighty Milwaukee firms dealing in iron, steel and heavy machinery. In 1913 their total output had been $105,622,059; this declined in 1914 to $83,605,985.[35]

On January 1, 1915, Otto Falk said that no one could predict how long the depression would continue, but real relief was dependent on three things: "First, the return to the market of the railroads and other large customers of the metal trades industry for at least a fairly normal volume of purchases. Second, a cessation of the enactment of laws adverse to the real business interests of the country and the political agitation accompanying the same. Third, the ending of the European War." With no lessening of the economic stringency, more and more manufacturers felt compelled to reduce wages. Allis-Chalmers held out as long as possible. Not until March 1, 1915, did General Falk reduce the wages of "all officers and employees from the President down by 10 percent." He had missed the Christmas Eve greeting to United States business from Charles M. Schwab, president of Bethlehem Steel, who has just returned from England. There was more than a bit of irony in his prediction that because of the war the United States was on the threshold of the "greatest period of prosperity it has seen in many years."[36]

When recovery from the depression arrived near the middle of 1915, it came in sweeping fashion. The largest cause was the discovery by the Allies, particularly the British, that they could not hope to win the war without buying great quantities of war materials and supplies abroad. Schwab's Christmas statement was the result of a large munitions contract awarded Bethlehem Steel by the British. Not until March did the subcontracts, which meant immediate prosperity, filter down to Allis-Chalmers. A subcontract from Schwab's company called for the machine work on two million forgings furnished by Bethlehem Steel at a price of $1.85 each, the work to extend into 1916. Although Allis-Chalmers had to spend about $230,000 for machinery and tools, much of it standard equipment, this cost had been taken into consideration in the contract. The first contract was expanded in May until it totalled about $3,760,000. Since further orders were probable, the Executive Committee developed a policy on war orders. It unanimously agreed "that the Company should endeavor to get orders of this character [the shell contract] provided the work is reasonably suited to the Company's manufacturing facilities and can be obtained at prices carrying good profits on definite terms."[37]

As president of Allis-Chalmers, Otto Falk came under immediate attack from the German-American Alliance which adopted the following resolution on March 20: "Shrapnel shells are manufactured by the Allis-Chalmers Company. We greatly regret that shells are being made for such purposes. Furthermore, we regret that a man in whose veins runs German blood is the head of a concern that makes weapons to be used to kill the Germans." General Falk issued an evasive statement that "We are filling an order for Bethlehem Steel Co. We do not know whom this is for. We do not know even whether it is for war purposes or not. This denial ought to satisfy every one once for all that we are not taking war orders." The Falks had been leaders in the Milwaukee German-American community as had the Vogels and Pabsts, who also had representatives on the Executive Committee. In 1902 Otto Falk had been on the Executive Committee and acted as Chief Marshal when Prince Henry, brother of the German Emperor, visited Milwaukee. As late as December, 1914, he had been one of the businessmen who brought to Milwaukee Dr. Bernhard Dernburg, former German Secretary of State for the Colonies, to speak on "the commercial relations of the United States and the European countries now at war." But despite his German background, Falk made his views quite clear in a statement on March 21, 1915: "So far as the Allis-Chalmers Manufacturing Company is concerned, it has stockholders in Germany, England, France and other countries engaged in the war. In performing my duties as President of the company I have heretofore acted and shall continue to administer its affairs in such manner as I believe to be for the best interests and consistent with its position as a representative American manufacturing concern which takes no part and has no prejudices for or against either side in the present European war." In short, General Falk was interested in the financial health of Allis-Chalmers. The profits from the shell contract were sufficient to restore half the salary reduction beginning July 1, 1915, and the other half in September.[38]

President Falk with Comptroller William A. Thompson.

About the end of April, General Samuel Pearson, an insurance man from Allentown, Pennsylvania, and a former supporter of the Boer cause in South Africa, sued Allis-Chalmers to prevent their sending munitions to England. He asserted that he had property in Germany and held German securities which were being "injured as the result of the 'conspiracy' to injure Germany." Pearson had strong support from the German population in Milwaukee. His suit threatened the prosperity of Allis-Chalmers as well as that of other companies, for if successful in Milwaukee he planned to start proceedings in Chicago, Pittsburgh and Dayton. An investigation conducted by Allis-Chalmers provided the basis for a personal repudiation of Pearson, and the court sustained a countersuit filed by the company. Whatever appeal Pearson might have had among the majority of Milwaukeeans seems to have vanished with the sinking of the *Lusitania* on May 7, 1915. The passions of the German-American population, however, still had to be reckoned with. Having discussed insurance against sabotage on November 22, the Executive Committee took out a six-month policy on December 28, but did not renew it because of the "general impression that danger in this respect appears more remote at the present time than when such insurance was originally placed." A strenuous campaign by the *Milwaukee Journal*, for which it won the Pulitzer Prize in 1919, and a changing pattern of thought generally turned the tide. But for some time Allis-Chalmers and Otto Falk were not highly esteemed by the German-American community.[39]

To stimulate shell production at Allis-Chalmers, a piece work system was introduced in the shrapnel department in late May, 1915. Although the number of pieces turned out was increased if a workman was to earn the same amount as under the twenty-five cent an hour rate, some men were said to be making as much as five dollars a day. The 500 men at work in this department was only half the number the company hoped to employ. One machinist reported: "It is a continuous circus out there. Lathes are being run by tailors, carpenters, shoemakers and all kinds of men." By the end of August Allis-Chalmers was turning out nearly 10,000 shells a day in their "closely guarded" shop.[40]

Whatever the background and experience of the men, they did put out shell casings at a steady pace. The contract with Bethlehem Steel was for one and three-quarter million 3.3 inch British shrapnel casings and one million 3.3 inch British high explosive casings. Once America entered the war, Allis-Chalmers produced one million 75 millimeter high explosive casings for the United States government. The company designed and manufactured a complete line of special single purpose machines for the manufacture of shell casings. The copper band-turning machine in particular was produced for other shell casing manufacturers not only in the United States but also in Italy and France.[41]

General Falk's efficiency earned the company a citation from the

Army Ordinance Department. The citation described how Allis-Chalmers, having no other space available, cleared off the south end of the ground floor of its pattern storage building to make a space 160 by 250 feet for the machining of 75 millimeter high explosive forged steel shells. The daily production, which ran from 7,500 to 9,550, was the second largest of any plant in the country, irrespective of size, and was declared by a visiting foreign commission of British, French, Italian and Belgian officers to be the highest in the world per square foot of space. This efficiency also earned the company a substantial profit, for these were not cost-plus contracts but flat price to all contractors. Since these prices were set high enough to provide a profit for all, efficient manufacturers such as Allis-Chalmers could make handsome profits.[42]

During 1915, when it became apparent that the war was not to be a short one, the orders from England and France increased. By early 1916 President Woodrow Wilson, while affirming the United States determination to remain at peace, was advocating preparedness and "a great navy second to none in the world." With American entrance into the war, the two principal considerations came to be the production and delivery of military goods. The government almost immediately inquired about Allis-Chalmers ability to produce forgings, marine turbines and engines for ships. They could produce the materials, but the critical problem became the

During World War I Allis-Chalmers supplied the Navy with slides and mounts for the 16-inch guns of battleships under construction. A treaty between Great Britain, Japan, and the United States resulted in the destruction of the ships before they were completed.

delivery, because in 1917 German submarines destroyed 6,618,623 tons of shipping. During the same year the entire production of new shipping in Britain came to only 1,163,474 tons, while the United States launched about 1,000,000 tons. All the other Allies combined with all the neutrals produced only 539,871 tons. In short, all of the building by the nations outside Germany and Austria came to only 2,703,345 tons—less than half the total destroyed by submarines. Clearly the key to allied success lay in building ships to deliver the necessary war supplies and controlling or eliminating the submarine menace.[43]

Ultimately the United States succeeded through organization, concentration of effort and inventiveness. The most ingenious response to the challenge was building ships of concrete. Daring in innovation but certainly more successful than concrete ships were the "fabricated" ships. Until this time the whole ship had been built in its yard, with each plate drilled and shaped to its own complex use. Some imaginative engineers thought of standardized parts made in distant plants and merely assembled at the shipyards. The idea worked. By the end of the war 85 percent of the hulls of ships and many of their other parts were made in distant steel-works. Special orders for marine and naval products were given to Allis-Chalmers because the West Allis Works was so well-equipped for such production. Hundreds of tons of steel plate were fabricated and drilled in the Allis-Chalmers shops for cargo vessels built at Hog Island by the Submarine Boat Corporation. The company made forgings for marine engines to be built by

Allis-Chalmers was called on to build marine engines for the Navy's "bridge of ships." The forty-nine engines from 1,400 to 2,800 horsepower that were ordered marked the end of an era for these large steam engines.

others as well as propeller and line shaftings. They worked on marine guns also—sixteen-inch gun turrets, sixteen-inch gun slides and mounts, and five- and six-inch gun barrels. Most of the gun slides and mounts, however, were never installed; the battleships for which they were intended were scrapped by the Washington Naval Conference following the war.[44]

Efficiency in the repetitive process of making shell casings was very profitable, as was the production of special marine orders, but most of the orders for war materiel came far closer to the regular Allis-Chalmers lines. Seventy-seven eighteen by thirty-six inch, 350-horsepower single cylinder Corliss engines were produced for the DuPont powder plants. On this order August Werner in the No. 2 erecting shop was able to maintain a production schedule of one engine a day. Known as "the man who always had an Ace in the hole," he was farsighted enough to keep parts on hand for a nearly complete engine, so that if anything happened he could still maintain his production record. Marine triple expansion engines operated on the same principle as the famous Reynolds-Corliss triple expansion pumping engine, long one of the company's standard products. Allis-Chalmers was therefore asked to produce twenty 1,400-horsepower marine engines for the wooden ships of the emergency fleet, together with nine 2,000-horsepower and twenty 2,800-horsepower engines for other cargo vessels.

The production schedule established by the Hanlon Dry Dock Company called for the keel of their first ship to be laid on July 19, 1918, the engine to be delivered no later than October 16, launching to take place on November 16, and final delivery of the completed vessel on February 16, 1919. The two Hanlon contracts indicate the rapid rise in cost of materials: six 2,000-horsepower engines on May 3, 1918, cost $565,000; six identical engines on November 25, 1918, allowing the same margin of profit, cost $600,000. A letter of February 1, 1919, from the Grant Smith-Porter Ship Company of Portland, Oregon, is indicative of the quality of the triple expansion engines delivered by Allis-Chalmers during the war. "During our construction for the Emergency Fleet Corporation, we have received engines from some eight different firms, and we wish to take this opportunity to compliment you on the quality and excellence of workmanship in your engines. Your firm is especially to be commended because you have not allowed the hurry and rush of war work to lower your standard of workmanship." These engines marked the end of an era. Although the last steam engine was not built until 1930, these were the last sizable steam engine orders filled by Allis-Chalmers.[45]

The largest single order filled by the company up to World War II was for $7,950,000 worth of marine main propulsion turbines and their fittings for thirty-four destroyers. These turbines were made to plans drawn by Warren Flanders of the Westinghouse Company, which recommended Allis-Chalmers for the job because of its reputation as a builder of fine machinery. Although

Allis-Chalmers was the smallest of the three major turbine builders of the day, it was actually best fitted for the job, for it had the most stable and experienced workers. Since each destroyer had two shafts, the company produced sixty-eight units, each with a high and low pressure turbine, for a total of 750,000 horsepower.

The marine propulsion turbines were installed in fourteen United States torpedo boat destroyers, numbers 185 on, built at Newport News, Virginia. Twenty identical destroyers, numbers 231 on, equipped with Allis-Chalmers turbines were built by the New York Shipbuilding Company. The *Clemson*, built at Newport News, set efficiency records for all ships of her class, and the *Brooks*, built at New York, set the speed record for her class.

Although the Allis-Chalmers turbines met the efficiency specifications and the ships met the speed requirements, naval inspection and acceptance was new to the company. Commander J. H. Rowan and forty-eight naval inspectors were stationed at the plant during the production of the steam turbines. To test these turbines, a specially built steam boiler subjected the cylinders to 500 pounds of pressure. The cylinders were tested by holding a mirror around the joints and other parts of the cylinder to check for telltale vapor. The turbines also had to pass acceptance tests at the shipyards, and final acceptance came fifteen months from 12 o'clock noon of the day the trial voyage was made. Only then did Allis-Chalmers receive final payment.[46]

Turbines, shells and engines are produced by men working on materials. During the war there was a greater and greater shortage of both. Production of goods and the number of workers rose greatly but as war prosperity came, Allis-Chalmers was competing for men and materials with hundreds of other producers across the land.

Allis-Chalmers, experienced in the building of steam turbines, received orders for propulsion turbines for thirty-four new destroyers. This view shows the last two leaving the West Allis Works.

Trying to hedge against possible future shortages, the company in August of 1915 purchased pig iron to the extent of 6,000 tons beyond existing commitments. But many other materials were far more difficult to obtain in the amounts sufficient for increasing production. Early in 1916 President Falk gave notice to salesmen and department managers that they must be increasingly careful about promising delivery dates, for the company was having trouble securing materials. By March, bars and plates that had a normal delivery time of thirty to sixty days were requiring a minimum of four months. Quotations on cold rolled steel with a shipping promise of a year was the best Allis-Chalmers could secure even before the United States entered the war. The company purchasing agent, Fred Haker, was often sent East to search out a billet of steel and bring it back in a baggage car.

Soaring costs of materials posed another problem. An indication of the incredible speed with which prices rose can be seen in the increase in some crucial materials from March 1915 to March 1916:

Copper	100%
Pig Iron	60
Tool Steel	600
Steel Castings	33
Forging Billets	150
Steel Plates	300
Electric Steel	150
Tin	50
Lead	150
Manganese	1000

Otto Falk had dealt successfully with many problems, but scarcity of materials and rapidly rising costs created "conditions most difficult to cope with." After the United States entered the war, he told the Executive Committee that it was very hard to figure ahead with any degree of certainty.[47]

Otto Falk had reason to worry, not only about the availability and cost of materials, but also about the availability of workers. Precise data on the number of employees had never concerned the company during the period 1901–1911. But such data was a matter of increasing concern by mid-1915 and, beginning with August 30, the General began to consistently bring such information to the attention of the Executive Committee. At that time 3,207 men were employed in West Allis and 507 in the Reliance manufacturing departments, 856 in the engineering and manufacturing departments of the Bullock Works, and 672 in administration, engineering, in the field or general offices, for a total of 5,242. Numbers increased steadily as the company moved into full production with war orders and an upswing in the American economy. On August 27, 1915, employees totaled 5,622; October 31, 6,290; November 30, 6,727; December 31, 7,298; and on January 31, 1916, 8,008. The peak for some time was reached on February 29, 1916, with 8,325 workers. Of this peak number, 5,515 were at West Allis, 807 at the Reliance Works, 1,200 at Bullock and 803 in the general offices. Save for July through September, 1916, when there was a machinists' strike, the number leveled off at around 7,800 workers until October 31, 1918, when it shot up to 9,251. The peak employment figures for February, 1916, and October, 1918, were caused by large shell orders.[48]

A serious labor shortage plagued Allis-Chalmers from 1916 through 1918. As early as October, 1916, general works manager Search reported that the "question of capacity was not regulated by the amount of machinery in the Company's works but by the number of men that can be obtained." The shortage worsened after the United States entered the war. Although American citizens were exempt from the draft for war work, this ruling did not apply to the citizens of foreign powers. Such men, and many were working at Allis-Chalmers, were subject to military service and many were drafted by their native countries. The company posted bond for employees who were citizens of "alien enemy countries" to help keep them on the job; these men could not travel and generally remained with the company throughout the war. American citizens unqualified for military service were free to move and many of them did. The company management resented advertisements in Milwaukee newspapers by firms outside the city offering inducements to workers to leave Milwaukee. They unsuccessfully took up the problem with the Metal Trades Association, hoping to

Allis-Chalmers supplied propulsion turbines for destroyers like this. Many were given to Great Britain before World War II in exchange for naval bases on British territory.

persuade the papers to discontinue the practice. By late 1917 the Executive Committee learned that the shortage of labor was serious and seemed difficult, if not impossible, to solve. More and more women were hired in 1918 to fill the labor gap. Although the company recognized that most of the shop work could not be handled by women, they resolved to hire as many women as possible for all positions that they could fill.[49]

Labor relations, too, caused wartime problems for General Falk. During the ten years following the molders' strike of 1906, Allis-Chalmers had no serious labor troubles. But in January, 1916, Otto Falk reported his anxiety to the Executive Committee "as to what might transpire during the year." He had good reason for anxiety. In the fall of 1915 the Machinists' Union had started an intensive campaign which increased their membership in Milwaukee from about 1,200 men to approximately 3,500 by the spring of 1916. They aimed for sufficient strength to demand an eight-hour day and complete unionization of the metal shops in Milwaukee. Anticipating union action, the Metal Trades Manufacturers in Milwaukee, including Allis-Chalmers, requested the Wisconsin Industrial Commission to conduct a survey to determine the working hours in competitive industries in the Midwest. The commission reported on June 30 its survey of 519 industries employing 93,540 men in machine shops, pattern shops and foundries. The survey found that 38 percent of the companies worked under fifty-five hours per week, 62 percent fifty-five hours or over per week. In the 519 companies, 50 percent of the employees worked fifty-four hours or under per week. On receiving the report, the Milwaukee Metal Trades Association on the same day passed a resolution to the effect that from July 1 the working hours in the

In 1918 Allis-Chalmers built the largest mixed pressure steam turbine generator unit in America—10,000 kilowatts, 1,500 rpm for the Calumet and Hecla Mining Company.

Milwaukee metal trades would be 52½ hours per week instead of 55. The hourly rate was to remain the same, but overtime was to begin after 52½ hours per week.[50]

This adjustment was insufficient, however, to satisfy union objectives. Banking on the shortage of workers, the urgency of the manufacturers to get out lucrative orders and their own strength, the machinists called a strike to begin on July 18. They planned to strike one Milwaukee shop each working day, hoping to encourage the first strikers by obtaining funds from the men still at work to pay the strike benefits. Allis-Chalmers, the largest employer in the area, was the objective for the first day. Faced with an ultimatum for an eight-hour day from a well organized union, the Metal Trades Association replied that they had only recently made a reduction in hours from 55 to 52½, and they would not entertain a proposition for further reduction nor submit the question to arbitration.[51]

At 10 a.m. on July 18 machinists at the West Allis and Reliance Works walked out. At West Allis the men were reported "light-hearted and orderly," and the streetcar company had twenty cars waiting to take them home. Although union organizer Emmett L. Adams boasted that not more than 50 machinists were left in the plant, Max Babb, speaking for the company, said that the figure was exaggerated. Babb claimed that of the 7,000 men employed by the company, between 2,300 and 2,400 were classed as machinists and of that number about 1,000 were on strike. By Wednesday the union was well ahead of its schedule of one plant per day, for Pawling & Harnischfeger, Nordberg, and Kearney & Trecker had all been struck. The union leaders announced that they were prepared for a "long, hard fight;" they would pay married men eight dollars a week and single men six dollars a week in strike benefits.[52]

On July 25, one week after the strike began, Governor Emanuel Lorenz Philipp reported his negotiations with both union and employers. "I did this," the Governor said, "because I believe industrial peace is in the interest of all concerned including the citizens of Milwaukee and the state generally." He said that the employers recognized the problem, but the union was trying to move too rapidly. The employers admitted that ultimately they would adopt the eight-hour day but, having just made one concession, they were not prepared for further concessions at that time. With twenty-one shops struck by the union, the Metal Trades Association held firm. The strike at Allis-Chalmers was concentrated at the West Allis Works where the number of workers dipped from 5555 men to 3225. But by August 14, Falk reported that approximately 700 men had returned to work and that they were returning at the rate of about 100 per week.[53]

On Saturday, September 23, the Machinists' Union met and officially voted to call off the strike at all plants on September 26. Falk reported that "the men returned without any concessions and upon exactly the same conditions existing when the strike was called." Except for a few men who had been particularly active in the strike, all machinists were back at work. In fact, by September, 100 more machinists were at work at West Allis and Reliance than when the strike began. The impact of the strike on Allis-Chalmers can be seen in the impairment of net earnings during the third quarter of the fiscal year.[54]

First quarter	$740,336.89
Second quarter	950,760.90
Third quarter	545,199.98
Fourth quarter	928,722.46

In the long run the workers benefited appreciably at Allis-Chalmers during the war years. What the machinists failed to gain through the strike of 1916, all workers at Allis-Chalmers were given on August 1, 1918, when the eight-hour day was put into effect to comply with stipulations in government contracts. In 1912 a law had made the eight-hour day mandatory for those handling government contracts. Although the law permitted the president to waive its provisions during time of war, the Naval Appropriations Act of March 4, 1917, provided that in this event time and a half must be paid for all work in excess of eight hours. Allis-Chalmers complied with the legal provisions. If the number of hours of work declined, the average earned rate per hour increased notably from January, 1913, to September, 1918.

Occupation	Jan. '13	Mar. '17	Sept. '18
Machinists	.2767	.3261	.4720
Molders Iron	.2960	.3829	.5653
Molders' Helpers	.1814	—	.3999
Blacksmiths	.3517	.4215	.5981
Blacksmith's Helpers	.2260	.2890	.4361
Patternmakers	.3074	.3568	.4885
Winders	.2006	.2557	.3844
Coremakers	.2325	.3356	.4813
Carpenters	.2712	.3135	.4581
Common Labor	.1951	.2538	.3908

Despite these wage increases, the company had "extreme difficulty" holding its employees because of wage inducements from other firms.[55]

In an attempt to hold the workers on the job, Otto Falk devised a bonus plan in late 1916 which was continued through the remainder of the war years. Employees who had worked for the company one year or more received a 10 percent bonus paid quarterly. Those working at Allis-Chalmers for a shorter period of time received a bonus based on a graduated scale. By making the full 10 percent applicable only to those who had worked a full year, he hoped to keep more men on the job longer. The labor turnover is indicated by Falk's projection that the proposed bonus would be equal to an additional 5 percent on the annual shop payroll. He hoped that the plan, by increasing labor retention, would cost the company less than the projected 5 percent. Falk believed that the expenditure would actually mean a savings to the company over the cost of losing employees and breaking in new men. A year later he was given power to grant an additional bonus to key men to retain their services.[56]

All manufacturers faced the same problems of shortage of labor and material. But on the other side of the coin, the war years proved to be a seller's market. President Falk reminded his salesmen that "orders that do not carry a profit are unattractive under existing circumstances." Lest the salesmen and buyers think that the company was seeking exorbitant profits, Falk assured them that while prices authorized by the company during the war years often appeared high in comparison with earlier prices, they actually carried "no more than a reasonable profit."[57]

Some businessmen have always found it expedient to fish in the troubled waters of wartime economy and come up with products offering the highest profit. This was not the philosophy of Otto Falk who sought to reduce business operations to a calculated certainty; war was the enemy of that objective. He took the shell contracts with a particular end in mind, that the quick profits from that source could support the company until the standard product lines could be made profitable enough to sustain the company. By the end of 1915 the war orders constituted less than 20 percent of the company's production. Save for the shell contract, the company refused to accept war orders which would require special equipment or organization, because the General did not want temporary business to interfere with regular business. Rather, he wanted Allis-Chalmers in a seller's market to take advantage of war orders in

The development of the multi-compartment grinding mill in the 1920s enabled cement and other companies to produce a finished product in one pass through the mill. Previously, they needed two separate mills for coarse and finish grinding.

regular products to improve and increase company equipment so that the firm would be better able to serve its customers when peace was restored.[58]

The company's standard lines prospered during the war. As the demand for metals and cement began to soar under pressure of modern warfare, the crushing, cement and mining department increased its sales of standard machines and in 1915 introduced the Compeb Mill which revolutionized grinding. Before this time almost all fine grinding installations in the cement industry consisted of preliminary ball mills followed by a battery of tube mills. The new design permitted both preliminary and final grinding in a single mill having two or more compartments and using only metallic media throughout. The Compeb Mill in large sizes replaced batteries of small preliminary grinders and tube mills as well as screens, elevators, and screw conveyors. This compact, self-contained and efficient machine made possible in one operation the efficient reduction of any grindable material from sizes of 1½ inch and under to any commercial fineness. Its popularity and success can be seen in the fact that in 1919 alone enough Compeb Mills were sold to cement plants to produce annually 6,000,000 barrels, an amount equal to 1/12 the total output of cement in the United States in 1918. With excellent equipment such as this, sales in crushing, cement and mining increased from $2,241,091 in 1915 to $4,796,961 in 1919.[59]

The acquisition of the flour mill division of Nordyke & Marmon Company meant that Allis-Chalmers equipment would produce 90 percent of the flour milled in the U.S.

Because the United States became the primary source of food for the Allies during the war, the government placed a high priority not only on the production of food but also on processing machinery. The milling industry was authorized to order all the equipment it wanted, and sometimes more than it needed, storing the equipment for future use. With bumper grain crops and great demand, Allis-Chalmers sales billed for grain milling products rose from $359,914 in 1915 to $1,241,919 in 1919.[60]

The demand for lumber also mounted during the war. Mills throughout the country were rebuilt, and old equipment was replaced to help fill government requirements. When the United States Signal Corps discovered that spruce was the strongest and toughest softwood for its weight and was ideal for airplanes, the demand was so great and so sudden that the Signal Corps formed a Spruce Production Division with headquarters in Portland, Oregon. In furnishing equipment for the production of spruce lumber, Allis-Chalmers led all other machinery manufacturers in the United States. When old spruce mills, remodeled and improved, still could not satisfy the demand, the company was called on to furnish two completely new lumber manufacturing plants at Port Angeles, Washington, and Toledo, Oregon. The contracts called for not only all the sawmill machinery but also a large amount of electrical equipment. Though the sales for wood processing machinery had declined during the hard times of 1913 and 1914 to a low point of $248,000 in 1915, they rose rapidly to $1,142,601 in 1919.[61]

Some of the earnings and profits during the war years were invested in the war effort. The personnel of the company subscribed to all the issues of United States Liberty Bonds as well as to the Red Cross, the Y.M.C.A., United War Chest and other relief funds. Moreover, the company itself subscribed, not only as a patriotic duty, but also to provide reserve funds that it had never had prior to the war. By the close of 1918 the company owned $1,686,350 par value of Liberty Bonds and $2 million of United States Treasury Certificates.[62]

The smokestacks of West Allis, Reliance and Norwood and the engines of Allis-Chalmers had helped achieve the final victory. As early as May 25, 1917, Secretary of War Newton D. Baker had seen the problem of war in an industrial society very clearly when he said, "War is no longer Samson with his shield and spear and sword and David with his sling; . . . it is the conflict of smokestacks now, the combat of the driving wheel and the engine." On November 11, 1918, thousands of Allis-Chalmers workers spontaneously left their work upon news that victory had at last been won. With pieces of tin, sheet steel—anything they could find that would make noise—they paraded for miles down Milwaukee's Grand Avenue. These men had provided the materials for war and for ultimate victory.[63]

Six and one-half years before this, Otto Falk had become president of the Allis-Chalmers Manufacturing Company. He had consolidated the plants and installed new equipment, won excellent rapport with company personnel, brought cooperation and loyalty to himself and the company, put the company's finances in order, and improved the quality of the products. One "final test" remained on Falk's agenda—"profits." The prospects for this, however, had not seemed particularly bright during the difficult years of 1913 and 1914. But by the end of 1915 the company was showing a fine profit. The progressive measure of success can be seen in the following figures:

Net results for first quarter, 1915	$ 8,914.99 deficit
Net results for second quarter	194,813.14 profit
Net results for third quarter	333,008.46 profit
Net results for fourth quarter	559,445.60 profit
Total	$1,078,352.21 profit

More important, the company was paying dividends. Although the accumulated arrearage on the preferred stock still amounted to 13 percent, it had made a start.[64]

In 1916, despite the strike, dividends of 9 percent were paid, and the prospects for the next year were even brighter. Net income for 1917 was $4,010,490.51, with dividends on the preferred stock of 10 percent. Over and above this, the book surplus after the deduction of the dividends amounted to nearly $5.5 million. The volume of business remaining on hand totaled $27 million, substantially in excess of the unfilled orders on hand at any time in the history of the company.[65]

Because of wartime conditions, costs and wages continued to rise but so did profits. After paying another 10 percent dividend in

During World War I, Allis-Chalmers equipment prepared large amounts of spruce lumber for airplanes in the sawmills of the west coast.

1918, 3 percent of which applied to arrearages, the net profit was more than $4.5 million, and the surplus had risen to nearly $8.5 million.[66]

With the war's end, 1919 showed a contraction of the company's business. Cancellation of government war contracts amounted to about $3 million. Because half of this sum had never been entered on the company's books and no work had been completed under those contracts, there was nothing to write off. Nonetheless, the net profit for the year still came to $3,599,713.56, and the net working capital had increased nearly $2.5 million over the previous year to a total of over $22 million. The balance sheet showed current and working assets of the company at $28,204,081.75. With the current liabilities at $6,141,198.44, the company was in excellent financial condition.[67]

When the dreary dividend record of the Allis-Chalmers Company from 1901 through 1911 is kept in mind, the most important part of the financial history of Allis-Chalmers from 1913 through 1919 is its record of dividend payments. During 1919 11 percent was paid as dividends on the preferred stock. Of this amount 7 percent made up the regular dividend and the remaining 4 percent paid up all the arrearages. For the first time in nearly twenty years the company was clear of financial obligations. General Falk had also achieved the final principle of "the true executive function. . . . *profits.*"[68]

Otto Falk called 1918 "the most eventful year in all history." He thanked the entire organization for the "earnest support and efficient service" which had brought maximum production and provided real assistance in winning the war. The next task was to convert from a war to a peace program. Falk had every confidence that this could be done successfully, and that the future held promise. If Otto Falk had great confidence in the future of Allis-Chalmers, the Board of Directors had every reason to trust him as president. In "appreciation of his services and his successful management," the board on March 6, 1919, voted him a gift of $50,000.[69]

Under Falk's dynamic leadership the company now faced a new decade. The bright prospects of 1913 had materialized, and it was time to look to the future. In his New Year's greeting for 1920, General Falk said, "The Company has endeavored to pursue a course which would enable it to meet present day problems and at the same time build for future permanent success." A conservative appraisal was characteristic of Otto Falk. But he had instilled a confidence and aggressiveness in the personnel of the company that can best be seen in a statement in the April, 1919, *Sales Bulletin:* "Speaking of being unprepared for peace, if we do as well in peace as in war without preparation, well, may the Lord have pity on our trade rivals." The challenge had been met, problems had been solved, and the future seemed assured.[70]

NOTES TO CHAPTER SIX

1 Unfortunately Otto Falk left no personal papers and virtually no executive correspondence. *Circular Letter* No. 16, July 1, 1912, No. 30, April 11, 1913. The *Milwaukee Sentinel* (evening edition) for April 16, 1913, notes that a bill to force Allis-Chalmers to reimburse all those who lost money through the reorganization was killed by the state Senate.

2 *Dictionary of Wisconsin Biography* (Madison, 1960), p. 126; John G. Gregory, *History of Milwaukee, Wisconsin* (Chicago, 1931), 4:464–466; *Power Review*, July, 1940, p. 5; Lyman Anson, "That is Why They Call Him 'Milwaukee's Foremost Citizen,' " *Electrical Manufacturing*, January, 1931, p. 31.

3 Anson, *Electrical Manufacturing*, p. 31. For the purchase of the Franz Falk Brewing Company by the Pabst Brewing Company, see Thomas C. Cochran, *The Pabst Brewing Company* (New York, 1948), pp. 83, 189–190.

4 *Milwaukee Sentinel*, January 5, February 7, 1911, October 27, 1937; *Milwaukee Journal*, June 6, 1907, October 27, 1937, Gregory, 4: 466–467.

5 *Milwaukee Journal*, July 28, 1906; Arthur Van Vlissingen, "50,000,000 New Dollars A Year," *Forbes*, June 1, 1938. n.p.; *Power Review*, July, 1940, pp. 4–5.

6 A-C Mfg. Co. Board of Directors, March 15, 1913; *Milwaukee Journal*, March 18, 1913.

7 A-C Mfg. Co. Board of Directors, October 7, 1915, May 5, 1916, December 2, 1927; A-C Mfg. Co. Executive Committee, March 17, 1913, August 2, 1915. The Executive Committee was a closely knit group. On December 10, 1901, General Falk had married Elizabeth A. Vogel, daughter of Fred Vogel, Jr. Their daughter, Elizabeth Louise, married into the Pabst family. When Gustave Pabst resigned from the board in October, 1915, he was replaced early the next year by Charles F. Pfister, a brother of Mrs. Fred Vogel. Otto Falk held directorships in the First National Bank (Fred Vogel, Jr., president), The Wisconsin Trust Company (Oliver C. Fuller, president), and the Hotel Pfister Company.

8 For the rise of the professional executive in the twentieth century and the declining power of boards of directors, see Thomas Cochran's *The American Business System* (Cambridge, 1957), pp. 11–12, 66–67; Otto H. Falk, "How a Change in Policy Saved Our Business," *System*, February, 1922, p. 136.

9 A-C Mfg. Co. Board of Directors, June 28, 1913; Neil M. Clark, "How General Falk Converted Bankruptcy into Profits," *Forbes*, February 15, 1926, p. 10.

10 A-C Mfg. Co. Board of Directors, Report by General Falk on the conditions of the company, March 24, 1913; Otto H. Falk to department managers, district managers, and department salesmen, April 28, 1913.

11 Clark, *Forbes*, pp. 10–11.

12 A-C Co. Executive Committee, April 25, 1904. As early as 1904 President Warren was recommending that the Fraser & Chalmers plant be closed. A-C Mfg. Co. Executive Committee, July 10, December 31, 1913; A-C Mfg. Co. Board of Directors, February 5, 1914; *First Annual Report*, April 6, 1914, p. 8; Charles Allendorf to author, August 10, 1959.

13 A-C Mfg. Co. Board of Directors, March 4, 1921; A-C Mfg. Co. Executive Committee, September 12, 1921, November 14, 1931, October 29, 1935, August 15, 1940; *The Times*, August 31, 1922. During 1935–1940 the last of the Reliance buildings were torn down because they "were in such condition as to be a danger to anyone who might be in them or to the public who might be passing by."

14 *Power Review*, July 1940, p. 5; Falk, *System*, p. 199. There was some resistance within Allis-Chalmers to the sale of the land on which the Reliance Works stood. Some held that it was an excellent location to serve the then projected St. Lawrence Seaway. C. A. McCormack to Alberta J. Price, July 19, 1945.

15 Falk, *System*, pp. 136–137, 199; Harold U. Faulkner, *The Decline of Laissez-Faire 1897–1917* (New York, 1951), pp. 121–122; Thomas C. Cochran and William Miller, *The Age of Enterprise* (New York, 1956) pp. 243–244.

16 Falk, *System*, p. 137; Anson, *Electrical Manufacturing*, p. 31. In his letter to departmental managers and salesmen of April 28, 1913, Falk wrote, "I am counting upon your hearty co-operation and am prepared to give you every assistance in my power. No one should hesitate to take up with me at any time anything pertaining to the welfare of the company. I shall welcome suggestions and appreciate any helpful information that is given to me." Also see *Power Review*, July, 1940, p. 5.

17 Falk, *System*, pp. 139, 199; Anson, *Electrical Manufacturing*, p. 31; Van Vlissingen, *Forbes*, n.p.; Clark, *Forbes*, p. 32. Many employees of Allis-

Chalmers have mentioned to the author that they took the General's invitation at face value and always felt free to go to him and were always well received.

18 R. S. Flesheim, "Memo on Graduate Training Course," October 1, 1947. Flesheim, a graduate of the University of Michigan, was a member of the class that began work in 1904. A-C Co. Executive Committee, April 5, 1906; Graduate Student Apprentice Program, July 6, 1908, July 18, 1908; A-C Mfg. Co. Executive Committee, November 13, 1916; Falk, *System*, p. 138.

19 Falk, *System*, p. 136; C. A. McCormack to Alberta J. Price, July 19, 1945; Charles Allendorf to author, August 10, 1959; Irving H. Reynolds to Alberta J. Price, August 31, 1945.

20 Falk, *System*, p. 137; J. L. Neenan to author, July 29, 1959.

21 Falk, *System*, p. 138; Clark, *Forbes*, p. 32.

22 A-C Mfg. Co. Board of Directors, July 3, October 1, 1913, October 7, 1915; A-C Mfg. Co. Executive Committee, June 16, 1915.

23 A-C Mfg. Co. Board of Directors, December 6, 1915; A-C Mfg. Co. Executive Committee, January 8, 1917.

24 Anson, *Electrical Engineering*, p. 32.

25 Charles Warner to author, July 27, 1960; Otto H. Falk and D. W. Call to C. V. Edwards, April 11, 1912. The proposal was accepted by Edwards in a letter dated April 16, 1912.

26 Falk, *System*, pp. 136–137; Charles Allendorf to author, August 10, 1959; *Circular Letter* No. 18, January 24, 1914. C. R. Beck expressed the general opinion of General Falk when he said, "A fine man and a damn good president." He also added, "Unlike the previous period, he watched where every dollar was going." Beck to author, July 24, 1959.

27 A-C Mfg. Co. Board of Directors, December 4, 1913, February 5, March 5, 1914. In the board meeting of December 12, 1916 action was taken to dissolve the Bullock Company.

28 Falk, *System*, p. 137; A-C Mfg. Co. Board of Directors, June 28, 1913, Report by Otto Falk to the board; Presidential Letter, May 20, 1913; Irving H. Reynolds to Alberta J. Price, August 31, 1945; Charles Allendorf to author, August 10, 1959. William Watson succeeded Search as Works Manager and served in that capacity until 1943.

29 Ernest Shaw to author, August 3, 1959.

30 *A-C Views*, September 27, 1954, pp. 4–5; *Sales Bulletin*, January, 1924, p. 287; A-C Mfg. Co. Executive Committee, April 10, 1916, for terms and conditions of a five-year contract with W. M. White.

31 *A-C Views*, September 27, 1954, pp. 2, 12; A-C Mfg. Co. Executive Committee, May 27, 1918; A-C Mfg. Co. Board of Directors, August 6, 1914.

32 *A-C Views*, October 27, 1954, p. 2; W. W. Nichols to Walter Geist, June 2, 1944.

33 *Milwaukee Sentinel*, November 25, 1915; *Allis-Chalmers Gas Engines & Generators, 1000 kva each and Larger*, MS copy, n.d. On November 29, 1915, it was announced to the Executive Committee that a contract for fourteen gas engine units costing $1,650,000 had been awarded to Allis-Chalmers by the Illinois Steel Company.

34 *First Annual Report*, April, 1914, p. 8; A-C Mfg. Co. Board of Directors, June 28, 1913. In his report to the board, Falk noted that the Westinghouse Company had shown a profit of only $38,961.90 for the year.

35 *Second Annual Report*, April 9, 1915, p. 7; A-C Mfg. Co. Board of Directors, March 5, 1914, February 4, 1915; *Circular Letter* No. 19, May 25, 1914; *Milwaukee Sentinel*, September 26, 1914, January 1, 1915.

36 A-C Mfg. Co. Executive Committee, March 2, 1915; *Circular Letter* No. 40, March 3, 1915; *Milwaukee Sentinel*, December 24, 1914, January 1, 1915.

37 A-C Mfg. Co. Executive Committee, March 24, April 9, 29, July 23, 1915.

38 *Milwaukee Journal*, March 4, 1902, March 20, 1915; *Milwaukee Sentinel*, December 11, 1914, March 21, June 16, 1916; Circular Letter No. 43, June 12, 1915.

39 *Milwaukee Sentinel*, May 4, 5, 1915; A-C Mfg. Co. Board of Directors, May 11, 1915; A-C Mfg. Co. Executive Committee, November 22, December 28, 1915, June 19, 1916. Still in *Milwaukee*, pp. 455–458, details the pro-German movement in Milwaukee during the early years of the war. Will C. Conrad, Kathleen Wilson and Dale Wilson, *The Milwaukee Journal: The First Eighty Years* (Madison, 1964), pp. 85–102, discuss the role of the *Journal* in combatting the pro-Germanism of the war years.

40 *Milwaukee Journal,* May 21, August 31, 1915, in Allis-Chalmers section, industry clipping files, Milwaukee County Historical Society. Real difficulty exists in determining the full role of Allis-Chalmers in war production. The reason for the lack of facts in some instances can be found in *Circular Letter* No. 90, April 20, 1918.

We have been cautioned by various Governmental Departments regarding information reaching the public concerning government orders and contracts which have been placed with us. You are therefore instructed in the future not to discuss or give out any information whatsoever regarding government work in our shops.

Effective May 1st and during the period of the war we will discontinue listing orders in our monthly sales bulletin.

Otto H. Falk, President

41 Charles Warner to author, July 27, 1960; A-C Mfg. Co. Executive Committee, September 20, 1915, December 11, 1917, February 4, 1918.

42 *History of the Chicago District, United States Army Ordinance,* United States Government, 1918, extract in Allis-Chalmers files; A-C Mfg. Co. Executive Committee, August 2, 16, 1915, July 24, 1916; A-C Mfg. Co. Board of Directors, August 6, 1920.

43 A-C Mfg. Co. Executive Committee, April 16, 1917; Mark Sullivan, *Our Times* (New York, 1932), 5:227–228, 390–399; *Milwaukee Sentinel,* October 20, 1915.

44 G. William Warner, "What Allis-Chalmers Did During the Last War," MS. c. 1943; J. J. Kern to author, June 15, 1961.

45 J. J. Kern to author, June 15, 1961; Malcolm C. Maloney to author, June 13, 1961; *Sales Bulletin,* February, 1919, p. 14. The last launching of the Hanlon Dry Dock Company was in 1921. But it is interesting to note that as of 1959 one of these ships equipped with an Allis-Chalmers engine was still in operation. It was originally commissioned as the *Medon,* but the name was later changed to the *Mary Olson.* The gross tonnage of the 320-foot vessel was listed at 3,474 tons. *Record of the American Bureau of Shipping* (New York, 1959), p. 964.

46 A-C Mfg. Co. Executive Committee, January 7, May 27, 1918; A-C Mfg. Co. Board of Directors, November 1, 1917; Warner, MS; R. C. Allen to author, July 3, 1961; *A-C Views,* September 27, 1954, pp. 6, 12. Also see Walter F. Peterson, "Allis-Chalmers: Great Lakes Arsenal of Democracy During World War I," *Inland Seas,* Winter, 1962, pp. 256–266. The original destroyer contract called for turbines for forty destroyers at a cost of $9,020,000; however, with the cessation of hostilities orders for six were cancelled. Some materials for these had been manufactured prior to cancellation of the contract, and the company shipped this equipment to the Norfolk Navy Yard as a gift. It is interesting to note that illustrations 21 and 211 for sections of the High Pressure Turbine End and the Low Pressure Turbine End found in *Naval Machinery* (Annapolis, 1937) as issued by The United States Naval Institute are of the Allis-Chalmers turbines.

47 A-C Mfg. Co. Executive Committee, August 9, 1915, April 30, 1917; *Sales Bulletin,* March, 1916, p. 1.

48 A-C Mfg. Co. Executive Committee, August 30, 1915, through December 9, 1918. Otto Falk reported monthly to the committee noting areas of increase or decrease.

49 A-C Mfg. Co. Executive Committee, October 23, 1916, August 13, October 15, November 22, 1917, June 3, 1918; *Circular Letter* No. 76, May 2, 1917.

50 A-C Mfg. Co. Executive Committee, January 27, June 19, July 5, 1916.

51 *Milwaukee Journal,* July 18, 1916.

52 A-C Mfg. Co. Executive Committee, July 17, 24, 1916; *Milwaukee Journal,* July 18, 19, 20, 21, 1916.

53 *Milwaukee Journal,* July 25, 26, 1916; A-C Mfg. Co. Executive Committee, July 31, August 14, 21, September 25, 1916. The Milwaukee branch of the National Metal Trades Association agreed that their various plants should render mutual assistance during the strike. When Bucyrus-Erie took in work from the strike-bound Allis-Chalmers plant, the Bucyrus workers went on strike as well. See Harold F. Williamson and Kenneth H. Myers II, *Designed for Digging* (Evanston, Illinois, 1955), p. 127.

54 *Circular Letter* No. 70, September 28, 1916; A-C Mfg. Co. Executive Committee, August 11, 1916; *Fourth Annual Report,* April 10, 1917, p. 8.

55 A-C Mfg. Co. Executive Committee, April 18, July 22, 26, 1918; "Allis-Chalmers Manufacturing Company Statement Showing Average Hourly Rates Paid Following Classifications. West Allis Works." n.d.

56 A-C Mfg. Co. Executive Committee, December 26, 1916; A-C Mfg. Co. Board of Directors, December 6, 1917: *Circular Letter* No. 87, January 9, 1918.

57 *Sales Bulletin*, March, 1916, p. 1.

58 *Sales Bulletins*, August, 1915, p. 2, December, 1915, p. 1; *Third Annual Report*, April 6, 1916, p. 8. The General was apparently successful in keeping special war orders to no more than 20 percent of production for the remainder of the war.

59 *Milwaukee Sentinel*, November 16, 1915; *Sales Bulletin*, January, 1920, p. 13; Memorandum to George Smith, April 14, 1947; Sales Billed (Corporate), Industrial Equipment Division, 1913–1954; C. R. Beck to author, July 24, 1959.

60 J. L. Neenan to author, July 29, 1959; Sales Billed (Corporate), Industrial Equipment Division, 1913–1954.

61 *Sales Bulletins*, September, 1918, p. 3–4, February, 1918, p. 17; Sales Billed (Corporate), Industrial Equipment Division, 1913–1954.

62 *Sixth Annual Report*, April 4, 1919, p. 10; A-C Mfg. Co. Board of Directors, June 7, August 2, 1917.

63 Warner, MS; Mark Sullivan, *Our Times*, 5:372.

64 A-C Mfg. Co. Executive Committee, June 28, 1913, July 23, 1915; A-C Mfg. Co. Board of Directors, December 6, 1915, February 3, 1916; *Third Annual Report*, April 6, 1916, p. 8.

65 A-C Mfg. Co. Board of Directors, February 7, 1917; *Fourth Annual Report*, April 10, 1917, p. 9; *Fifth Annual Report*, April 6, 1918, pp. 8–9.

66 *Sixth Annual Report*, April 4, 1919, pp. 10–11.

67 *Seventh Annual Report*, April 9, 1920, p. 10; A-C Mfg. Co. Executive Committee, November 18, 1918; A-C Mfg. Co. Board of Directors, December 5, 1918.

68 *Seventh Annual Report*, April 9, 1920, p. 10.

69 *Sales Bulletin*, December, 1918, p. 1; A-C Mfg. Co. Board of Directors, March 6, 1919.

70 *Sales Bulletin*, April, 1919, p. 15, December, 1919, p. 15.

William Monroe White
Chief Engineer

CHAPTER SEVEN

THE TWENTIES: DECADE OF PROSPERITY

PEACE IS THE GREAT AND INEVITABLE result of war. Otto Falk, who had not allowed Allis-Chalmers to become solely involved in war production, looked forward to peace. Throughout the decade of the 1920s the directing and pervasive force behind Allis-Chalmers was more than ever that of Otto Falk. With all arrearages paid and with an efficient plant and able staff at his disposal, he had an opportunity to develop the company along new lines, especially in the electrical area. The processing machinery field faced problems in some areas and opportunity in others. Power generation, however, both steam and hydroelectric, moved into a new era of prosperity. But the 1920s was also a period of severe competition which thwarted some efforts and demanded new approaches in others.

The twenties opened inauspiciously with a depression. Orders booked by Allis-Chalmers declined steadily through the early months of 1920, hitting bottom in August with only $850,000. With the end of the war the number of company employees had leveled off at around 7000 men through 1919 and 1920. But during 1920 about 200 men were laid off each month until a low point of 4,480 men was reached in February of 1922.[1]

Economic dislocation was also a result of war, and General Falk initially had the responsibility for dealing with the problem of decreasing business. Wartime bonuses were withdrawn and wage reductions imposed. Falk told the Executive Committee that from December, 1920, to September, 1921, production had decreased by 58 percent, but factory expense had been reduced by 57 percent to offset it. He had anticipated the course of the depression and reduced the entire situation to its proper perspective. In December, 1920 he wrote:

> Our Government, our institutions and our industries are sound; and our people are intelligent. After a brief period of readjustment, this country, restored to a normal basis, will resume active business, expand and grow, and enjoy prosperity and life as no other people may hope to do for generations to come.

The depression was a time for "every man to be up and doing." Allis-Chalmers weathered a short period of readjustment and by 1922 prosperity had returned.[2]

A graph of the period indicates that the recovery began in 1922, developed into a business boom which reached a peak in mid-1923, then dipped slightly in 1924 only to rise again in 1925 and 1926. Following a slight dip toward the end of 1927, the line moved up through 1928 to the peak of prosperity in 1929 before plunging into the abyss of 1930 and 1931.

Year	Indus. Prod.	Wholesale Prices	National Income (Billions)	Real Income Per Capita (1929 prices)
1921	58	97.6	$59.6	$522
1922	73	96.7	60.7	553
1923	88	100.6	71.6	634
1924	82	98.1	72.1	633
1925	90	103.5	76.0	644
1926	96	100.0	81.6	678
1927	95	95.4	80.1	674
1928	99	96.7	81.7	676
1929	110	95.3	87.2	716

This pattern was seen quite distinctly in Milwaukee. From 1922 to 1928 the value of the city's manufactured products rose from $789,519,605 to $1,053,472,000. During 1929 manufactures gained $100 million. The metal trades accounted for much of this, increasing 68 percent in dollar value between 1923 and 1929 and increasing in volume of production from 37 percent to 48 percent of Milwaukee's total manufactures.[3]

In the 1920s considerable interest arose in plants to treat wood for preservation. This shows a tie treating plant built for the Atchison, Topeka and Santa Fe Railroad.

Allis-Chalmers accounted for a substantial part of the industrial growth of Milwaukee. Company sales for 1913, the year Otto Falk was elected president, totaled $11,127,621. During the peak war year of 1918 this figure tripled to $35,031,234. With cancellation of war contracts and the depression, sales steadily declined until 1922, when they stood at $20,794,046. From that low point a steady increase took place so that 1928, with sales of $36,294,562, surpassed 1918, and 1929 set a record of $45,302,356. The net profit on billings for the twelve-year period from 1919 through 1930 indicates a remarkably consistent pattern:

1919	11.9%
1920	11.3
1921	9.0
1922	10.6
1923	10.6
1924	11.6
1925	11.8
1926	11.7
1927	9.5
1928	8.1
1929	9.6
1930	8.7

The twenties may have been "the era of disillusionment" for some, but the unparalleled prosperity of the company was not disillusioning to Otto Falk. He recognized that he was "living in a time different from any within our experience," but this was what he had been waiting for.[4]

As president of Allis-Chalmers during World War I, Otto Falk had proven his judgment, foresight and leadership. As his abilities showed themselves more and more he was given larger powers. On March 1, 1917, the Board of Directors gave the Executive Committee, "acting through the President," the power to make donations to charitable and worthy causes. Although this was an insignificant authority at that time, Otto Falk used it increasingly in the prosperous twenties. In 1920 when the office of chairman of the board was abolished, Falk became the presiding officer at board meetings. After 1920 more and more items on the agenda of the board "were left to the President with power." Allis-Chalmers became the visible industrial image of Otto Falk.[5]

General Falk recognized that the future of Allis-Chalmers, or any other industry, was based on constant improvement of the product lines. The sound financial situation achieved during the war provided the basis for the substantial profits of the twenties. These profits, in turn, were increasingly invested in research and development as the decade progressed. In 1922 and 1923 Allis-Chalmers, along with other industries, contributed to research on metal fatigue at the University of Illinois under the Engineering Foundation and the National Research Council. A new and modern chemical laboratory was built at West Allis in 1924 because "the work performed by this department is exceedingly intricate and important and it is greatly to the Company's interest." By 1928 considerable improvement had been made in the use of welded steel construction instead of cast iron and steel, which helped to reduce the investment in patterns. A new hydraulic turbine testing pit and cavitation laboratory, one of the most completely equipped in the country, was put into service at the West Allis Works during the summer of 1930. By 1929 the annual expenditures for experimental and development work had reached $1,317,957.57. This cost was equal to $1.15 a share. The *Chicago Journal of Commerce* wrote, "The outlay has been justified by the position in which the company finds itself in the fore ranks of manufacturers of heavy machinery and in electrical and mechanical equipment, a position which has been attained through keeping abreast of the latest discoveries in its lines of endeavor."[6]

Research provided the basis for improvement; standardization brought greater economy of operation. During the early twenties the Department of Commerce under the aggressive leadership of Herbert Hoover was trying to eliminate the existing industrial chaos through standardization. This effort was not lost on Allis-Chalmers. General Falk organized the Engineers Standardization Committee to introduce uniform practices and procedures in 1921, to keep engineers in touch with developments in inter-departmental apparatus and to minimize expense. Such action was essential for, by 1930, the company was divided into eight separate departments, each with its own manager and engineering specialists; these were the electrical department, steam turbine depart-

The twenties also saw the modernization of sugar mills, particularly in Cuba. Allis-Chalmers supplied steam turbine generator units for electrical power and motors for driving the cane crushing rolls.

ment, hydraulic turbine and centrifugal pump department, engine department, pumping engine department, crushing, cement and mining machinery department, milling machinery department, and tractor department. Standardization in a company so large and so diverse was necessary for orderly development.[7]

Between 1923 and 1929, manufacturing output per manhour in the country rose almost 32 percent. While much of this gain on the national scene represented gains in mass production industries, the results of research and standardization can be seen in the increasing efficiency of Allis-Chalmers operations.

Year	Inventory Turnover in Net Sales	Sales Per Dollar of Plant	Ratio of Sales to Working Capital	Ratio of Sales to Total Assets	% Income to Invested Capital
1921	1.97	0.78	1.03	.41	4.07
1922	2.08	.66	.87	.35	4.05
1923	2.06	.84	1.06	.42	4.93
1924	2.30	.89	1.12	.44	5.75
1925	2.35	.93	1.10	.45	6.02
1926	2.35	1.00	1.12	.47	6.24
1927	2.61	.98	1.57	.54	7.85

This steadily increasing efficiency was a tribute to the administration of the company.[8]

Under Falk's leadership, Allis-Chalmers had become a largely self-contained unit in terms of personnel. He had brought the stability, security and opportunity necessary to hold men. Trade apprenticeships as patternmakers, molders, machinists, woodworkers, tool and die makers, electrical mechanics, blacksmiths, and boilermakers were open to boys who had completed at least

Allis-Chalmers has long been a leader in the training of trade apprentices. This group was training in the main foundry.

the eighth grade. Apprentices were also encouraged to complete high school by attending continuation schools. As early as 1916, 210 boys were continuing their high school education while under apprenticeship contract. Of these, 68 were in school in West Allis and 142 in Milwaukee. Drafting apprenticeships were available to boys who had elementary drafting training in high school.

Training was also carried on in cooperation with engineering schools. Arrangements were made with Marquette University for engineering students to alternate periods of study at the school with periods of work in the Allis-Chalmers shops, giving the students practical experience as well as academic knowledge. This system was later enlarged to include other engineering schools. The graduate student system, established in 1904 and given hearty support by Falk during the war years, continued through the twenties as an important part of the training program. Between twenty and twenty-five young men were taken in each year for the two-year program. The result of all these training programs was that between 1918 and 1928, with one exception, every executive position in the shop was filled by men trained in the plant.[9]

Plant facilities received the constant attention of Otto Falk during his twenty-seven years with Allis-Chalmers. Contending with neither war nor depression in the twenties, he kept all the works, and particularly the West Allis Works, in first-class condition. Consolidation of manufacturing facilities in Milwaukee had been achieved in 1922–23 with the closing of the antiquated Reliance Works. Expansion was provided for by the purchase of land to the east of the West Allis plant. Throughout the decade, additions were made almost yearly to the West Allis Works to keep

At its installation in 1928, this forty-foot boring mill was the largest machine tool of its kind in the United States.

pace with expanding business, to enlarge and improve existing facilities and to add completely new ones. Typical of these improvements was the erection in 1928 of a great forty-foot boring mill costing $175,000. The mill, largest of its kind in the United States, provided unexcelled facilities for machining the large component parts needed for machines of ever-increasing size. This was only a part of the 1928 program of capital expenditures which totalled $1,790,588. A new shop completed in 1930 was equipped with 60- and 100-ton cranes and a large vacuum tank for drying out power transformers. Some conception of the size of the great West Allis Works in 1930 can be gained from the fact that if all its buildings had been placed end to end, they would have extended two and three-quarter miles. Other significant measurements were:

Total floor area of plant, square feet	2,901,639
Total ground area, acres	155.58
Plant boiler horsepower	10,045
Miles of railway track	21
Number of cranes	162
Heaviest casting produced (tons)	120

The *Chicago Journal of Commerce* stated, "It is doubtful whether any other plant is better tooled and equipped to build heavy machinery."[10]

With these facilities as the backbone of its operations, the company could then launch a program to expand its product lines and facilities. In this, Allis-Chalmers was in the mainstream of American business in a period of industrial mergers. During the years 1919 to 1928 nearly 6000 independent mining and manufacturing enterprises were absorbed by other companies. In part this trend stemmed from the belief that a large concern with a varied business would be more stable than one devoted to fewer lines. To some extent it came from a desire to economize operation. If at one and the same time some competition could be eliminated, and one salesman could represent two or more lines of goods, the cost of selling would decline and profits would increase. Advances in technology also made it easier for companies to spread out. Communication, the nerve system of business, was constantly improving. Innovations in accounting techniques now allowed management to maintain a head office in a metropolitan center such as Milwaukee, and from there efficiently supervise plants in areas of lower cost, closer to raw materials.

Operating an efficient and consistently profitable concern, the General was prepared in the twenties to extend the scope of operations. In 1924 Allis-Chalmers bought from the Worthington Pump and Machinery Corporation its line of mining, crushing, cement and creosoting machinery. In 1926 Falk engineered the purchase of the Hoar Shovel Company and the Midwest Engineering Company. Later the same year he bought the flour milling

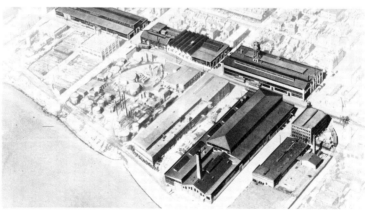

The company continued to expand. Additional tractor manufacturing facilities were constructed at the West Allis Works, and the Norwood, Ohio, plant was added in 1904. In the 1920s, Allis-Chalmers acquired plants in La Crosse, Wisconsin, Springfield, Illinois, and Pittsburgh, Pennsylvania.

business of Nordyke & Marmon Company and the following year purchased the Pittsburgh Transformer Company. The growing importance of the tractor division was evident in the purchase of the Monarch Tractor Company in 1928 and the La Crosse Plow Company in 1929. In 1930 Falk added the Stearns Motor Company to his list of acquisitions. By 1930 Allis-Chalmers had, besides its main plant in West Allis, plants in Norwood, Ohio, Pittsburgh, Pennsylvania, Springfield, Illinois, and La Crosse, Wisconsin. All told, these plants covered 333 acres and had a combined floor area of 4,831,561 square feet. Although Allis-Chalmers and its management expanded during the twenties, the individual plants were kept at the size best suited to their particular operations.[11]

The processing machinery field included sawmilling, flour milling, crushing, cement and mining equipment. In this field, Allis-Chalmers maintained its position as an important producer and in some of these areas became the leading manufacturer.

By 1920 Allis-Chalmers had been building sawmills for sixty-one years and bandmills for thirty-five. Its position in the field had been established by building more mills than any other manufacturer and by providing some of the best machinery. It was also the only company that could offer a complete plant, including all electrical equipment and the power plant.

In 1920 there was every reason to anticipate a building boom which would presumably call for more sawmilling machinery. In 1918 only 20,000 new houses had been built in the United States and in 1919, 70,000. The average number of American families for every 100 dwellings had increased from 110.5 families in 1890 to 121 families by 1920. In view of these statistics and in anticipation of increased business, Allis-Chalmers brought Ernest Shaw back to the company in 1920 to redesign all their sawmill machinery. Shaw's type "C" mill and other sawmill refinements offered some of the most modern and efficient equipment on the market.[12]

From 1919 through 1924 wood processing billings averaged about $1 million a year. In 1925, however, sales dropped by half, a situation that continued for the rest of the decade. Sawmilling was the one department that failed to grow during this generally prosperous decade. The decline in sales resulted from the fact that by 1924–1925 competitors in the industry had picked up the ideas introduced by Allis-Chalmers. But it stemmed also from a change in departmental managers. With the death of John F. Harrison in 1919, Allan Hall became departmental manager. Since both flour milling and sawmilling were under the same department head, the success of either product line depended on the interests of the departmental manager. Harrison had been a sawmill man for thirty-six years, while Hall was a flour mill man with little interest in sawmilling. Other companies developed lightweight, portable mills which met the changing requirements of the lumber industry. But perhaps the chief reason for declining sales was that sawmilling technology had reached something of a dead end.

There was constant refinement of band saws, edgers and trimmers and development of horizontal and vertical re-saws, but there was no development which would make all previous methods obsolete and open the door to sales of new machinery. Once Allis-Chalmers mills were installed, they required only replacement parts. The result was that the sawmilling department was languishing well before the depression virtually finished it.[13]

The Reliance Works and the Allis name had been associated with flour milling since 1847. With the introduction of the roller process in the late nineteenth century, the Allis Company led the industry. In the course of time, constant improvements were made in flour milling machinery. They included purifiers, separators, graders, scourers, finishers, bolters, dusters and the placing of roller mills in batteries, and all of these contributed to the production of better flour at lower cost. During the twenties plant cleanliness was increasingly stressed. Toward the end of the decade laboratories were being installed so that grain could be analyzed and blended before milling to assure uniform quality of flour.[14]

Under pressure of war, sales of Allis-Chalmers flour milling equipment soared to $1,581,349 in 1920. The following year sales were reduced by half, and a constant decline set in until by 1926 they were about $318,000. The loss of foreign markets as well as the changing diet of the American people contributed to this decline. Before the war five companies in the United States sold complete mills. All these companies were affected by the deterioration of the market, and by the mid-twenties they had been reduced to three: Allis-Chalmers, Nordyke & Marmon Company of Indianapolis, Indiana, and the Wolf Company of Chambersburg, Pennsylvania.[15]

Nordyke & Marmon, having branched out into the manufacture of the Marmon Motor Car, felt its future lay in that direction. By

Rapid growth in the use of automobiles and trucks greatly increased the demand for petroleum and for economical pipeline transportation to refineries. Allis-Chalmers supplied considerable equipment for the pipelines, including the diesel engines which drove plunger pumps.

1926 they were prepared to discontinue flour mill and transmission machinery lines and were seeking a buyer. Allis-Chalmers paid $350,000 for the milling machinery business and inventory of Nordyke & Marmon. The Executive Committee reasoned that "by combining the two lines the Company would not only eliminate the competition but it would also round out its work in the milling machinery shop which has been comparatively slack for some time."[16]

The addition of Nordyke & Marmon equipment to Allis-Chalmers meant the union of two of the oldest and best flour milling lines in the country. The Nordyke & Marmon Company had been founded in 1851 by Ellis Nordyke and his son Addison for the manufacture and erection of flour mills. The addition of the unusual mechanical ability of Daniel W. Marmon strengthened and expanded the company to the point that Richmond, Indiana, proved to be too small to satisfy the demands for labor and materials, and in the early seventies the concern was moved to Indianapolis, where it prospered. The Nordyke line was broader than the Allis line and though its machines were more complicated, the name had an equally fine reputation. The purchase of Nordyke & Marmon brought to Allis-Chalmers the Nordyke square sifter, which had been constantly improved over twenty-five years and was better than the comparable Allis-Chalmers model. An excellent flaking mill for the preparation of cereals also came from Nordyke. During 1927, while Allis-Chalmers was offering the combined lines and also filling Nordyke & Marmon's current orders, sales shot up to $1 million, but they declined to about $700,000 for the remainder of the decade.[17]

It has been claimed that Allis-Chalmers sales after the purchase of the Nordyke line accounted for nearly 90 percent of the industry and constituted a virtual monopoly in flour milling. Even a monopoly, however, could not bring about a continued sizable increase in sales. The truth was that with declining demand, the United States had more flour mills than it needed. As early as 1923, if all mills had been run at capacity they would have produced nearly three times the national demand. Since both the Allis-Chalmers and the Nordyke machines were very well constructed and no inventions had made existing machines obsolete, there was no great demand for new machines. The monopoly amounted to little more than one of supplying replacement parts. Grain milling machinery simply provided a stable but unspectacular source of sales through the entire period.[18]

The decade of the twenties was the heyday of the crushing, cement and mining machinery department. The average annual sales of around $2 million from 1913 through 1916 doubled to about $4 million from 1917 through 1923. Sales rose to a peak of $8 million in 1926 and continued at roughly this high level through the remainder of the twenties.[19]

The famous Gates crushers, brought to Allis-Chalmers in the merger of 1901, had long constituted the standard line of the firm. By the twenties this design was outdated by newer crushers. Traylor Engineering Company of Allentown, Pennsylvania, Kennedy-Van Saun Engineering Company of New York, Smith Engineering Works of Milwaukee, Austin Manufacturing Company of Chicago, and Chalmers and Williams of Chicago offered ever keener competition in the crusher field. In an attempt to reestablish its supremacy, Allis-Chalmers in 1924 purchased the crushing, cement, mining and creosoting lines from the Power & Mining Division in Cudahy, Wisconsin, of the Worthington Pump & Machinery Corporation, Harrison, New Jersey.[20]

The Power & Mining Machinery Company was founded at the

Another world's record was achieved when Allis-Chalmers built two sixty-inch all-steel gyratory crushers for crushing copper ore in the Andes Mountains of Chile. Sectionalized for easier transportation, each complete crusher weighed a million pounds.

beginning of the century by Harry Holthoff and William D. Gray, both former engineers with Allis-Chalmers. Their McCully crushers introduced in 1911 were superior to the Gates and Blake crushers produced by Allis-Chalmers. Otto Falk was most interested in the McCully crusher and was willing to pay $1,125,000 for the rights to this machine as well as the patents on mining, cement and timber-processing equipment. He was not interested in the Cudahy plant, since these lines could be produced in the West Allis Works. With its excellent facilities, a fine sales organization and the addition of an outstanding line of crushers, the company for a time pulled ahead of its competition.[21]

Shortly after the acquisition of the McCully line, the company was offered the patents for a reduction crusher developed by Will Symons, a Chicago engineer. But Otto Falk thought that $2 million was too much for the patents to this secondary, high speed crusher, and they were taken by the Nordberg Company of Milwaukee. In time, however, the Symons Cone Crusher virtually killed the sales of the McCully fine reduction crusher and seriously cut into all the smaller size crusher sales. Moreover, this introduced the Nordberg Company into the crusher field, and it eventually brought out an entire line of crushers. Nordberg, along with other competitors, now brought stronger competition than ever before into this field, steadily reducing sales of Allis-Chalmers equipment.[22]

In an attempt to meet this problem, the company brought out the Newhouse crusher, which was not particularly successful. The

In 1926 Allis-Chalmers bought the right to manufacture the Hoar Shovel, operated by compressed air and used for underground mining and excavation. Few were sold, however, due to the development of the engine-driven front end loader.

Newhouse oil-adjusted crusher, at the end of the decade, was even less successful because it was too expensive and too complicated to offer real competition to the Symons Cone Crusher.[23]

Although Allis-Chalmers may have lost business in small crushers, it remained the principal producer of large crushers. Because of its extensive engineering experience, the company could handle specialized orders, including those for foreign installation. The shipment of two 500-ton Superior McCully gyratory crushers to Chile in 1926 is an example. Designed and built for the Chile Exploration Company, a subsidiary of Anaconda Copper Mining Company, these machines had to be specially constructed. On November 8 the Executive Committee adjourned to inspect the crusher that had been erected in the shop, "it being the largest and most powerful gyratory crusher ever built."[24] The crushers had received openings of 60 by 150 inches, and practically the entire machine was constructed of cast steel for excessively hard ore. The steel castings and forgings for the crushers were among the largest ever made in the United States. To meet the special transportation problems, the machines were built in sections. The lower main frame of each crusher was made in two parts, with a combined weight of 88,000 pounds. The top shell was in two horizontal sections, with the lower ring weighing 114,900 pounds and the upper part halved in two sections, each weighing 92,000 pounds. All of these sections had to be handled with a special crane on the steamer Chilcop, special lighters were made to receive them at the port of Mejillones, and special cars were built in Chile to transport the parts by a narrow gauge mountain railroad to an altitude of 9,500 feet. This was a specialized job from beginning to end.

The 1926 Chilean crushers were no more than preparation for the installation of four sixty-inch all steel gyratory crushers in the same country three years later. Each of these crushers weighed one million pounds. To be transported on the narrow gauge railroads, no single piece could weigh more than sixty tons. When installed, each crusher could handle 2,500 tons of copper ore per hour. The imposing sight of twenty-five railroad cars transporting this shipment from West Allis to New York merely indicated the size of order and degree of specialization that the Allis-Chalmers Manufacturing Company was prepared to undertake in the twenties.[25]

To strengthen the mining line, Allis-Chalmers in 1926 bought from the Hoar Shovel Company of Duluth, Minnesota, the right to manufacture a small loading shovel. A decade before this a practical miner had invented this shovel for use in the iron mines of Michigan. The Hoar Shovel was an extremely compact machine that could be successfully operated in spaces as small as seven feet high by seven feet wide and could revolve a full 360 degrees. It was designed primarily for use in underground mucking and surface excavation in very restricted quarters. By 1926 about 140 of these were in use in most of the mining districts of the United States and in a number of foreign countries. Although Allis-Chalmers did not anticipate a large market for these shovels, company officials did expect favorable possibilities for future growth. Moreover, the company risked nothing, for its only commitment was a royalty of 5 percent of the selling price on all shovels manufactured and sold under the agreement.

Relatively few Hoar Shovels were ever built by Allis-Chalmers. In part this was a result of extremely keen competition. The three principal competitors—the St. Louis Power Shovel Company, the Myers-Whaley Company of Knoxville, and the Nordberg Company of Milwaukee—usually had direct personal representation on tunnel jobs requiring this kind of shovel and spent much more time and effort in securing orders. But the main reason for lack of sales was that the rights to the Hoar Shovel were acquired only very shortly before the development of the tractor shovel, or front-end loader, which largely terminated the manufacture of the earlier machine.[26]

Some persons estimate that Allis-Chalmers did as much as 50 percent of the total mining and crushing business in the United States during the twenties. However, as the decade progressed it was the cement business that acquired an increasing importance. The Compeb Mill, which had been introduced in the last years of the previous decade, now came into its own. During the period 1920–1929, Allis-Chalmers probably did as much as 75 percent of the domestic business and certainly was the prominent American concern in the foreign cement business. The increasing use of cement in poured concrete buildings at the same time that the United States began to pull its ever growing number of automo-

This laboratory tested samples from cement and mining companies under a variety of conditions and with different types of equipment.

biles out of the mud by placing them on paved roads helped to produce this demand. Then, too, American builders were finally disabused of the notion that German cement was superior to American portland cement. Suspicious of the superiority of European cement, Ray C. Newhouse bought a complete set of German testing equipment and installed it in the Allis-Chalmers laboratory next to American testing machinery. Samples of cement, both domestic and foreign, were tested by the two methods and it was found that the technique used by the German testing apparatus and not the cement itself was responsible for the apparent higher strength. As a matter of fact, Newhouse proved that the strength of American cement was equal to or even greater than the German cement which had previously been imported in the United States.

In 1928 six new portland cement plants were built in the United States, five of them completely equipped with Allis-Chalmers products. In foreign trade during the decade, Japan bought more than twenty cement plants. The Asano Cement Company of Japan alone purchased about $9 million worth of company equipment. Edward Latrobe Bateman, the aggressive cement machinery salesman in foreign areas, also sold a great deal of equipment in other parts of the Far East and in South Africa.[27]

The expansion of the electric power industry during the 1920s was truly amazing. From 1922 to 1930 the capacity of electric generating stations in the United States grew from twenty-two million horsepower to forty-three million, a rate of growth made possible through revolutionary technology. Among the new devel-

In the early 1900s the U.S. Government built a plant at Muscle Shoals, Alabama to extract nitrogen from the air for use in munitions or as fertilizer. The plant was equipped with three Allis-Chalmers rotary kilns and three coolers.

opments were improvements in design that greatly reduced the cost of generating power, cheaper and more effective long-distance transmission of power, and the interconnection of stations serving local regions in order to even distribution of loads. The cost of these changes averaged about $750 million a year during the mid-twenties. The Allis-Chalmers share of this impressive total rose from $10,677,039 in 1920 to $16,213,048 in 1929.[28]

The trend to larger turbines made the twenties a record-breaking decade for Allis-Chalmers. Hardly a year passed without new hydroelectric records. In 1922 an installation in the Henry Ford and Son, Green Island Plant, included the largest fixed blade propeller runners ever. The following year the largest combined hydroelectric unit, 83,000 hp, 65,000 kva, 213 foot head, 107 rpm, was installed in the Niagara Falls Power Company's Schoellkopf plant. This generator developed 607 kva per revolution, 10 percent greater than that of any other water wheel generator built until 1935. In 1924 the installation of the largest capacity single jet impulse turbines in the Western States Gas and Electric Company's Eldorado plant were just part of forty main units aggregating 408,000 horsepower installed during the year. During the following year, the largest capacity double overhung Francis turbines were installed and the largest capacity double jet impulse turbines to date. In 1926 four 54,000 horsepower turbines installed in the Susquehanna Power Company's Conowingo plant included the largest butterfly valves to that date, each twenty-seven feet in diameter. Two years later the largest propeller turbine plant in the

These 2,200 horsepower hydroelectric units, installed in the Green Island power plant of the Ford Motor Company in 1922, are unique in that they produce both alternating and direct current from the same generator.

United States and the largest capacity hydroelectric unit with full automatic control were installed in the Louisville Hydro-Electric Company's Falls of the Ohio plant. Although such detail may seem tedious, it does convey the impression of increasing size and continuity of development in hydroelectric units.[29]

By 1930 Allis-Chalmers had installed hydraulic turbines aggregating nearly six million horsepower. Most of these were built for power plants in the United States and Canada, but installations were also made in several of the South American countries, Japan, Spain, Mexico and the Philippines.[30]

Enormous progress had taken place in turbine design and efficiency since Allis-Chalmers had entered the field in 1904. The rapidity of change can be seen in power development on Wisconsin's Peshtigo River. In 1909 a firm of consulting engineers designed and started construction of a plant at Johnson's Falls. Work was suspended, however, and the project postponed until 1922 when construction was resumed. By 1922 progress in turbine design and the vertical generator required a powerhouse only

This runner was installed as part of the world's largest combined hydroelectric unit at Niagara Falls in 1919.

one-fourth as large as the horizontal turbine called for in the plans of 1909. This also points up the growing interrelation between the hydraulic turbine department and the electrical department of Allis-Chalmers. Such contact was also true with the steam turbine department. As the twenties progressed, the two turbine departments and the electrical department became more closely related, all to the benefit of the company.[31]

When the Niagara Falls Power Company conducted the opening ceremonies for its new hydroelectric installation on December 20, 1919, Allis-Chalmers was proud that of the three units ordered from each of three companies, its unit was the only one ready for operation. Throughout the history of this unit, as was true of its other units, Allis-Chalmers fully lived up to its slogan of "continuous operation, greater dependability and freedom from shutdown." The success of the company in this field came largely from excellent leadership and the quality of the men in the department.

Four years later, that Niagara Falls power plant was enlarged with the installation of this Allis-Chalmers unit, rated initially at 70,000 horsepower and later uprated to 75,000 horsepower.

In 1920 the Committee on Awards and Prizes of the American Society of Mechanical Engineers voted unanimously that "the annual Award for Life Membership for the best paper presented for the year" be awarded to Forrest Nagler for his paper on "A New Type of Hydraulic-Turbine Runner," which appeared in the 1919 volume of the *Transactions* of the society. The decade closed with more honors for the department. William M. White, manager and chief engineer, was awarded the honorary degree of Doctor of Science by his alma mater, Tulane University. This honor, the first of its kind ever conferred by Tulane, recognized White's scientific and practical achievements in hydraulics.[32]

Improvement of design and constant development always required testing facilities. The first hydraulic testing was initiated in 1915 at a temporary outdoor flume set up in the powerhouse spray pond at the West Allis Works. In 1920 deep excavations were made for underground flumes in part of the No. 2 erection shop. The first test in Pit No. 1 was made in March, 1921. In 1930, this pit was completely rebuilt and modernized. Continued testing resulted in improved turbine design and produced greater efficiency. In 1926, White received a letter from James Rickey, hydraulic engineer for the Aluminum Company of America, submitting his log of operation of 31,000-hp unit 3 at their Narrows plant (Cheoah) at Badin, N.C., for that October. The efficiency of this one of three identical Allis-Chalmers turbines was better than 99.9 percent. He concluded: "I did not pick out the above month as an especially good

The turbines at the Conowingo plant on the Susquehanna River, installed in 1928, established a record for physical size. They were equipped with twenty-seven foot diameter butterfly valves, the largest ever built.

one but just the general run of this plant. From this log you will note that the delays were all electrical. Consequently your turbines are 0.56% better than Ivory Soap!"[33]

In the twenties the tendency in hydroelectric development was to build larger units because of the interconnection of the various power systems. Further reasons for enlargement were the multiple use of such units to deliver the maximum amount of primary power and the maximum amount of kilowatt hours during high flow periods and to use hydroelectric units to carry the peak loads of the system. This expansion of service made it commercially advisable to develop streams to larger capacities and also to install larger machines.[34]

The hydraulic turbine department contributed steady sales to the company during the twenties. The first million dollar year for the department had been 1917. From that year through 1931 hydraulic turbine sales ranged from one to two million dollars yearly. Although this was not the largest department nor the greatest source of profits for the company, its records for size, consistent high quality and unexcelled performance helped enhance the general reputation of Allis-Chalmers.[35]

Steam turbine construction was another major segment of Allis-Chalmers business during these years. The first Allis-Chalmers steam turbines, built in 1904, had been based on English design. In fact, a close connection with England continued through the twenties. Sir Charles Parsons, inventor of the reaction steam turbine, the type manufactured by Allis-Chalmers, visited the United

Night view of the Narrows Development, Tallahassee Power Company, Badin, North Carolina; the power house contained three Allis-Chalmers 31,000 horsepower hydraulic turbines.

States in 1924. As a guest of President Falk he toured the West Allis Works and discussed design with the engineers of the steam turbine department. Five years later Lord Falmouth, Director of C. A. Parsons & Co., Ltd., visited the West Allis plant. In line with the policy of continuing cooperation between the two companies in engineering matters, he too discussed recent developments in engineering and design with company engineers.[36]

The company did more than just keep posted on English developments in the steam turbine field. Arthur C. Flory, departmental manager from 1919 to 1943, provided the internal leadership for significant advances in the field. This department developed an esprit de corps which was of greater importance than Flory alone. It was teamwork on which success was built. Louis L. Chapotel, who began in the service section of the department in 1907, put it this way: "All Management, Sales, Engineers, Shop Organization, Service Section working as a team made it possible to attain the success that has been achieved."[37]

With excellent leadership and the department working as a team, the company was prepared to build larger and larger turbines. In fact, it had to develop larger turbines or become a minor figure in the steam turbine business. Machines up to 12,500 kilowatts had been successfully built by Allis-Chalmers, but competitors were building larger ones, and the company had to keep pace. Because blading troubles on large machines were common at that time, the company's competitors had incurred considerable expense and some loss of prestige. In view of this, Allis-Chalmers carefully studied the technical and metallurgical problems before entering large turbine production. Even then Falk and the Executive Committee felt it wise to proceed step by step, building on experience "rather than attempt to take on at one time contracts for a number of units of various large sizes never before built by the Company." Operating on this basis, the company saved itself both expense and embarrassment.[38]

On occasion, Allis-Chalmers received specifications, particularly from cities, calling for forfeiture on steam consumption, shipment or both, on steam turbine contracts. Otto Falk felt rather strongly that a contract calling for forfeiture without an equivalent bonus was not equitable. In this, Allis-Chalmers completely agreed with its chief competitors. As a rule, forfeiture, forfeiture and bonus, and bonus contracts were not good business propositions. All too often they proved to be a source of argument and disagreement even with old customers. For good and valid reason, Allis-Chalmers tried to discourage this type of contract whenever possible.[39]

In the 1920s larger capacity units were required for the increased use of electric light and power. To provide the necessary power, increasingly higher pressures and temperatures were called for. A good example of this trend is the Waukegan Station of the present Public Service Company Division of Commonwealth Edison Company during the period 1923 to 1931. The original proposal called for the installation of two 20,000 kilowatt and four 30,000 kilowatt units. Before construction began, the requirements were increased to an initial unit of 25,000 kilowatts. The second unit called for 35,000 kilowatts, and the third was to develop 50,000 kilowatts. The fourth and fifth units at Waukegan were the first Allis-Chalmers reheat turbines. Work on the first of these units, which was rated at 65,000 kilowatts, was begun in 1930. The second unit, started in 1931, had a rating of 115,000 kilowatts. In this power plant it was originally proposed to have six turbines with a capacity of 16,000 kilowatts. As finally completed, however, it had five Allis-Chalmers units with a total capacity of 290,000 kilowatts and a "topping" turbine made by another company. This is indicative of the growing requirements and the active response of the company.[40]

The workmanship on large turbine units had to approach perfection, and the company worked to achieve that goal as a matter of pride as well as of necessity. When the first turbine in the Waukegan station was installed, the erection engineers found that at 10,000 kilowatts it ran like a "sewing machine." But when another 1,000 kilowatts was added it shook the power house until they thought it would collapse. The engineers tore the turbine down

The trend toward ever larger steam turbine generator units is demonstrated by the Waukegan station of the present Commonwealth Edison Company. The unit shown above, installed in 1923, had a rating of 25,000 kilowatts. In succeeding years, units of 35,000, 50,000, 65,000, and 115,000 kilowatts were installed; all were built by Allis-Chalmers.

and put it together six times before Charles Robertson found that a slight defect allowed cold water to drip on the balance piston. When that was fixed it ran at full load "like a watch."[41]

The crucial year in the Waukegan contracts and in the entrance of the company into the large turbine field was 1925. With 25,000 and 35,000 kilowatt units installed, the order for the 50,000 kilowatt turbine depended on the satisfactory performance of the first two units. Since this was part of the rapidly expanding electrical empire of Samuel Insull, additional orders from this source were also dependent on quality performance. Securing the 50,000 kilowatt turbine contract really marked the entrance of Allis-Chalmers into the large steam turbine market. Since each of these turbines operated with condensers of the company's manufacture, this also meant sales and profits beyond the turbines alone.[42]

These power units on which so much depended for the expanding American economy were truly "tireless monsters," a term used for Allis-Chalmers steam turbines by the Southern Indiana Gas and Electric Company. One Allis-Chalmers 7,500 kilowatt turbo-generator unit was placed in operation at their power station at Evansville, Indiana, on January 23, 1923. From then until January 3, 1926, 1,089 days later, it ran 24,631 hours and 48 minutes out of a possible 26,064 hours, or 94.5 percent of the time. Through performance such as this the Allis-Chalmers steam turbine earned a splendid reputation for efficiency and reliability. The fact that many of the products of the department were for customers already

In 1925 an agreement was made with A. Reyrolle & Co. Ltd. of England for the right to manufacture and sell the first Reyrolle metal-clad switchgear in North America. The first installations were in the Waukegan station of the present Commonwealth Edison Company.

using company equipment was an indication of the operators' confidence in the company.[43]

It was not only in electrical generating equipment that Allis-Chalmers was important in the electrical manufacturing field. The company was a very important producer of smaller electrical equipment such as motors, switchgear, and transformers. Small Allis-Chalmers motors were manufactured in Cincinnati at the old Bullock Works, now known as the Norwood Works. The volume of business from this source remained very steady throughout the twenties. In 1920, for example, sales of electrical equipment from Norwood totaled $4,824,104; in 1929 they were $4,078,456. While the dollar volume remained almost the same, the number of employees decreased from about 1,200 in 1920 to an average of 900 in 1929, indicating greater efficiency.[44]

With the expanding use of electricity for industrial purposes, the Norwood Works pioneered in a number of fields, one of them being synchronous motors. In 1919 the first of these motors was supplied to the Goodrich Rubber Company for rubber mill drives. Two years later synchronous motors with dynamic braking were supplied to the same mills. The first successful application of this type of motor to main roll drives in steel mills was made in 1924. To satisfy the growing demand for motors which could operate successfully over long periods of time in dusty and dirty places the company developed a line of totally enclosed motors. A new technique patented by the company in the construction of large direct current motors was the "Frog Leg Winding," which rapidly received wide recognition from engineers throughout the country. This winding gave complete equalization of voltage on the motor armature without the use of complicated cross connectors. However, Norwood Works did not share in the expanding sales of smaller size electrical motors during the decade despite its improvements in the product.[45]

Large electric motors and generators were produced at the West Allis Works. Sales of this large electrical equipment increased between 1920 and 1929 from $5,852,935 to $7,582,895. Allis-Chalmers was called on to install generators of record-breaking sizes during the twenties. The hydroelectrical unit installed at Niagara Falls in 1923 was of world record proportions. The generator capacity of 75,000 kva was the largest combined unit until the company built the 82,500 kva units for the Hoover Dam in 1952. The 115,000 kilowatt turbo-generator placed in operation in the Waukegan station in 1931 was the largest 1800 rpm generator then built. The forging for the rotor, 169,000 pounds, was also the largest up to that time.[46]

Competition with the great electrical giants, General Electric and Westinghouse, seems to have prevented any great internal growth in the electrical department during the 1920s. Means had to be found for the growth that Otto Falk desired outside the development of standard electrical products. This could be done either by buying rights to a new product or by purchasing a plant producing electrical lines that would complement existing products. Because the company was in sound financial condition in the twenties, Otto Falk could move ahead on both these fronts. In 1925 it was announced that the company had entered into an agreement with A. Reyrolle & Company, Ltd., Hetburn-on-Tyne, England, for the manufacture and sale of metal-clad switchgear in the United States and its possessions, Cuba and Mexico.[47]

After nearly two years of careful investigation of various types of European switchgear and the potential domestic market, the decision was made in favor of the Reyrolle type. Alphonse Reyrolle had established his first workshop in 1886 and shortly after the turn of the century began to develop and manufacture metal-clad switchgear. Progress was very slow, but adherence to sound engineering principles gradually brought the Reyrolle Company to the top as builders of reliable, mistakeproof switchgear. The name Reyrolle, in the minds of British engineers, became synonymous with "safe and reliable" equipment. The Barking Station in London, the ultimate in English power station construction at the time and officially opened by King George V on May 26, 1925, was equipped throughout with Reyrolle metal-clad switchgear.[48]

In April, 1925, Robert B. Williamson, Herbert W. Cheney, and W. C. Furnas went to England to complete arrangements for the manufacture of Reyrolle switchgear in the United States. They returned in May with "loads of drawings." A space on the fifth floor of the main building at West Allis was cleared for the switchgear department, with Cheney as engineer-in-charge. The first order came in September when the New York Edison Company purchased equipment worth $145,000. Cheney died the following March and was succeeded by Henry V. Nye. The first really

This is another type of Reyrolle switchgear. These fourteen units were installed in a substation of the Public Service Company of Northern Illinois; each unit was rated 37,000 volts, one million kva.

significant installation of the "Armorclad" units took place in 1926 at the Waukegan station where 25,000 and 35,000 Allis-Chalmers steam turbine-generator units were in service. For these early installations very little original designing was done, the English switchgear drawings being used as they stood. Gradually, American conditions and requirements dictated changes. Labor costs forced the abandonment of the English plan of assembling the unit for plant testing, dismantling it for shipping, and then reassembling it at its destination. Allis-Chalmers engineers redesigned the frame with two pressed steel halves bolted together, so that after the units were assembled in the shops, the equipment could be split up into a number of shipping groups, each one completely assembled. The English design of electrically operated breaker closing mechanisms was found too elaborate and too delicate for American practices. Early in 1927 a completely new line of solenoid-operating mechanisms was designed by company engineers. In 1928, Allis-Chalmers shipped its first completely American-designed switchgear; a foreign innovation had been completely redesigned to fit American needs and practices.

The principal asset of this equipment and its chief selling point was its virtual guarantee for the lives of the operators. To be safe, a switchgear had to function without any possibility, by skilled operator or common laborer, of accidental contact with a line conductor or arcing distance from it. This was accomplished through Reyrolle equipment. Moreover, the plant was most compact, continuity of service was established and maintenance was minimal.[49]

These indoor units, installed in a station of the Waukegan Generating Company, were rated 3,000 and 3,400 amperes, 15,000 volts, and 1,500,000 kva.

The increasing acceptance of this equipment in the United States is evident in the mounting sales through 1930, when the impact of the depression began to be felt. In 1925 two contracts covering 52 units were received. Contracts increased to 8 in 1927 and to 48 in 1929 for a total of 307 units in that year. Installed in power stations, substations, office buildings and industry, the new Allis-Chalmers switchgear offered far greater operating safety, dependability, and ease of installation than the common open type. Switchgear also provided a new source of income greater than some of the company's old established lines such as sawmilling. By 1927 sales had already passed the half million dollar mark and they went on to reach a peak of over one million dollars in 1929.[50]

Otto Falk made a different approach toward expanding the electrical department in May, 1927, when Allis-Chalmers purchased the entire capital stock of the Pittsburgh Transformer Company, Pittsburgh, Pennsylvania, for approximately $4.5 million. This was the largest company in the country which manufactured transformers exclusively, and it was very well known in the electrical field. From May, 1927, until March, 1928, it operated as an independent subsidiary; on that date, the Pittsburgh Transformer Company was taken over completely by Allis-Chalmers and operated as the Pittsburgh Works.[51]

The Pittsburgh company had been building small transformers since 1897, when it was founded by Frederick C. Sutter. In 1906, Robert V. Bingay was brought into the partnership, and under his dynamic leadership the firm expanded rapidly. Incorporated in 1910, the company grew enormously during World War I. New buildings were erected with the most modern and efficient techniques of production. When Allis-Chalmers bought the company, it employed around 500 people, a figure which remained rather constant until the end of the decade. In response to slightly

This was one of the last steam engines built by Allis-Chalmers. It provided power for a sugar plantation in Hawaii.

increased demand, the works were further expanded in 1928 after being taken over by Allis-Chalmers.[52]

The purchase of the Pittsburgh Transformer Company by Allis-Chalmers precipitated an investigation by the Federal Trade Commission. The basic question was whether this was a violation of Section 7 of the Clayton Act prohibiting one corporation from acquiring all or any part of the stock of another corporation where the effect of such acquisition would be to substantially lessen competition between the two, to restrain commerce in any section or community, or to tend to create a monopoly of any line of commerce. The investigation indicated that Allis-Chalmers and Pittsburgh Transformer together accounted for about 10.6 percent of the transformer business of the country. The commission also found that the Pittsburgh Transformer Company was largely confined to small transformers while the Allis-Chalmers lines consisted, for the most part, of the larger sizes. The lines of the two companies largely complemented each other. After a hearing the Board of Review dropped the matter.[53]

The manufacture of Pittsburgh types of distribution and power transformers was continued at the Pittsburgh Works in sizes up to and including 5,000 kva single phase and 15,000 kva three phase at 44,000 volts, and including also those transformers up to 4,000 kva single phase and 12,000 kva three phase, at 66,000 volts, and all transformers for electric furnace work. The larger transformers and those for higher voltages were manufactured at the West Allis Works. The sales offices of the Pittsburgh Transformer Company were consolidated with, and operated as part of, the Allis-Chalmers district office organization, with branches in all principal cities. Integrating this line into the larger organization increased the firm's facilities and advanced its manufacturing methods and extensive service organization; Allis-Chalmers was now equipped to meet all transformer requirements.[54]

In the Pittsburgh Transformer Company, Allis-Chalmers brought into its expanding organization a highly profitable line. From the acquisition of the company in 1927 through 1931, sales never fell below $2 million per year, and in 1928, 1929 and 1930, they were well over $3 million annually. The electrical line entered into in 1904 with the acquisition of the Norwood Works had grown by 1929 until it accounted for one-half of all sales of the industries group and one-third of the total sales of the company.[55]

The electrical business was profitable to Allis-Chalmers, but it involved the company in competition with the giants of the electrical industry. Constant demands were made to contribute to such worthy causes as the "Better Home Lighting Activity," sponsored by the National Electric Light Association, or to join such organizations as the Society for Electrical Development. The argument was that General Electric and Westinghouse were contributing the lion's share of the funds, and something was expected of Allis-Chalmers. The share of the electric market held by the company in 1920 is seen in terms of a drive conducted by the National Electric Light Association of New York City to cover a deficit of $50,000. The association's officers levied an assessment of $25,000 on General Electric, $15,000 on Westinghouse, and the remaining $10,000 to be contributed in equal parts by the Wagner Electric Company, the Western Electric Company, and Allis-Chalmers. Economy-minded Otto Falk recommended to the Executive Committee that the company's contribution be $1,500, since he questioned both the activities of the committee that made the assessment and the formula on which it was based.[56]

The electrical department, more than any other, involved the company in a variety of financial arrangements to hold or secure electrical contracts. In 1923 Allis-Chalmers was compelled to buy $10,000 par value of 7 percent cumulative, preferred stock in the Jersey Central Power and Light Corporation to secure a contract from A. E. Fitkin and Company, New York, which owned most of the stock. The Fitkin Company had purchased several turbines from Allis-Chalmers, who had never purchased any stock in a Fitkin-controlled company. The Fitkin Company simply told Allis-Chalmers that they would get the order for a steam turbine for the Kansas Power Company of Concordia, Kansas, if they purchased $10,000 worth of the Jersey Central stock. Allis-Chalmers purchased the stock and was then awarded the contract. In 1924 the company purchased 1,000 shares of stock in the Southern California Edison Company, which had purchased $750,000 in Allis-Chalmers products in the previous four years and was ready to place some large orders. Since "certain prominent competitors" had purchased large blocks of stock in the California concern, Otto Falk thought it prudent to invest in the stock and, he hoped, in the firm's good will. During the twenties Falk was always attentive to the good will of the powerful Insull interests. In 1925 the Executive Committee authorized the purchase of stock in the Wisconsin River Power Company. In 1923 and 1924 Allis-Chalmers made contributions to St. Luke's Hospital in Chicago because it was one of Samuel Insull's favorite projects.[57]

The decade of the 1920s was an immensely profitable and extremely productive period in the history of the company. Certainly uninterrupted production contributed to the achievement of these two goals; there was no work stoppage from strikes at the West Allis Works during the entire decade. Although most of the machinists in Cincinnati were on strike from May through August of 1920, the men at the Norwood Works of Allis-Chalmers remained on the job throughout that period, with only a few interruptions. The only strike on record in any of the company plants during this decade was at the Pittsburgh Works in the summer of 1929, a comparatively peaceful period generally on the national labor scene. From 1922 through 1925 the average number of labor disputes per year was only 37 percent of those in the years 1916 to 1921, and only 48 percent as many workers were involved. Milwaukee was particularly free from strikes. Between 1927 and the end of 1930 seven strikes involved only 647 workers. Although the decade may have been peaceful, it was certainly not happy for organized labor. Membership in the American trade unions declined from an estimated 5,110,800 at its peak in 1920 to no more than 3,444,000 in 1929. But this decline in membership does not measure the accompanying loss of vigor, morale and influence. This change in labor came not only from decreasing union membership and less vigorous leadership but also from the improved position of workers.[58]

In an attempt to prevent organization of their workers by national unions and possible resultant strikes, many large industries during the twenties organized their own company unions. By the end of the decade 432 major companies had their own unions covering 1,369,076 workers. Allis-Chalmers declined to organize its own union in the twenties; in fact, there is no evidence that the matter was ever discussed. Otto Falk depended on his personal relationship with the employees and the company sponsored welfare programs to keep the employees contented and unorganized. The company clubhouse and recreational programs had been part of this program for two decades. The old Allis Mutual Aid Society had been incorporated into Allis-Chalmers at an early date, and in 1925 a new shop hospital opened at the West Allis Works. The lower floor of this brick building was a general service shop

This two-story hospital, completed in 1925, was equipped to care for all but the most serious injuries and illnesses. A company ambulance was also available to move patients to larger hospitals.

hospital, with three nurses on duty during the day. The upper floor had special rooms and a laboratory for the regular physician and rooms for an eye, ear, nose and throat specialist and a dentist. The 1924 figures indicate that the old facilities were widely used, for the doctors gave 3,724 consultations or treatments and the nurses treated 53,000 cases during the same period. The new hospital was expected to be used even more and to provide better service. Perhaps less dramatic but of great benefit to the employees was the commissary department. Shoes and overalls were sold to the workers at cost. By 1913 this department had thirteen stations in the plant for the distribution at low prices of milk, ice cream, coffee and soup. This department also sold thousands of plugs and pails of tobacco. Changing patterns of employment and the prosperity of the twenties combined to make possible a two week paid vacation for all those who had been employed a year or more.[59]

Throughout the twenties the company continued its "open shop" policy in industrial relations. This policy simply meant that workers had "the right to work when they please, for whom they please, and on whatever terms are mutually agreed upon between employee and employer and without interference or discrimination on the part of others." This long-standing policy of the company was popularly known as "the American plan of employment." This policy was tested on April 20, 1929, when the workers at the Pittsburgh Transformer plant went out on strike. Their only demand was union recognition, and consistent with its policy, Allis-Chalmers refused to grant this recognition. The union was free to exist—the company never used "yellow-dog contracts"—but Otto Falk refused to bargain with the union. The strike ended in mid-July when the workers returned without union recognition.[60]

The hospital was staffed with doctors and nurses who were available around the clock. These nurses stood in front of the hospital that was replaced by the 1925 structure.

W. W. Nichols, assistant to the president, gave a speech at Tacoma, Washington, early in April, 1926 which he submitted before publication to vice president Max Babb for a legal opinion. Babb's response provides insight into the attitude of the Allis-Chalmers management toward organized labor. Several references in Nichols's speech "to the patriotism, intelligence, etc. shown by the labor organizations and leaders of the country" moved Babb to warn Nichols on April 15 that, "In view of our open shop position, if you make any reference of this kind, it should be in probably only one place rather than in several and in quite moderate terms." By April 20, Babb had thought further and wrote Nichols, "I also think you had better omit all reference to labor leaders, labor organizations, etc. as that might get us into trouble with some of our own people."[61]

If Allis-Chalmers did not recognize labor unions, neither did it participate in the trade associations prevalent in the twenties. Herbert Hoover, as secretary of commerce, was one of the leading sponsors of these cooperative industrial movements which were designed to limit competition. Following the Supreme Court decision in the Maple Flooring Manufacturers case in 1925, these activities were considered quite legal although they were actually in restraint of trade. The first statement of company policy in this matter had been made by President Call as early as 1911: "I desire to reaffirm what is now and has been our position in the past in this respect, i.e., that our Company at no time and under no circumstances will be a party, directly or indirectly, to any agreement or understanding of any kind which seeks to restrain trade, control prices, distribute work, or to do any act of doubtful propriety." Call concluded his statement: "Our Company has always been absolutely free and independent in the direction referred to above, and it will be our unswerving purpose to keep it so."[62]

W. W. Nichols on September 24, 1913, explained to the Executive Committee a proposed Machinery Builders Society which was to "promote the standard of manufacture of the various lines of machinery concerned and a more friendly relationship among its members." Although it took no action, the committee favored a policy which would encourage friendly relations and cooperation with competitors, "but only to such extent as was entirely proper and clearly within the law and not open to any possible criticism." This discussion was apparently a test of sentiment on behalf of John H. McClement, chairman of the Board of Directors. On December 29, 1914, a letter from McClement to the Executive Committee was read to that group by its chairman, Fred Vogel, Jr. McClement hoped that the "ruinous competition" among the three largest pump builders—Bethlehem Steel, International Steam Pump, and Allis-Chalmers—could be replaced by close cooperation. President Eugene Gifford Grace of Bethlehem Steel was explicit in his response to McClement's query. He was willing to make verbal agreements on what bids should be made and to divide the contracts equitably among the bidders. Grace added, "I do not know whether this course is legal or not, but it is sound business and common sense." International Steam Pump was willing to cooperate to a degree but was far more guarded and concerned with legality. The Executive Committee of Allis-Chalmers, while aware of the severe competition in the pump division, referred the matter to the corporate general counsel.[63]

Max Pam, general counsel, with his partner in the Chicago firm of Pam and Hurd rendered a decision on February 3, 1915. Pam replied to the question of whether Allis-Chalmers could legally enter into an agreement jointly or separately with Bethlehem Steel and International Steam Pump Company "to prevent ruinous competition among and between the contracting parties" particularly in the regulation and control of bidding. His conclusion was that: "It is our opinion, unhesitatingly expressed, that no such agreement could be entered into which would have legality or validity and which would not subject the Allis-Chalmers Manufacturing Company to serious legal obligations and consequences and its officers and directors to the possibility of civil and criminal liability." On the next day the board entered the following in the *Minutes:* "IT WAS FURTHER RESOLVED that the conclusions of General Counsel as expressed in said opinion be followed and observed in the conduct of the business of the Company." A copy of the opinion was sent to each member of the board and the president was advised that all officers and agents of the company be instructed to follow and observe the conclusions of the general counsel. This decision guided company policy on cooperation through the Falk administration.[64]

Rather than depend on such artificial means as allocation of markets and orders through trade associations, Allis-Chalmers operated competitively in the market, selling its products on the basis of quality, integrity and service rather than price. Herman Schifflin pointed out the fact that the company often got an order not because it was the lowest bidder, but rather because the name Allis-Chalmers meant "a conscientious determination to supply its customers machinery of such design and manufacture and materials that it will do the work which it is represented to do with full measure of efficiency." The high quality insisted upon by General Falk was linked to a reputation for integrity in business. When Irving Reynolds retired in 1935, Harold W. Story of the legal department remarked, "I don't know you very well."' To this Reynolds replied, "I have handled $100,000,000 worth of business for Allis-Chalmers as assistant chief engineer and chief engineer without ever consulting the legal department." This statement attests to the integrity of contracts entered into by the company and to the quality of its products.[65]

Selling the name Allis-Chalmers meant convincing the prospective purchaser that it was in his best interest to deal with the company because he would receive better service from Allis-

Chalmers than from any other company. Whether it was Saturday or Sunday, a holiday or during the night, "service to our customer was the first consideration," claimed Abe Goldberg, who was in charge of crushing and cement machinery repairs during the twenties. As illustration: when a thirty-inch Superior gyratory crusher operated by the Hagersville Quarries, Limited of Hagersville, Ontario, was found to be unfit for use because of a damaged mantle on a Sunday in 1927, Goldberg received word by long distance telephone at 1:30 that afternoon. William Gorman, general foreman of No. 3 Shop, volunteered to go to the plant to prepare the 6000 pound mantle and self-tightening nut for shipment. When the Chicago and North Western Railroad declined to promise shipment, the Milwaukee Road was appealed to and agreed to do its best to handle it on their 5:55 train. Abe Goldberg then accompanied the shipment to Chicago and saw it loaded in a Michigan Central express "messenger" car leaving at 1:50 A.M. Monday. Clearing the customs at St. Thomas, Ontario, sometime Monday night, the express car arrived at the customer's plant at 5:15 A.M. Tuesday enabling the plant to resume operations by Tuesday night.[66]

Not only a company's reputation but also its future orders are built on service of this kind. Herman Schifflin, manager of the crushing, cement and mining machinery department, received a letter of appreciation from James Franceschini, president of Hagersville Quarries, for the excellent service provided by Goldberg and the company. Franceschini concluded, "I can assure you that I, personally, appreciate very much the attention which we received, and as we are very large operators of stone quarries throughout the Province of Ontario, we shall not forget the service which we received when in the market for additional equipment such as you manufacture." Otto Falk saw to it that quality and service went hand in hand.[67]

Municipal water and sewage systems were modernized in the 1920s. These centrifugal pumps, driven by synchronous motors, were installed in Minneapolis's Fridley station.

For a long time the relations of Allis-Chalmers and the public were relatively unorganized. What contact there was originally came mostly through advertising. Arthur Warren headed a large advertising department in 1905 while his brother was president of the company. This department declined during the years before reorganization and was abolished by General Falk as an economy measure. The purchasing department then handled advertising with the engineers of the various products writing their own copy. Arthur K. Birch, advertising manager from 1920 to 1937, gradually rebuilt the department, but even in 1927 it consisted of only three men, plus two photographers, an artist and four women.[68]

Although a number of large companies, including General Electric and Westinghouse, had established "public relations departments" by 1921, Allis-Chalmers had not. In part, this was probably because the company, especially under Otto Falk, had made an excellent name for quality, integrity and service. Perhaps to a greater extent such a department was thought unnecessary because of satisfactory relations with labor and an immense local prestige. But the main reason is probably found in the dual role of General Falk, who was not just the president of the company; to the public he *was* Allis-Chalmers. His public image was already clear before he came to the company. His prestige continued to grow following his connection with Allis-Chalmers in 1912, and in 1924 he was honored as "Milwaukee's Foremost Citizen." Edward Bernays, one of the early leaders in the field of public relations, wrote, "The personality of the president may be a matter of

These centrifugal pumps were placed in the North Side Treatment Works of the Sanitary District of Chicago.

importance, for he perhaps dramatizes the whole concern in the public mind." General Otto H. Falk did that.[69]

Out of revulsion to the war a new spirit had been born in America even before Warren G. Harding introduced the era of "normalcy." The spirit grew under Calvin Coolidge, who declared that "the business of America is business." Business in general and businessmen in particular enjoyed a public approval and adulation unequalled in American history. Otto Falk responded to this change in spirit. During the period from 1912 to 1919 he had been busy putting Allis-Chalmers back on a solid base. The soldier, the social leader of Milwaukee had subordinated all other interests to those of the company. But in the twenties, a period of substantial profits and constant expansion, a different Otto Falk emerged. He had never enjoyed himself so much. The story of how General Falk saved Allis-Chalmers appeared in any number of periodicals. The titles changed but the theme remained the same. Neil M. Clark wrote about "How General Falk Converted Bankruptcy into Profits" for *Forbes* in 1926. "How a Change in Policy Saved Our Business" appeared in *System* in 1922 over Falk's name. When such free publicity was available and when the company had such a newsworthy president, no great need for a public relations department existed.[70]

Otto Falk's military experience also provided generous copy. *Personal Efficiency* magazine in 1926 carried an article on how "Soldierly Qualities Win in Business." The increased respect given the salesman during this period can be seen in an editorial interview with General Falk by Sam Spalding in "Soldier and Seller: Both Belong to Glorious Professions, says the President of Allis-Chalmers Manufacturing Company." Another interview with General Falk resulted in "Barrage of Advertising Clears Path for Sales Engineers." In this article Falk was quoted as saying that

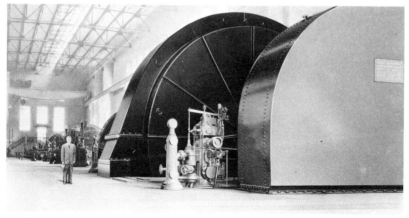

This interior view of the San Francisquito power house of the Bureau of Power and Light, City of Los Angeles, shows the largest impulse type hydroelectric unit ever built. It was rated 32,000 horsepower, 25,000 kva and operated at 143 rpm.

he had great respect for the power of good advertising but, being a military man, he was inclined to liken most of the usages of commercial and industrial practices to those in military circles. "Following this reasoning, and endeavoring to compare industrial selling with the attack of a military force in war time, I usually think of advertising as being the artillery of the industrial commander." But pursuing this metaphor, the General pointed out that battles are not won by superiority of artillery but by the infantry. Although Allis-Chalmers did a good deal of advertising in the twenties, he thought it had advertised more in its early days. "I believe that some of its troubles were due to the fact that it was relying too much on the advertising—the artillery—and too little on its sales engineers—the infantry." The General concluded the article by stating, "Thus far, I believe advertising has acquitted itself well. I believe, with the unremitting attention it is receiving, it will be an even more important weapon in the future. Perhaps in the end, it will be considered the machine gun of industry, working with the troops on the front line." Few companies had interesting and colorful presidents prepared to deliver such prose.[71]

In a decade when the worship of business almost became a national religion and when success in business was the culminating achievement of life, the voluble and quotable General Falk was happy to provide the formula for success in such statements as "The Decalog of Cooperativeness." Another widely quoted decalog was "Soldierly Qualities Win in Business." These qualities were "Courage, Loyalty, Energy, Ideals of Service, Health and

This bucket wheel and shaft assembly of an impulse type hydraulic turbine, suspended from two cranes, weighed 235,000 pounds. It was installed in the Big Creek 2A Development of the Southern California Edison Company.

Strength, Leadership, Obedience and Thoroughness, Organization, Strategy, and Training." He expanded on each of these qualities for the edification of his readers. The decalog was sometimes stretched to "Twelve Principles of Policy." These principles were "Cut the frills, Pay your patriotic debts, Soldierly qualities win in business, To fail, loaf, Study and understudy, There is no substitute for common sense, A sick man can't win, Most troubles are only human, There are lots of things that cost too much, The restless fellow fails where the steady worker wins, Have confidence in good men, and A wise man respects his job." He amplified each principle as he did with number five: "Study and understudy. After you have really mastered your job and know more about it than anyone else, study the job of the man ahead. Be an understudy. Be prepared for opportunity." Here was a man in tune with the times: he knew what the public wanted and he gave it to them.[72]

Otto Falk was an excellent ambassador of good will for Allis-Chalmers. Since foreign orders ran from $2 to 5 million a year, or 9 to 17 percent of total orders during the period 1920–1925, his trip to Europe in 1923 was important to Falk and to the company. The General returned from his four months in Europe with observations which indicate some of his personal views. He found that the businessmen of the world were tired of the chaos caused by the politicians and were ready to step in and correct the situation. In all Europe, he thought, it would be difficult to "muster a corporal's guard" that was still favorable to government ownership of public utilities. Europe's disillusionment should, he felt, stand as an object lesson to the United States. His personal philosophy is reflected in his conclusion: "World conditions will depend upon the acceptance of Christian principles by the peoples of all nations. Until those principles are accepted in their true meaning and spirit there is little hope for world peace."[73]

As steel mills were enlarged or modernized in the 1920s, Allis-Chalmers supplied equipment like this 5,000 horsepower, 595 rpm wound rotor induction motor.

Otto Falk and Allis-Chalmers showed a growing concern for community relations in the 1920s. It has been noted how the Allis interests had alienated many in the West Allis area shortly after the turn of the century. Although the administrations of Warren and Whiteside moved to remedy the situation, the rapprochement was not complete. Until the end of the war Otto Falk was preoccupied by the war effort and by his struggles to place the company on a sound financial basis. But during the profitable twenties Falk and Allis-Chalmers could be more concerned with community relations than before. The company subscribed $1,000 to the building funds of both the West Allis Baptist Church and the Holy Assumption Church of West Allis. The records of Lake Park Lutheran Church of Milwaukee indicate that Otto Falk, a good Episcopalian, gave $500 and that the Allis-Chalmers Manufacturing Company gave $1,000 to its building. Since this gift does not appear in the *Minutes*, it is entirely possible that there were many more unre-

The last vertical triple-expansion pumping engine built by Allis-Chalmers was this million gallon per day unit for Milwaukee's Riverside pumping station. It and three other Allis-Chalmers pumping engines were dismantled in the 1950s and replaced with smaller motor-driven centrifugal pumping units.

corded gifts. Because the company's Aid Society sent most of its patients to Trinity Hospital, the Executive Committee appropriated funds to help make up the hospital's annual deficit. One of the few split votes recorded in the *Minutes* of the Executive Committee concerned a request for $5,000 over a five-year period to support the Christian colleges of Wisconsin. Falk, Oliver C. Fuller and James D. Mortimer voted "yea." The one "nay" was cast by Charles F. Pfister because he favored a larger subscription. Milwaukee-Downer College, Milwaukee School of Engineering and Marquette University all received financial support from the company. Marquette, with gifts of $50,000 in the twenties, got the lion's share. In 1920 the Executive Committee expressed its concern over a housing shortage in West Allis and underwrote a Milwaukee Association of Commerce plan for its employees who wished to build homes in that area. Over the decade the company's support of the Centralized Budget of Philanthropies tripled from $4,000 in 1921 to $12,000 in 1929. By the mid-twenties, West Allis was proud to be the home of the largest industrial plant in Wisconsin, one of the largest plants in the world devoted to the building of heavy machinery.[74]

These increasing contributions to charity and education, as well as research and expansion of plant facilities, were all made possible by the high and even pattern of profits during the twenties. One of the company's advantages was the comparative absence of seasonality in revenues. From March through December of 1923 the orders ranged only from $11.5 to $13.0 million per month. In 1925 the range for the entire year was only $9.7 to $10.8 million. This stability came in part from the variety of products in the company line. During the mid-twenties the complete list numbered well over 250, generally grouped under the following headings: cement machinery, condensers, crushing machinery, engines, forgings, flour mill machinery, hoisting machinery, hydraulic machinery, mining and metallurgical machinery, powdered coal plants, power transmission machinery, pumping machinery, sawmill machinery, small power shovels, timber-preserving equipment, tractors, both steam and water turbines, and about fifty forms of electrical apparatus. This great diversity contributed to the stability of company earnings.[75]

General Falk's leadership combined with wartime prosperity had provided the company with a sound financial base by the end of the war. With all arrearages on preferred stock paid up in 1919, the question became the payment of dividends on the common stock. On March 1, 1920, the Executive Committee carefully weighed the question and took a conservative position. In view of the projected expenditure for plant improvement and the uncertainty of business conditions, the matter was deferred until the June meeting; at that time the decision was yes. For the last three quarters of 1920 the company paid a dividend of 1 percent per quarter, the first common stock dividends ever paid by Allis-

Chalmers. In 1925 these dividends were raised to 1.5 percent per quarter. But Otto Falk never allowed the payment of dividends to dominate all other company concerns. Ample amounts were used for research, expansion, and to build up a surplus account.[76]

The General preferred to operate with a minimum of fixed costs. For this reason, and also for reasons of economy, in May, 1927, an issue of $15,000,000, ten-year, 5 percent gold debentures was sold, the proceeds being applied toward the redemption on July 1, 1927, of the entire issue of $16,500,000 of 7 percent preferred stock. The old Allis-Chalmers Company had fallen so far behind in its cumulative preferred dividends that they could never have been paid. The prosperity of World War I enabled the Allis-Chalmers Manufacturing Company to pay up its cumulative preferred dividends. By retiring the preferred stock, fixed charges could be eliminated. It will also be recalled that it was the mortgage of 1906, secured by the company property, which sent the earlier company into receivership. The debentures of May, 1927, were a direct obligation of the company and not secured by mortgage. Another of the earlier pitfalls was thus avoided. The difference between the 5 percent debenture issue of $15 million and the $16.5 million 7 percent stock was taken from the working capital. The General also shrewdly provided a substantial saving to the company through this stock retirement program. The interest charges of $750,000 were substantially less than the charges of $1,155,000 for preferred dividends, offering a saving to the company of roughly $400,000 per year.[77]

At their annual meeting on May 3, 1928, the stockholders authorized an increase in the capital stock of the company from 260,000 to 500,000 shares of common stock, with a par value of $100 each. On January 11, 1929, the Board of Directors authorized the issuance of 26,000 shares of this increased capitalization, pro rata, to the

As the world's mines probed deeper into the earth, powerful hoists were needed to lift men and materials to the surface. This 58,000 pound rope pull hoist could lift copper ore from a mile underground at 2,250 feet per minute, faster than most passenger elevators.

holders of record on January 25, 1929, at the rate of one share for each ten shares outstanding at the price of $140 per share. These subscriptions were paid in full on February 20, 1929. Proceeds from the sale of this stock, sold well below market price, were used for plant extensions.[78]

The stockholders, at a special meeting held September 20, 1929, authorized an amendment to the certificate of incorporation to provide for the exchange of 500,000 shares of common stock of a par value of $100 each for 2,000,000 shares without par value, and the issuance to stockholders of 1,144,000 shares of the new stock in lieu of the 286,000 shares of common stock then outstanding, each stockholder receiving four shares of the new stock for each share of the old. On September 27, 1929, the Board of Directors authorized the issuance of 114,000 additional shares of common stock, pro rata, to the holders of record of the outstanding stock on October 10, 1929, at the rate of one share for each ten outstanding, for subscription and purchase by them at the price of $60 per share. The income from this action, taken just before the stock market crash, was used for the expansion of the company.[79]

The trend of Allis-Chalmers stocks during the bull market of the twenties was steadily upward. Common stock rose from a low of 78¼ in 1926 to a high of 330 in 1929. The new common stock ranged from 35¼ to 75½ in 1929. Allis-Chalmers common stock consistently gained in favor among investors during the twenties. The increasing number of holders is indicated in the following table.

	1929	1928	1927	1926	1925	1924
Common	4,284	4,056	3,594	2,964	2,543	2,434
Preferred	*___	*___	*___	3,232	3,368	3,554

* Redeemed July 1, 1927.

That Allis-Chalmers stock held up very well during the first year of the depression is seen in the fact that the market had ranged from a low of 49¼ to a high of 68 during 1930. This was not lost on the general investor, for with more stock on the market and the soundness of the company proven in the early depression shakedown, the number of investors increased from 4,284 at the end of 1929 to 9,525 at the end of 1930.[80]

Had General Otto H. Falk known that October, 1929, would bring the great depression, he could hardly have put the company in better shape. Through increased diversification of products he had broadened the base of operations, bringing in new lines such as tractor and Texrope drives. Both of these were to prove very profitable. He had prepared a cushion for the lean years ahead by building up a good-sized surplus, prevented the company from being embarrassed by unpaid, cumulative dividends by retiring the preferred stock, and broadened support for the company by increasing the number of stockholders. Allis-Chalmers stood ready to weather the storm ahead.

NOTES TO CHAPTER SEVEN

1 A-C Mfg. Co. Board of Directors and A-C Mfg. Co. Executive Committee, August, 1920-December, 1922, *passim*. On February 28, 1922, 2,976 men were employed at West Allis, 166 at Reliance, 526 at Bullock, and 812 in the general offices. *Sales Bulletin*, December, 1920, p. 75; *Eighth Annual Report*, April 9, 1921, p. 11; *Ninth Annual Report*, April 8, 1922, p. 9.

2 A-C Mfg. Co. Executive Committee, February 21, 28, March 21, October 10, November 28, 1921; *Sales Bulletin*, December, 1920, p. 77; *Circular Letter* No. 124, December 23, 1920.

3 George Soule, *Prosperity Decade; From War to Depression: 1917–1929* (New York, 1947), p. 108. Soule's table is based on *Federal Reserve Bulletin* (October, 1945), p. 1049, for industrial and wholesale prices; Simon Kuznets, *National Income and Its Composition, 1919–1938* (New York, 1941), pp. 137, 153, for national income and real income per capita. Bayrd Still, *Milwaukee: The History of A City* (Madison, 1948), pp. 478, 484–485.

4 *Sales Billed (Consolidated)* for the years 1913–1957; "Net Profit on Billings," memorandum, n.d.; *Sales Bulletin*, December 1921, p. 123.

5 A-C Mfg. Co. Board of Directors, March 1, 1917, May 7, 1920, August 5, 1921, July 7, 1922.

6 A-C Mfg. Co. Executive Committee, November 27, 1922, May 12, 1924; *Fifteenth Annual Report*, March 10, 1928; *Eighteenth Annual Report*, March 14, 1931; *Chicago Journal of Commerce*, March 28, 1930.

7 *Works and Products Bulletin* No. 146, 1930, p. 1; *A-C Views*, September 27, 1954, p. 9.

8 *Chicago Journal of Commerce*, December 5, 1928.

9 A-C Mfg. Co. Executive Committee, November 13, 1916; General Otto H. Falk, "Training Men to Assume Responsibility," *Machinery*, July, 1928, n.p.; General Otto H. Falk, "The Need for Specialized Knowledge is Increasing," *Milwaukee Journal*, August 28, 1927. The reason for the one exception in filling top shop positions was that the company acquired a new product and a man experienced in its manufacture came with Allis-Chalmers to continue the same duties he had filled with the previous concern.

10 A-C Mfg. Co. Board of Directors, August 8, 1924, May 8, 1925, September 13, 1926, June 10, 1927, September 2, 1927; *Sixteenth Annual Report*, March 16, 1929; *Works and Products Bulletin* No. 146, 1930, p. 1; *Chicago Journal of Commerce*, March 27, 1930.

11 *Works and Products Bulletin* No. 146, 1930, p. 1; *Chicago Journal of Commerce*, March 28, 1930.

12 *Sales Bulletin*, April, 1920, p. 19, September, 1924, p. 197; Ernest Shaw to author, August 3, 1959.

13 A-C Mfg. Co. Executive Committee. When Harrison died he was the oldest department manager in the service of the company. *Seventeenth Annual Report*, March 14, 1930. In 1929 the company acquired the right to manufacture the McMillan Defiberizing Machine, which made wood fibre from waste. Although considerable demand was expected for this machine, it went into production unhappily just as the depression was beginning and did nothing to bolster the sales of the department. Ernest Shaw to author, August 3, 1959; *Sales Billed* (Corporate) Industrial Equipment Division.

14 *Sales Bulletin*, January, 1929, pp. 632–633.

15 *Sales Billed* (Corporate) Industrial Equipment Division; J. L. Neenan to author, July 29, 1959.

16 A-C Mfg. Co., Executive Committee, November 22, 1926; A-C Mfg. Co., Board of Directors, 1926; *Circular Letter*, No. 182, December 7, 1926; J. L. Neenan to author, July 29, 1959.

17 *Sales Bulletin*, January 1927, pp. 361–2, contains a reprint of an article published in the *American Miller*, February 1, 1926, on the 75th anniversary of Nordyke & Marmon; *Sales Bulletin*, February, 1927, p. 371; *Sales Billed* (Corporate) Industrial Equipment Division; J. L. Neenan to author, July 29, 1959.

18 John Storck and Walter Dorwin Teague, *Flour For Man's Bread: A History of Milling* (Minneapolis, 1952), pp. 282–283. The authors point out that the number of flour mills in the United States had declined from over 15,000 in 1909 to about 4,000 in 1927. During the first quarter of the century the production of flour in pounds per man-hour had doubled from 200 to 400. During the period 1919 to 1927 the number of workers in flour mills had declined from 45,000 to

30,000. *Sales Bulletin*, May, 1928, pp. 521–522; J. L. Neenan to author, July 29, 1959.

19 *Sales Billed* (Corporate) Industrial Equipment Division.

20 *Sales Bulletin*, August, 1924, p. 161; *Twelfth Annual Report*, April 10, 1925.

21 A-C Mfg. Co. Executive Committee, April 10, 1924; A-C Mfg. Co. Board of Directors, May 9, 1924; *Sales Bulletin*, April, 1924, pp. 223–224; *Circular Letter* No. 154, April 29, 1924. Besides the Gates Gyratory Crusher, Allis-Chalmers had gained the Blake jaw crusher from Fraser & Chalmers in the merger of 1901. The Blake crushers were redesigned and improved in 1907. From 1915 to 1924 the Blake all-steel crusher was manufactured in three sizes: 48 x 36; 60 x 48; and 36 x 24. Meanwhile the Power and Mining Company was building the Superior line of jaw crushers, starting in 1911 with a 42 x 40 which was followed by a 36 x 24 and five smaller sizes down to 15 x 10. When Allis-Chalmers acquired the McCully gyratory and the Superior jaw crushers in 1924, the Gates gyratory and the Allis-Chalmers jaw crushers were discontinued except for the large all-steel jaw crushers. Lashway notes, "Allis-Chalmers Crushers," pp. 6–7.

22 C. R. Beck to author, July 24, 1959; *Sales Bulletin*, January-March, 1931, p. 503. The *Sales Bulletins* from 1929 on indicate the concern over the sales of the Symons Cone Crusher. The attempts to refute the Symons claims are not at all convincing.

23 Lashway notes, "Allis-Chalmers Crushers," pp. 5–6; Abe Goldberg to author, August 28, 1959.

24 *Sales Bulletin*, September, 1926, p. 243; "Allis-Chalmers Crushers," pp. 4-5; A-C Mfg. Co. Executive Committee, November 8, 1926.

25 Lashway notes, "Allis-Chalmers Crushers," pp. 6–7.

26 A-C Mfg. Co. Executive Committee, February 26, 1926; *Sales Bulletin*, March, 1926, p. 207; May, 1926, p. 211, May, 1927, p. 276.

27 C. R. Beck to author, July 24, 1959; J. N. Friedbacher to author, August 5, 1959; Memorandum to George Smith, April 14, 1947; *Sales Bulletin*, March, 1929, p. 365, April, 1929, p. 371; *Fifteenth Annual Report*, March 10, 1928; *Sixteenth Annual Report*, March 16, 1929.

28 *Sales Billed* (Corporate), 1913–1954.

29 "Hydraulic Turbine Records, 1920–1928," memorandum, n.d.; *Sales Bulletin*, December, 1924, p. 367.

30 *Chicago Journal of Commerce*, March 27, 1930.

31 Edward Uehling, "American Hydro Power in the 20th Century," reprinted from *Allis-Chalmers Electrical Review*, Third and Fourth Quarters, 1955, p. 10.

32 A-C Mfg. Co. Executive Committee, December 22, 1919; *Sales Bulletin*, December, 1921, p. 149, January, 1926, p. 457, June, 1930, p. 950.

33 *A-C Views*, September 27, 1954, pp. 5–11; *Sales Bulletin*, December, 1926, p. 516.

34 C. R. Martin, "Size of Hydro-Electric Machinery is Becoming Limited by Mechanical Restrictions," *Sales Bulletin*, August, 1926, p. 495.

35 *Sales Billed* (Corporate), Industrial Equipment Division. In 1922 the Executive Committee discussed the formation of a Hydraulic Patent Holding Company in conjunction with Wm. Cramp and Sons Ship and Engine Building Co. This proposal was discussed at different times from June until November, but there is no indication in the records that the agreement was ever consummated. A-C Mfg. Co. Executive Committee, June 26, September 18, 1922; A-C Mfg. Co. Board of Directors, July 7, November 10, 1922.

36 *Sales Bulletin*, October, 1924, p. 279, October, 1929, p. 379.

37 *A-C Views*, September, 1954, p. 6. In the same issue of *A-C Views*, James Wilson, who started with Allis-Chalmers as a steam turbine engineer in May, 1905, said: "It all boils down to teamwork and I maintain the steam turbine department has always been a shining example of teamwork."

38 A-C Mfg. Co. Executive Committee, January 7, 1924; *Sales Bulletin*, January-March, 1931, p. 719. In 1923 the method of casting the blades was abandoned. Paul Dimberg, who was with the steam turbine section from 1906 on, wrote, "As the large turbine business increased it was necessary to develop blading for higher stresses and in 1923 we adopted a nickel chrome alloy steel which was formed and rolled into our blade sections. This material was also rolled in bar stock from which our individual machined blade was used at the low pressure end of the turbine where the stresses and top speed were high." *A-C Views*, September 27, 1954, p. 6.

39 *Sales Bulletin*, June, 1922, p. 87.

40 *Sales Bulletin*, December, 1924, p. 187, May, 1930, p. 705; *A-C Views*, Sep-

tember 27, 1954, p. 12; Hans P. Dahlstrand, "Development of Steam Driven Power Generation," MS dated May 6, 1947, p. 6.

41 *A-C Views*, September 27, 1954, p. 8. An advertisement by the Standard Oil Company (Indiana) in 1923 used an illustration of an Allis-Chalmers turbine to typify the dependability, simplicity and freedom from breakdown that could be achieved "if it is properly lubricated." The company took great pride in this unsolicited advertisement for its product. *Sales Bulletin*, June, 1923, p. 127.

42 A-C Mfg. Co. Executive Committee, July 27, 1925. The entrance of Allis-Chalmers into the large steam turbine field did not mean that it had become the principal producer. The company had been producing units of 10,000 kw or larger since 1918. Since that time they had built 35 turbines of 10,000 to 12,500 kw for a total kilowatt rating of 374,500. They had also built seven units of 15,000 kw capacity, eight units with capacities ranging from 20,000 to 50,000 kw. The total kilowatt capacity in all fifty units of 10,000 to 50,000 kw aggregated 709,500 kw. *Sales Bulletin*, February, 1928, p. 323.

43 *Sales Bulletin*, February, 1926, p. 257; *Steam Turbine and Alternator Units*, Bulletin No. 1122, 1922.

44 Steam Turbine and Alternator Units, Bulletin No. 1122, 1922; A-C Mfg. Co. Executive Committee, 1920–1929, *passim*.

45 *Sales Bulletin*, October, 1926, p. 417, January, 1926, p. 437. By the twenties the Louis Allis Company of Milwaukee was in active competition with Allis-Chalmers in the manufacture of motors. The *Sales Bulletin* for December, 1922, p. 235, expressed concern over the confusion in use of the word "Allis."

46 *Twelfth Annual Report*, April 10, 1925, p. 9; *A-C Views*, October 27, 1954, pp. 2–6; "A-C Electrical Generators, 1920–1930," undated memorandum.

47 *Thirteenth Annual Report*, April 9, 1926, pp. 11–12; *Sales Bulletin*, December, 1925, p. 424.

48 *Sales Bulletin*, December, 1925, p. 424; *A-C Views*, October 27, 1954, p. 2.

49 R. Stowe, "Beginning of Allis-Chalmers Switchgear Department," MS, October 4, 1957; *Works and Products Bulletin*, No. 146, 1930, p. 1. It is noted in the Minutes of the Board of Directors, May 7, 1926, that approximately $350,000 was to be spent in development and adaptation of Reyrolle switchgear.

50 Stowe, MS, October 4, 1957; J. H. Michael, "History of Allis-Chalmers Switchgear Development Work," MS, October 2, 1957; *Sales Billed* (Corporate), 1913–1954.

51 A-C Mfg. Co. Board of Directors, May 6, 1927, February 3, 1928; *Circular Letter*, No. 185, May 16, 1927; *A-C Views*, October 27, 1954, p. 3.

52 J. C. Shupe, "Brief History of Pittsburgh Transformer Company," MS, March 7, 1939; J. C. Shupe, "Historical Sketch—Allis-Chalmers Pittsburgh Works," MS, n.d.; William Feather, *R. V. Bingay; Builder of Transformers & Transformer of Men*, August 29, 1925, pp. 1–8, *passim*; A-C Mfg. Co. Board of Directors, October 5, 1928; A-C Mfg. Co., Executive Committee, February 24, 1928.

53 A-C Mfg. Co. Board of Directors, May 4, 1928.

54 *Sales Bulletin*, February, 1928, p. 604; J. C. Shupe, "Historical Sketch—Allis-Chalmers Pittsburgh Works," MS, n.d.

55 *Sales Billed* (Corporate), 1913–1954.

56 A-C Mfg. Co. Executive Committee, December 20, 1920, July 14, 1924, September 15, 1925; A-C Mfg. Co. Board of Directors, August 8, 1924.

57 A-C Mfg. Co. Executive Committee, December 13, 1920, August 20, 1923, May 19, 1924; A-C Mfg. Co. Board of Directors, August 8, 1924.

58 A-C Mfg. Co. Executive Committee, August 2, 1890. Although the company was no longer a member of the Metal Trades Association, and had not been since the first decade of the century, it contributed $1,000 to help fight the strike. Soule, pp. 202, 225–227; Still, p. 497.

59 *Circular Letter* No. 163, March 19, 1925; *Sales Bulletin*, December, 1925, p. 372–3, April, 1926, p. 400, May, 1926, p. 410, June, 1926, pp. 419–420, July, 1926, p. 429, August, 1926, pp. 440–41, September, 1926, p. 449, October, 1926, p. 458; Soule, p. 223.

60 The statement on the open shop is part of a resolution adopted at a meeting of the National Conference of State Manufacturers Associations meeting in Chicago on January 21, 1921, as quoted in Soule, pp. 200–201; *Circular Letter* No. 204, July 17, 1929; A-C Mfg. Co. Executive Committee, July 8, 15, 1929.

61 Max W. Babb to W. W. Nichols, April 15, 20, 1926.

62 *Circular Letter* No. 191, August 9, 1911.

63 A-C Mfg. Co. Executive Committee, September 24, 1913, December 29, 1914.

64 A-C Mfg. Co. Board of Directors, February 3, 4, 1915.

65 Irving H. Reynolds to Alberta J. Price, August 31, 1945; *Sales Bulletin*, January, 1923, p. 167. Abraham Goldberg said that "a machine in operation was regarded as an Allis-Chalmers machine even though it was paid for. Allis-Chalmers felt it was still responsible for keeping it operating." Abraham Goldberg to author, August 28, 1959.

66 *Sales Bulletin*, October, 1927, pp. 299–301; Abraham Goldberg, "1903–1958" MS, pp. 13–14.

67 James Franceschini, president, Hagersville Quarries, to Mr. H. Schifflin, manager, crushing, cement and mining machinery department, September 16, 1927.

68 Earl Lashway to author, August 30, 1960.

69 Thomas C. Cochran & William C. Miller, *The Age of Enterprise* (New York, 1956), pp. 310–311, quotes Bernays. Bernays worked to improve the public image of both Westinghouse and General Electric during the twenties. Eric F. Goldman, *Two Way Street* (New York, 1948), p. 20.

70 Neil M. Clark, "How General Falk Converted Bankruptcy Into Profits," *Forbes*, February 15, 1926, pp. 9–11, 32, 52; Otto H. Falk, "How a Change in Policy Saved Our Business," *System*, February, 1922, pp. 135–139, 199.

71 "Soldierly Qualities Win in Business," An Interview with General Otto Herbert Falk, by the Editor of *Personal Efficiency*, March, 1926, pp. 144–150; "Soldier and Seller," An Editorial Interview with General Otto H. Falk, by Sam Spalding, *Sales Tales*, October, 1929, pp. 185, 229–232; "Barrage of Advertising Clears Path for Sales Engineers," An Interview with General Otto H. Falk, *Class & Industrial Marketing*, September, 1927, pp. 19–20.

72 General Otto H. Falk, "Make Yourself Liked in Business," *System*, December, 1930, p. 468; "Soldierly Qualities Win in Business," *Personal Efficiency*, March, 1926, pp. 147–148.

73 The Board of Directors appropriated $10,000 to defray the expenses of Falk's trip to Europe in 1923, and $5,000 to defray the expenses of a trip to the "Southern and Western offices of the Company, and the Company's agent in Cuba" in 1927. A-C Mfg. Co. Executive Committee, November 16, 1925; A-C Mfg. Co. Board of Directors, January 5, 1923, February 4, 1927; *Sales Bulletin*, July, 1923, pp. 186–187.

74 A-C Mfg. Co. Executive Committee, October 8, 1917, August 1, 1918, June 2, November 17, 1919, January 19, March 15, June 14, 1920, October 25, 1921, August 21, 1922, November 26, 1923, August 8, 1928, June 7, September 27, 1929; Records of Lake Park Lutheran Church, Milwaukee; J. H. Burbach, *Historical Review of West Allis*, 1927 ed., pp. 73–83, *passim.*

75 A-C Mfg. Co. Board of Directors, March-December, 1923, January-December, 1925, *passim; Chicago Journal of Commerce*, December 5, 1928.

76 A-C Mfg. Co. Executive Committee, March 1, 1920. By late 1921 the company had received a number of inquiries concerning the purchase of Allis-Chalmers stock from an investment point of view. On November 21, 1921, the Executive Committee took the position "that in making reply to such inquiries the officers of the Company should give such facts and information as to the Company's actual condition as were asked for and it was proper to furnish, but should refrain from advising with reference to the purchase of the stock or expressing an opinion on the question of a continuance of such dividends."

77 A-C Mfg. Co. Board of Directors, May 6, 1927; A-C Mfg. Co. Executive Committee, May 23, 1927; *Fifteenth Annual Report*, March 10, 1928, pp. 10–11; *Chicago Journal of Commerce*, December 5, 1928, March 28, 1930. This money was also used, in small part, to retire the $936,000 in bonds of the Bullock Electric Manufacturing Company, which was secured by a mortgage against Allis-Chalmers. As early as September 23, 1927, Otto Falk was using surplus funds to buy up the debentures. A-C Mfg. Co. Board of Directors, September 23, November 1, 1927, July 13, 1928, January 11, 1929.

78 A-C Mfg. Co. Board of Directors, April 5, 1928; *Sixteenth Annual Report*, March 16, 1929, p. 8.

79 A-C Mfg. Co. Board of Directors, January 11, August 16, September 27, 1929; *Seventeenth Annual Report*, March 14, 1930, p. 5.

80 *Eighteenth Annual Report*, March 14, 1931, p. 7; *Chicago Journal of Commerce*, March 28, 1930.

Harry Merritt
Tractor Manager

CHAPTER EIGHT

THE FARM TRACTOR: A SUCCESSFUL EXPERIMENT

THE STOCK MARKET CRASH OF 1929 marked the end of the prosperous twenties and the beginning of the depression-ridden thirties. Nearly all facets of American life and particularly the business community were seriously affected. Although Allis-Chalmers was no exception, the effects of the depression on the company were cushioned by the rapid development of the tractor department in the late 1920s. Increasing farm equipment sales, which accounted for more than half the Allis-Chalmers production by 1935, provided the basis for company expansion and reasonable economic stability in the thirties. Since it served as a bridge between the two decades and was so important to the company, the development of the Allis-Chalmers tractor and farm equipment business demands individual treatment.

As president of Allis-Chalmers, Otto Falk in 1913 realized that the nature of the company placed it at something of a disadvantage. Unlikely as it may seem, it lacked diversification. Within the company great diversity did exist, but it was nearly all in equipment for heavy industry such as engines and turbines or crushing, mining, and cement-making machinery. No one product could sustain the company if sales to heavy industry declined. Moreover, nearly all the heavy equipment was hidden away from the public in factories, power stations and the like. Because of its products, Allis-Chalmers was known to few Americans outside the heavy industries. What the company needed was a product that would attract wide publicity. After a great deal of serious thinking Otto Falk decided that his goal could be achieved through the production of farm equipment, particularly tractors. As it turned out, diversification into farm equipment was one of General Falk's most important contributions to the company. Allis-Chalmers succeeded in a field where Henry Ford had failed and where General Motors had lost $33 million. Most important, the tractor division was to carry the company through the bitter years of the Great Depression.[1]

The mechanization of American agriculture came from the importance of commercial agriculture, the size of the American farm, and the scarcity of farm labor. General Falk saw the changing pattern in agriculture. The old Falk Brewery had as a neighbor a farm machinery plant which had turned out more and more farm implements. Falk himself farmed as a hobby, proud of his Waukesha County farm. He must have been aware that horse- or mule-powered agriculture had reached the peak of its development by the turn of the century. Steam power had already come to many large farms in the last quarter of the nineteenth century; it was, however, merely a supplement to horsepower and not a replacement. As a source of belt power, the steam engine was an important contribution to agriculture, but for drawbar work it had serious limitations. The mammoth steam-powered implements were practical only on large western farms and could not profitably be used on smaller eastern farms. Then, at the turn of the century, the internal combustion engine began a new agricultural revolution.

The first gas traction engines were huge, awkward, and difficult to start, drive, steer and maintain. Despite these drawbacks, eleven companies were making them by 1906. Charles W. Hart of Iowa and Charles H. Parr of Wisconsin built some of the first successful gas traction engines. To distinguish their machines from competi-

Before the days of the combine, the All-Crop harvester and the tractor, men experimented with better ways to cut grain. This crude machine was the forerunner of the reaper or binder.

tive steam traction engines and to avoid the unwieldiness of the term *gasoline traction engine,* they coined the word *tractor.* It turned out to be the perfect word to describe and to advertise this new form of power. By 1910 there were 1,000 tractors on American farms and only four years later in 1914, 17,000.[2]

The swift progress of tractor engineering and the remarkable popularity of tractor use was not lost on Otto Falk. At a meeting of the Executive Committee on August 15, 1913, he reported the recent visit from Walter Goeldner, a representative of Motoculture, Ltd., of Basel, Switzerland. He sent Max Patitz to Europe to investigate agricultural developments in general and the Motocul-

The first motive power on farms was provided by unwieldy steam powered traction engines such as this Rumely unit. Commonly used to power threshing machines like that pictured below, the engine pulled the machine to the field and was then connected to its driving pulley with a leather belt.

ture "motor driven Rotary Plow" in particular. Patitz brought back the rotary plow and an Austrian engineer named Haderer to demonstrate it. On the last day of 1913 the Executive Committee authorized a contract calling for payment of $10,000 in cash and a royalty of 6 percent on all rotary plows sold in the United States. At the beginning of 1914, Allis-Chalmers was one of fifty United States concerns entering the field of agricultural equipment. Rotary plows were tested by Allis-Chalmers on the Pabst farm near Milwaukee, in Virginia and in Colorado during 1914 and early 1915 before Falk considered it "reasonably safe and advisable to

(Top) The Oil-Pull tractor developed and built by Rumely was a significant advancement over the traction engine. Smaller than the traction engine, it could move at greater speeds but was still too heavy and slow for most farm jobs.

(Bottom) Allis-Chalmers first venture into the production of farm equipment was this rotary tiller. Developed in Austria, it proved unacceptable to American farmers and only a few were sold.

offer them for sale to the general public." Unlike the plow which had to be followed by a series of other tools to make a seed bed, the rotary tiller in one operation pulverized the soil to the full width and desired depth as the wire tools or claws cut the soil. Despite the numerous advantages claimed for the rotary tiller, it was never widely accepted by the American farmer. Only a limited number were manufactured by Allis-Chalmers on special orders, with no sales recorded after 1916.[3]

Otto Falk had a specific interest in developing an Allis-Chalmers tractor. He declined a proposition from Lyons, Knoll and Hartsough of Minneapolis for the manufacture and sale of a "Bull Tractor" because he intended to put an Allis-Chalmers farm product on the market rather than join a syndicate. Falk proposed committing the company on a rather broad front with the rotary tiller, a small tractor, and a tractor truck. By 1914 the motor truck had gained little acceptance outside urban areas, since it lacked tractive power on the generally poor roads. C. Edwin Search, works manager of the West Allis plant, proposed to Falk that the company produce a truck with a conventional body but propelled by crawler-type treads instead of rear tires. The increased traction would enable this truck to negotiate any road in any weather over most terrain. The market for such a truck was thought to be sizable. The General's response was immediate; he ordered work on the project to begin at once.

Work on the tractor truck was directed by Ray C. Newhouse, chief engineer of the crushing and cement department. Although

Allis-Chalmers ventured again into motive power when it designed and built this tractor-truck, which today would be called a half-track.

certain patent rights pertaining to crawler tractors were purchased from the Monarch Tractor Company of Watertown, Wisconsin, Search added further refinements which, according to the patent application, would provide "an endless track mechanism for tractors which is simple in construction and efficient in operation."[4]

By early 1915 the first machine had been completed and "given considerable test work and so far has given a most excellent account of itself." Newhouse planned to ask about $5,000 for the "5-ton Tractor Truck." An illustrated brochure showed the machine at work hauling timber and pipe and negotiating snow drifts with a full load. But however versatile and useful this machine was, there was almost no market for it in the United States. Domestic rejection came at the very moment of foreign opportunity, for World War I had engulfed Europe. The truck was demonstrated before the military authorities of the major allied powers, but only Russia submitted an order for ten machines at $5,500 each. Unhappily, these were assembled in Russia but never used. The last of the tractor trucks served for a number of years after the war in and around the West Allis Works for plowing snow and hauling castings. This prototype of the "half-track" of World War II was an excellent machine, born of fertile imaginations nearly a generation too early.[5]

The first conventional tractor, called by Falk the "small tractor," was an unwieldy tricycle type with one speed forward and one reverse. Rated at 10-18 (drawbar and belt horse-power), it went into production in November, 1914, and was placed on the market in 1915. Weighing 4,000 pounds, it sold for $1,950 and was "the only tractor that has a one piece steel heat-treated frame—no rivets

Allis-Chalmers built the first farm tractors in one of the West Allis shops. This view shows some of the 20–35 horsepower tractors being assembled. The 20 indicated the pulling horsepower, the 35 the horsepower at the belt.

to work loose—will not sag under the heaviest strains." This tractor could plow twenty acres a day on an average fifteen gallons of gasoline. When the farmers had trouble starting it in the morning, they had to "warm up the gasoline in the priming can in hot water and heat the spark plugs." From then on they ordinarily had little difficulty. Largely an experimental model, the small tractor never achieved wide sales to American farmers.[6]

As early as April, 1915, Allis-Chalmers engineers were developing a smaller tractor than the 10-18. Falk hoped to place it on the market in the near future, but for unexplained reasons it was not produced in any quantity until 1918. The principle of the 6-12 was radically different from that of the 10-18 and most other tractors of the day. It had two steel driving wheels in the front, pivoted by a turning mechanism at the center. The operator sat at the end of a long pole on lighter wheels at the rear. Without the sulky, the tractor could be attached to any two-row, horse-drawn implement, thus saving the farmer money. When company engineers found that a single tractor had insufficient power for some jobs, they devised a way of putting two of them together, one running forward and the other in reverse.

The 6-12, which Allis-Chalmers held to be "the greatest tractor for general farm use" on the market, was assembled in the gallery of No. 2 Shop. Each machine was built on a sawhorse, and a crew worked hard to assemble one tractor a day. It was equipped with Le Roi engines, with wheels and castings made at the West Allis Works. After the cleats had been bolted on the wheels, the tractor

The first tractor built by the company in 1914 was this rather awkward tricycle type, rated 10 horsepower at the drawbar and 18 at the belt.

was swung down by crane to the test rack in the No. 1 Erecting Shop, where the brakes were tested and the tractor was painted by hand. The price of the 6-12 was $850, F.O.B. Milwaukee. The 700 manufactured proved to be an overestimation of the market, and about 200 had to be sold at greatly reduced prices.[7]

Although the rotary tiller had been a failure and neither the 10-18 nor the 6-12 tractor much of a success, still General Falk was encouraged by the promise tractors held. As tractor engineers simplified design somewhat, reduced the weight and provided greater ease in operation, the 6-12 and the 10-18 looked slightly more attractive. The Allis-Chalmers tractor from the outset had been the product of several engineers working without any overall direction. With progress in production and sales, the company in 1917 set up a separate tractor department with Fred W. Kamm as manager. Beginning in December, 1918, the tractor department was given space in the company *Sales Bulletin*.[8]

General Falk decided that the new tractor department should have its own shop if it was to produce efficiently enough to be competitive. Acting on detailed recommendations from C. Edwin Search, the Executive Committee on January 6, 1919, authorized $1,368,000 to remodel and equip a large building which had been built during the war for pattern storage. This became Tractor Shop No. 1, designed to produce ten tractors daily of each size: the medium size 10-18, which had been on the market for several years, the rather new 6-12 general purpose tractor, and a new 15-30 prototype.[9]

The next tractor design had two large pulling wheels in front, two small ones in the rear. The small wheels could be removed and the tractor mounted directly on the equipment it was to operate.

Spurred on by the enthusiasm of Otto Falk, the Executive Committee in October, 1919, authorized $800,000 for a new foundry measuring 500 by 140 feet. This foundry, next to the No. 1 Tractor Shop, was completed and in operation in less than a year. The castings could then move efficiently from the north end of the foundry east into the tractor shop. The capacity of the foundry was 100 engine and transmission castings a day. With this new foundry, Allis-Chalmers had the potential to become one of the world's largest manufacturers of farm tractors.[10]

When production began on the 15-30, tests showed that it was more powerful than its rating, and it became known as the 18-30. Built for operators who demanded a three- or four-plow tractor, this conventional four-wheel tractor had two speeds forward (2.3 and 2.8 miles per hour), one speed reverse, and a drawbar pull of from 3,000 to 4,000 pounds. Its list price was $2,100 F.O.B. Milwaukee. When the official tractor test Number 83 was conducted by the University of Nebraska on the 18-30, the actual rating was 22 drawbar horsepower and 38 belt horsepower. Again its rating was raised, this time to 20-35. While many tractor manufacturers consistently overrated their machines, Allis-Chalmers preferred to underrate theirs, allowing a liberal reserve behind the guaranteed power rating.[11]

Early in 1921 the 12-20 tractor was introduced, apparently to replace the old tricycle type 10-18 and to satisfy demand for greater power than was available in the average two-plow type. This machine was able to handle a three-bottom plow under normal conditions and a two- bottom plow under "any and all conditions." Built along the same lines as the 20-35, it had a drawbar pull of 2,000 pounds at a top speed of 3.25 miles per hour. This tractor's rating was also raised, to 15-25.[12]

The 15-25 and the larger 20-35 were the principal Allis-Chalmers tractors during the first half of the 1920s, all other models having been discontinued. Falk hoped that these tractors with their green motors, yellow striped transmissions and bright red wheels would capture a larger share of the tractor market. Certainly the company had the manufacturing facilities for such a goal.[13]

General Falk had persuaded the Executive Committee and the Board of Directors to manufacture tractors. The minutes of both groups indicate that they spent more time discussing tractor business than all other products of the company put together. But the tractor department was the only department that steadily lost money. This venture had incurred heavy expense in Milwaukee, only to tie up funds away from home in disconcerting ways. For example, on March 28, 1918, the S. J. Tabor Company of Fargo, North Dakota, the company's tractor representatives in that area, appealed to the Executive Committee for financial assistance. They had found that it was the practice of other tractor manufacturers to deposit sums of money in interest-bearing certificates of deposit with banks in the towns where they had their best distributors. This deposit was made with the understanding that the money was to be available to the manufacturer's jobbers in paying for their machines and for discounting dealer's paper. To keep the Tabor Company contented, the Executive Committee authorized the deposit of $20,000 for nine months in the Merchants National Bank of Fargo. But this was only the first such request, for subsequent deposits of from $3,000 to $40,000 had to be made in other banks in North Dakota, South Dakota, Minnesota, Iowa and Illinois.[14]

Among other problems which plagued Falk and the Executive Committee was credit. Allis-Chalmers had always sold for cash; farmers have always been short of cash. The Committee's first decision was to follow its regular practice. But the company's dealers reminded Falk and the Committee that other tractor manufacturers took farmers' notes in partial payment on agricultural machinery. Pressure mounted as the financial condition of farmers on the plains, where most tractors were sold, worsened during the depression of 1920. To meet their competition in a declining market, the Executive Committee after December 20, 1920, allowed a discount on dealers' notes up to 75 to 80 percent of the billing price, provided the dealers' and farmers' notes were recommended by a national bank of recognized standing and secured by a chattel mortgage on the machine. Even this did not solve the problem, for the price of raw materials to the company constantly increased at the same time that the price of farm products decreased.[15]

The manufacture of tractors had been the General's pet project, but things were simply not going well. The substantial investment to that date had resulted only in consistent losses. On January 17, 1921, Otto Falk made an extended report to the Executive Committee on the present state and future prospects of the tractor department. He also explored with the Committee how the shop and foundry could be used for the manufacture of other products should the company decide to discontinue tractors. The Executive Committee requested Falk to make the same statement to the Board of Directors in the near future for a determination of policy.[16]

Since the company was making substantial profits despite the losses of the tractor department, the decision was to move ahead with an advertising campaign and a change in personnel. Early in 1921 Burge M. Seymour replaced Fred W. Kamm as departmental manager. At the same time the company launched the most intensive advertising campaign in its history. The advertising firm of Klau-Van Pietersom-Dunlap of Milwaukee bought double-page spreads in thirty-seven publications, full page advertisements in the farm papers and two and four page color advertisements in the trade papers. The cooperation of the dealers and distributors was carefully cultivated. Many were brought by special train to Milwaukee for an entire week of "instruction, lectures, service schools, demonstrations, sight-seeing trips and inspection trips." They were then sent home to contact all prospective tractor buyers.[17]

General Falk was apparently delighted in 1922 with the first use of the radio as a means of publicizing Allis-Chalmers tractors. A broadcast by Sears and Chaffee Supply Company of Great Bend, Kansas, who were counted "among the most progressive tractor dealers in the west and southwest," attracted a good deal of attention and comment. Lee Chaffee narrated the following story:[18]

> I am iron—my teeth are steel—my joints turn on balls, and rollers; my heart is lubricated under high pressure. I believe and assert I am the fittest of any kind in my size. If I stand idle, I go hungry. I prefer to work, and am not afraid of long hours. Give me four plows, and I will turn twelve acres of stubble in ten hours. In the hottest days of harvest, I will pull two binders and the amount we will reap will depend only upon you, who is [*sic*] my master. I will turn your 28″ or 32″ separator and make ample and abundant work for eight good teams, and if you take a liking to me, I will promise to do this work satisfactorily for you for the next ten years.
>
> By men who demanded and had the right to know the capacity and endurance of my strength, I was given the highest attainable honors and awards.

The first really successful Allis-Chalmers tractor was the 20–35. Originally called the 15–30, after tests at the University of Nebraska it was rated 18–30. The actual test showed 22 horsepower at the drawbar and 38 at the belt, so the tractor was finally rated 20–35.

For forty-one hours they tried to abuse me, but they found no weakness in me, and really, I enjoyed it all, as it was my second official test conducted by the Agricultural Engineers at the University of Nebraska. I am mighty proud of my name—the A-C 20-35.

I am yours for service, if you will but say the word.

Cordially and Sincerely,
AN ALLIS-CHALMERS TRACTOR

The time was ripe for this kind of aggressive sales campaign. A survey indicated that in the eight leading tractor states—Illinois, Iowa, Kansas, Minnesota, California, North and South Dakota, and Nebraska—there were only 106.8 tractors for each one thousand farms on January 1, 1920. An enormous market existed for the tractor manufacturer and salesman.[19]

But by the mid-twenties, despite the advertising campaign, the Allis-Chalmers tractor had not captured the market as General Falk and the company had hoped. They had produced a fine tractor and built an excellent shop and foundry, but the cold fact was that they had actually lost ground. The prospect list for 1925 came to only 12,000 names. By March of that year the 20-35s were coming out of the shop at the rate of only five a day. During the whole of that year only one new dealership had been established. Obviously the facilities for tractor production greatly exceeded the demand for Allis-Chalmers tractors. The tractor department launched and promoted by General Falk was in a critical state. The questions were simply these: if Allis-Chalmers was to stay in the tractor business, how could it move ahead, or if it was going to get out of the business, how could it get out gracefully? The time for decision had arrived.[20]

January 1, 1926, stands as one of the most important dates in the history of the tractor department. Effective on that date was the appointment of Harry C. Merritt to the position of manager of the tractor department. Born on August 28, 1881, at Vermont, Illinois, Harry Merritt had been a farm machinery salesman since he was seventeen. He had been with the Holt Manufacturing Company before joining the southwestern division of Allis-Chalmers, where he rose to the position of manager of the Wichita branch tractor office. He brought a new confidence and imagination to the department, and he had singleness of purpose. It is said that he never gave anyone his undivided attention unless the subject was tractors. He probably disrupted more weekends than any other Allis-Chalmers man of the time, for he was always calling his men on a Saturday or Sunday to say, "I'm going down to the office today. You wouldn't by any chance be planning to go down, would you?" They always did for he was "the sort of man you'd go through hell for."[21]

After the fashion of E. P. Allis, General Falk had been looking for the right man to head a department. For over a decade without the right man at its helm the tractor department had been losing money. With reluctance he decided that it was time to give up the venture and he proposed to do it as rapidly and as expeditiously as possible. He chose Harry Merritt, a human dynamo who loved tractors, to apply the *coup de grace*. To dispose of existing stock as rapidly as possible, Merritt reduced the price of the 20-35 from $1,950 to $1,295. This tractor now sold as it had never sold before. Obviously, there was a market for an Allis-Chalmers tractor of that rating and that quality at that price. Merritt requested a reprieve. Was it possible, he asked, to produce a quality tractor at the lower price and still show a profit?

With his engineers he disassembled an entire tractor and spread it out on the floor. They studied it part by part, and wherever possible substituted a lighter part of the same strength and quality for the old one, especially in the motor and transmission. Merritt received a special dispensation from the General to buy tractor parts outside the company if other concerns could provide the same quality at a lower price. Then he and his staff stripped all nonessential apparatus from the old model. Under Merritt's direction, the engineers then concentrated on the general appearance to produce what Merritt called tractor "sex appeal." In the end they succeeded in creating a trim, snappy-looking tractor that could do everything its predecessors had done and more, yet was nearly half a ton lighter and sold at the attractive price of $1,295.[22]

The 1927 model of the 20-35, which came out of these efforts, "smooth and clean-cut in appearance, is now offered to the farmer for a price which gives more horsepower per dollar than any other tractor on the market." Refinements had been made to "provide greater comfort for the operator and to make the tractor lighter, more compact and easier to handle." The greatest engineering feature was a motor that was fully sealed against dirt and grit, with oil and gas filters as standard equipment. To bring out this tractor at the lowest possible price, Allis-Chalmers discontinued the 15-25 and concentrated all efforts on producing the 20-35 as economically as possible. Tractor sales from 1921 through 1928 show the impact of Harry Merritt and his redesigning of the 20-35. In 1921 only 104 tractors were sold, with 181 the following year; by 1923 sales reached 334, with a slight increase to 416 in 1924. The critical year 1925 had 635 sales and with 1926 given over largely to reorganization and redesigning, this number increased slightly to 682. The 20-35 introduced in 1927 sold a record number of 1,814 and this more than doubled to 4,867 in 1928. Imagination and hard work had produced amazing results.[23]

The policy of producing the best possible equipment for sale at the lowest possible price in large part accounted for the success of the Allis-Chalmers tractor after 1926. Engineers were sometimes almost reluctant to suggest an idea for improvement because Merritt would invariably ask them to begin and would then hound them until it was perfected. This policy brought constant change and improvement. As the tractor department began to make money, more was available for research and development. A good example of this constant development can be seen in the "WC" tractor, which from the first of its kind to the last one off the line, incorporated 33,000 changes.[24]

According to Harry Merritt, the pricing of a tractor was always the last step before marketing. "First we study the needs of the market. Then we develop the product that best fits those needs. After that we figure costs and allow ourselves a fair profit." But a "fair profit" to Allis-Chalmers proved to be less than that of major competitors. A report of the Federal Trade Commission indicated that in the 1930s the average Allis-Chalmers profit on new farm machines and implements in some years was less than 4 percent, compared to 6.7 percent for International Harvester, and a whopping 18.0 percent for Deere and Company. Obviously, a willingness to take a smaller return had a good deal to do with the success of Allis-Chalmers in building a lower-priced tractor and in its rapid emergence as the third largest manufacturer of farm equipment.[25]

But quality and price were not quite enough. Allis-Chalmers equipment was in no way distinctive; there was no way of spotting an Allis-Chalmers tractor, for at a distance it looked like any other and was equally drab. In 1929 Merritt went to California and saw acres of brilliant orange wild poppies, their color making them visible for miles. Merritt brought some of the poppies to Milwaukee and asked experts of the Pittsburgh Plate Glass Company to duplicate the color. They developed "Persian Orange," a paint first used on the United Tractor. Soon all Allis-Chalmers agricultural implements were the same bright orange. Competitors ridiculed the flashy color, some thought it in poor taste, but it gave distinction to Allis-Chalmers equipment and helped it sell as it had never sold before. Merritt's judgment and foresight were once again important in the promotion of the company's tractors.[26]

Another extensive advertising campaign was launched to sell the tractor—a campaign that gained momentum as sales increased. Advertisements were run in all the farm journals and some popular magazines. Tractor schools taught farmers the care of tractors, discussed their potential uses, and pointed out the excellence of Allis-Chalmers tractors. After two years of experimenting with radio advertising, the company in October, 1929, announced a "radio program which provides coverage from coast to coast and is planned to reach virtually every farm home." The program with Pat Barnes, one of the most popular announcers of the day, featured a "snappy orchestra, a stirring band, vocal artists, famous humorists, something to please every taste." The company hoped that the vast unseen audience would catch "the tempo of progress" from the programs.[27]

The enthusiasm of Harry Merritt carried over into the sales force. By 1927 ten men of a field force of twenty-two each succeeded in selling more than $100,000 worth of tractors, thus winning admission to the tractor division $100,000 Club. In the heyday of the secret society, Merritt established the Ancient and Honorable Order of Fles (Friendship, Loyalty, Economy and Service) to build the morale of the tractor sales force and weld the tractor division together as an effective group. Only members of the tractor division or those directly connected with it were eligible for membership. Execution of the application, with the required approval and acceptance, entitled the applicant to receive the "Slippery Truncheon." It was fully understood that all prospective members had to be "enthused to the degree of performance and quick results." As of May, 1929, Merritt held the office of President Emeritus and was supported by the Exalted Cyclops, Illustrious Conservator, Worthy Interrogator, Eminent Registrar, Trusty Invigorator, Ceremonial Master and Honorary Distinguished Inquisitor. The entire field organization of the division was made up of members of the order who assembled each year for the annual "Pow-Wow," where applicants received the chip of the "Wahhoo" and were inducted into membership. This society may sound amusing but it was effective.[28]

In looking over government reports, Merritt found that in 1925 over 44,000 tractors had been exported by United States firms. "There is no good reason why Allis-Chalmers should not get a share of this business," he said. In 1927 he saw to it that a distributor was appointed for Western Canada, and a sizable volume of sales was secured from Saskatchewan and Alberta. But a man or a firm had to work hard to even begin to please Harry Merritt. In 1929 a more dynamic distributor, the Cockshutt Plow Company, Ltd., of Brantford, Ontario, was appointed for Canada, and it produced even greater sales.

In 1927 B. F. Avery & Sons of Louisville, Kentucky, undertook the distribution of Allis-Chalmers tractors in Argentina. During the same year two 20-35 tractors were exhibited at the Trans-Caucasian Agricultural Exposition at Tiflis, Russia, and tested under the supervision of the Tiflis Agricultural Academy. The fact that the Allis-Chalmers tractors finished first in the plowing contest considerably ahead of the nearest competitor and had the best fuel economy was not lost on the new Soviet state, which under its first Five-Year Plan was rapidly moving toward mechanization. The report from Jack R. Sullivan of the tractor division that "with a fair break it will not be long before Ivan Ivanovitch will be turning over the dirt with an A-C" was soon realized. In 1930 a $3 million order was placed with Allis-Chalmers for 1,850 tractors to be shipped immediately to Russia by the Amtorg Trading Corporation, the purchasing bureau of the U.S.S.R. These Persian Orange tractors were to be used in the wheat regions of Russia's state farms. It is interesting to contemplate General Falk, the conservative American industrialist, meeting with Messrs. Kolomoitzeff and Dimitrieff of the Amtorg Corporation to conclude arrangements.[29]

By 1930 the tractor operation had grown too large to be managed completely by one man, even Harry Merritt. Though an excellent salesman, his primary interest was in the tractor as a piece of engineering. In 1930 Merritt brought W. A. (Bill) Roberts to Milwaukee as agricultural sales manager, and the next year Roberts was made general sales manager. Merritt knew that what the tractor division needed was a well-organized sales force. He could supply the spirit but he did not have time to provide the necessary organization and proper supervision. Roberts at the age of thirty-three had the necessary experience as sales representative in both the United States and in Canada. The company had a well-engineered product unable to sell itself; it needed an organized sales force, and this was Roberts's assignment.

The tractor division, reorganized and renamed in 1930, now had the perfect team—Merritt handling the engineering and Roberts the sales. From 1930 on, almost all credit for tractor publicity went to Bill Roberts, who looked over every word that went out from the department. He was also principally responsible for the development of the company's far-flung sales and dealer organization. His keen eye searched out able young men such as Robert S. Stevenson and Willis Scholl for rapid advancement. Always direct and intense, he had the drive and determination to sell tractors. Here was the man who in 1941 would succeed Harry Merritt as manager of the tractor division and who, in time, would become president of the company.[30]

Tractor "sex appeal," Persian Orange, and the Order of Fles were admittedly gimmicks to induce the farmer at home and abroad to buy Allis-Chalmers tractors and to stimulate salesmen to sell them. But sell them they did. Instead of closing out the tractor department, Harry Merritt developed a product and an organization that turned losses into profits and a department into a division. The writer of the following words in the *Sales Bulletin* of December, 1925, had no idea how prophetic they were: "For the first time within the history of our Tractor Division, we have just cause to feel sanguine regarding the financial results to be obtained in the year immediately ahead of us." In 1926 tractor sales had reached a peak of $1,800,000, and the division was still in the red. But in the next four years until the beginning of the depression, tractor sales began to climb impressively as did the percentage of tractor sales to those of the company as a whole.

Year	A-C Total Sales	Tractor & Farm Equipment Sales	% of Total	A-C Total Net Income
1927	$33,350,000	$ 2,260,000	6.8	$3,182,000
1928	36,295,000	5,000,000	13.8	2,934,000
1929	45,300,000	10,890,000	24.0	4,331,000
1930	41,470,000	12,415,000	29.9	3,605,000

Harry Merritt in only five short years had established the tractor division as the principal unit of the company.[31]

Tractor sales and profits provided a basis for continuing expansion through acquisition. In the *Nineteenth Annual Report* for the year ending December 31, 1931, Otto Falk wrote, "Due largely to existing conditions, the Company has been able to acquire several carefully chosen properties and products at exceptionally advantageous terms." This had been going on in the tractor division since 1928, when the Monarch Tractor Corporation of Springfield, Illinois, was acquired. Organized in 1913 at Watertown, Wisconsin, as a local stock company manufacturing crawler-type tractors, it had grown rapidly during the war and was reorganized in 1919 as Monarch Tractors. Inc. In 1925 a new company, Monarch Tractor Corporation, was formed at Springfield, Illinois, to produce two crawler tractor models. In 1928 the Monarch plant, consisting of modern assembly and machine shops on six acres, employed about 200 men. Although Monarch had sales for the previous year of $1,400,000 Allis-Chalmers was able to purchase the company for $500,000 in cash. With the addition of the Monarch six-ton and ten-ton tractors to the Allis-Chalmers 20-35, the company was able to present to the public an excellent range of equipment from which to choose for almost any purpose.[32]

The two crawler tractors produced by Monarch in 1927 had much greater pulling capacity and economy than any other tractor of comparable size, according to the University of Nebraska engineering tests. Allis-Chalmers had purchased not only an excellent crawler-type tractor to add to its line but, more importantly, had secured valuable license arrangements so that the company could use the same patents that the Caterpillar Tractor Company worked under. Since sales of crawler tractors during the first year after purchase amounted to $2,300,000, Falk proposed the immediate construction of a $350,000 addition to the plant. By 1936, with eight different models of crawler tractors adapted for service in every

Two Monarch tractors at work on Staten Island.

type of agricultural and industrial use, the sales reached $9 million.[33]

In spite of the excellence of its tractors and the leadership of Harry Merritt, Allis-Chalmers in the late twenties faced an enormous handicap in its competition with International Harvester, Deere, and Case. These companies offered a complete line of farm implements as well as tractors. Merritt realized that he simply could not bring about the success he hoped for by selling tractors alone. Ten or fifteen years earlier that would have been possible, but not after a wave of consolidations between tractor and implement manufacturers. Allis-Chalmers needed to acquire an established name in implements, with expertise in production and a well-suited plant.

Otto Falk solved this problem in 1929 by purchasing the La Crosse Plow Company, which traced its history to 1860. Under the direction of Albert Hirshheimer, the company had established a reputation for building a fine line of "La Crosse Made" implements, including plows, harrows, disks, drills and cultivators. It had a reputation as a pioneer in the business, having marketed the first practical balanced frame horse-lift cultivator, the first power-lift grain drill, the first power-lift tractor plow, and other advanced implements. Despite its well-known products and a fine location on major railroad lines and near cheap barge transportation to the South on the Mississippi River, the La Crosse Plow Company saw

Harry Merritt, tractor manager (left), explains the features of the crawler tractor to sales manager W. A. (Bill) Roberts on tractor.

its business decline to the point where it either had to consolidate or go out of business. Aware of this, Otto Falk informed the Executive Committee on August 5, 1929 that he thought it could "be purchased on what is believed to be a very low basis." Allis-Chalmers paid only $275,000 for the La Crosse Plow Company.

New life was infused into the nearly defunct company, which had only seventy-four employees in August, 1929. The ten acres of buildings were modernized and expanded nearly threefold during the next twenty-five years. The products which were at first marketed as the Allis-Chalmers La Crosse Line of Farm Implements, were distributed through the international sales organization of the company. All of the La Crosse activities—manufacturing, engineering, purchasing, accounting and selling—were coordinated with the executive departments at the home office in Milwaukee.[34] With the addition of La Crosse products Allis-Chalmers had joined the ranks of the "full line" companies. It could now meet its competition on more even terms than ever before, precisely what Harry Merritt had in mind.

To broaden the product line of the tractor division, the company in 1930 purchased the rights to the grader line of the Ryan Manufacturing Corporation of Hegvisch, Illinois. Allis-Chalmers could then offer power-controlled leaning wheel graders, a motor patrol grader powered by the model "U" tractor, a leaning wheel grader with hand control, and a line of small graders for maintenance work. The next year the company acquired rights to the elevating grader manufactured by the Hy-Way Machine and Manufacturing Company of Omaha, Nebraska. In both instances the company purchased only the right to the product, the graders being manufactured in the existing plants.[35]

In the late twenties Henry Ford, more and more disappointed with the sale of the Fordson tractor, finally transferred assembly operations out of the country. The companies that had manufactured equipment for the Fordson, left quite high and dry, near the end of 1928 formed the United Tractor and Equipment Distributors' Association. Allis-Chalmers was contacted and agreed to build a "United Tractor" which would sell for less than $1,000, which their engineers immediately set to work designing. Although space and equipment at the West Allis Works could have accommodated the work, the desire to reduce costs and prices led the company to build again. "We decided that only the most modern production equipment would be used," Otto Falk wrote in 1930. "Today every piece of equipment in our *new* tractor shop is itself brand *new*! By an investment of $2,000,000 in plant and equipment, we have staked our faith not only in the future industrial and agricultural prosperity of the country, but also in what modern equipment will do." This new No. 2 Tractor Shop was equipped throughout with semi-automatic and automatic machinery.

The United Tractor, sold through the Equipment Distributors' Association, and the Model "U", its counterpart, sold through the regular Allis-Chalmers distributors, did fairly well both in agriculture and in industry. A wide variety of industrial equipment including full swing cranes, ditchers, drag line equipment, winches, hoists and side booms was adapted for use with the "U." Fleets of model "U" tractors were purchased by the state highway departments of New York, Mississippi, Alabama and Indiana. Company engineers put a tricycle front on this tractor, moved the rear wheels out and thus converted it to a cultivator which they called the "UC." But Allis-Chalmers did not pick up as much business as was anticipated and could not take over the Fordson business.[36]

The reason for the failure to capture the Fordson business, and in fact part of the problem of the tractor division from the outset, was that Allis-Chalmers simply did not have an adequate distribution system. The bulk of its business had always been heavy machinery, which largely sold itself. Buyers generally came to the company rather than the company going to the buyers; parts and repairs were handled through the factory. But tractors proved a different matter. The farmer wanted personal contact with a local dealer rather than the impersonality of a regional distributor or the factory. He wanted and needed parts, equipment and repair facilities near home. These needs demanded a radical departure from normal company practice, and even Bill Roberts could not do this immediately nation-wide.[37]

Grading was done like this before the advent of tractors and motorized graders.

Setting up an adequate system of distribution would cost time and money. In 1931, however, Allis-Chalmers was able to secure all this and more through the purchase of the Advance-Rumely Corporation of La Porte, Indiana. Established in 1852 in La Porte by a young German immigrant named Meinrad Rumely, the firm grew rapidly. To meet the growing demand for agricultural equipment in the nineteenth century it consolidated with Gaar-Scott and Company, Richmond, Indiana, and the Advance Thresher Company of Stillwater, Minnesota. In 1915 the merged companies were reorganized as the Advance-Rumely Company. At the time of its purchase by Allis-Chalmers, it was one of the great leaders in the farm equipment industry with a world-wide business.[38]

Internal problems and competition in the industry gave Advance-Rumely financial difficulties even before the depression. In 1930 it had sustained a net loss of $1,213, 605, compared with a loss of $395,503 in 1929 and $340,666 in 1927. It had last shown a profit in 1928. But whereas Allis-Chalmers had only five branch houses in 1931 and a relatively small number of dealers, Advance-Rumely had twenty-four branch houses and about 2,500 dealers. These branch houses were "in the selected hearts of the agricultural sections which makes it possible to carry our full line of agricultural products . . . for the important purpose of rendering a first class service to owners and operators." This excellent distribution system together with Advance-Rumely's well known harvesters and threshing machinery was bought for $4.5 million. This acquisition brought 854 new employees to Allis-Chalmers from the La Porte Works, including both general office personnel and those engaged in distribution.[39]

In 1930 Allis-Chalmers purchased the Ryan Grader Company and
began to develop graders to meet every condition of operation.

In 1931, Allis-Chalmers built a five-story office building at West Allis to accommodate the increasing business of the tractor division. But Allis-Chalmers with all other tractor producers faced the cold fact that the tractor had not been generally accepted for farm work and was not yet replacing the horse on the farm as the automobile and the truck had replaced him on the highway. This was because of the cost, limitations, and inefficiency of the wheel equipment used. Although the crawler or track-type tractor with its continuous tread was developed soon after the first wheel tractors appeared and had much more efficient traction, its higher cost offset the improvement in traction and therefore offered no satisfactory solution for the farmer. The cheaper steel wheels with lugs, on the other hand, damaged meadows, orchards, and barnyards; the sign "Tractors with Lugs Prohibited" appeared on most well-surfaced roads.[40]

Thousands of dollars were spent by manufacturers in experimenting with lugs of different designs, shapes, sizes and spacings to make the tractor more efficient and more useful. The sheer inefficiency of the lug-type wheel is indicated in the old tractor ratings of 10-18 and 20-35. Tests proved that the tractive efficiency—the ratio between the power delivered at the drawbar and the power produced by the motor when used under field conditions—varied from as low as 40 percent to a high of about 65 percent. Very simply, power was necessary to push the lugs in and to pull them out. The result was that, even on level ground, the tractor was constantly compelled to climb a rather steep grade. As the speed was increased, more of the total horsepower was required to merely move the tractor; at higher speeds it tended to approach the total output of the engine and leave little power for useful work. The consequence of this was that conventional tractor work was slow, inefficient and expensive.[41]

Engineers had flirted with the idea of putting rubber tires on tractors. Experiments were conducted with both hard rubber tires and high pressure pneumatic tires similar to those on trucks. But when attempts were made to plow with this equipment, the tractors could perform only under the most favorable ground conditions and were absolutely useless on wet ground. In 1930 Allis-Chalmers arrived at a partial solution when it sold Model "U" tractors with interchangeable wheels—steel wheels with lugs for farm work and rubber tires for the highway. Obviously, though, this was no real solution to the problem, and it was expensive as well. Research was continued and in 1930, Allis-Chalmers and the Goodrich Tire and Rubber Company developed the "O" pressure tire. Although this tire had no air pressure, the casing was so heavy it approached the weight of a solid rubber tire. The tractor engineering staff finally solved the problem—a low pressure tire with a flexible casing allowed the tread to spread out and distribute the load for necessary traction. Harry Merritt was reported to have said, "Go ahead—we can put it across." The low pressure tire was more than a solution to a problem; it was a significant breakthrough for the industry. As *Farm Implement News* put it, "Just about the time this industry seems to have dropped into a rut and reached a static point with no outstanding developments in sight, something arises to change its course. Rubber may be the pivot of the next turn."[42]

Company engineers used two 48 by 12 Firestone airplane tires for the initial test. These were put on a Model "U" tractor belonging to Albert Schroeder of Waukesha, Wisconsin, whose farm was chosen for the experiment because its 155 acres had a wide variety of soils and slopes. Schroeder had given his permission for the experiment, but he had his misgivings and insisted that the company bring along another tractor with five-inch lugs on its steel rims, in case the rubber-tired one would not do the work. Although it was clearly marked on the wide of the tires, "Keep Inflated to 70 Pounds," the air pressure was reduced to 15 pounds. This test, in April, 1932, was so successful that Albert Schroeder used this air-tired tractor until he traded it for a new Allis-Chalmers model in 1946.[43]

The company set up a program of continuous testing. Six Allis-Chalmers tractors with air tires were put on selected farms so that the tests could be conducted under a wide variety of work and conditions. The reports were uniformly enthusiastic. All those involved in the tests agreed with Albert Schroeder: rubber actually seemed tougher than steel, the tractors rode more comfortably, the air-tires were easier on the tools used, provided greater fuel economy, presented a greater tractive surface, and most important, they permitted greater speed in the fields.[44]

The first public demonstration of the air-tired tractor took place near Dodge City, Kansas, on Labor Day, 1932. After the demonstration, a wire to the home office announced that a farmer near Garden City, Kansas, had paid cash for the tractor and driven it home. On October 13, 1932, Allis-Chalmers made the first announcement of air-tires as standard equipment for the Model "U," declaring it "the most important development in tractor design in 10 years." This air-tired farm tractor had four speeds: 2-1/3, 3-1/3 and 5 miles per hour plowing speeds, and a 15 mile per hour road speed.[45]

Engineers pointed out that the real success of the air tire came from its large area of contact with the ground, permitted by low air pressure. A tire with twelve pounds pressure operating under load would flatten out to make contact with an area approximately 19½ inches long and 9½ inches wide for a total of 150 square inches. Moreover, the thin, flexible construction permitted the tire to bend and conform to the irregularities of the ground and thus exert a uniform pressure and tractive effort over the entire area of contact. More importantly, the traction developed by a pneumatic tire was found to have a direct relation to the load it carried, for as the load increased, the tire flattened out to make a larger area of contact with

(Top) In addition to road and construction work, crawler tractors were used to pull logs out of the woods.

(Bottom) The versatility and easy handling of the Model B tractor, designed for smaller farms, made them adaptable to many farms. One key to their good performance was their air tires.

the ground and, of course, greater traction. In October, 1932, *Farm Implement News* reported that tests under plowed surface conditions demonstrated that low pressure rubber tires consumed only half as much power to propel the tractor as did steel wheels with lugs. Therefore, under those conditions a tractor with a twenty-five horsepower engine and rubber tires could deliver just as much power at the drawbar as another tractor of thirty-five horsepower with steel wheels. The smaller engine would consume about a gallon less fuel each hour. Moreover, in the future the tractor would not be barred from the highways. The farm tractor was no longer limited in its operation but had become a general utility machine to be used wherever power was required.[46]

As a creature of habit the American farmer has resisted change as much as, and perhaps more, than others. During the summer of 1932, hundreds of farmers went to demonstrations with the expectation of seeing the air-tired tractor fail. They believed that this tire would slip just as badly as truck tires in wet weather, and that barnyard conditions would cause the rubber to rot. Some were still not convinced even when the Allis-Chalmers HTP 314 plow, drawn by a Model "U" air-tired tractor, won first place in the Manufacturers and Dealers Class at the 1933 Wheatland Plowing Match near Plainfield, Illinois. Never before had Allis-Chalmers won, but the air-tired tractor came home with a wide margin to spare.[47]

Not only the farmer needed further persuasion. One frantic competitor issued a nine-page "White Paper" with "proof" that the Allis-Chalmers claims were false. The president of one company issued a public statement that "whenever a machine is so designed or equipped that it can be stalled by a bucket of water, it's a mighty long way from perfection." But the editor of another farm publication recognized the achievement of Allis-Chalmers: "The two

Farmers were slow to accept the All-Crop Harvester. Demonstrations like this soon made them popular.

great turns in the road of farm machinery development came with the introduction of the internal combustion engine and the rubber tire. One company, alone and derided, can take credit for the latter." But one word of praise could not combat the enormous odds to convince the customer of the soundness of this innovation.[48]

The tractor division realized that it would have to do something spectacular to break down resistance to and build public acceptance for air tires. Because most tractors were unable to go more than five miles an hour, W. Elzey Brown of the advertising department thought of speed racing. Special high speed gears were installed in stock models of the Model "U" tractor, and the first public speed test was made at the State Fair race track at West Allis on June 18, 1933. Frank Brisko, a famous local driver who had recently competed in the Memorial Day classic at the Indianapolis Speedway, was the star in the auto races. The spectators could hardly believe their eyes when an Allis-Chalmers tractor, which had been plowing in the infield of the race track, was unhooked from the plow, turned over to Frank Brisko and then driven against time at 35.4 miles per hour. The effect was sensational. During the remainder of the summer, similar tractor races were advertised as attractions at many state fairs. The high point in this series of demonstrations occurred at Dallas, Texas, on September 17, 1933, when Barney Oldfield, the veteran automobile racer, drove a Model "U" air-tired tractor over a measured mile course at 64.28 miles an hour. The first man to drive an automobile at more than 60 miles an hour had also become the first man to drive a tractor at

In order to demonstrate the exceptional performance of air-tired tractors, they were demonstrated and raced by famous drivers at state fairs.

more than a mile a minute. Because this event was sanctioned by the American Automobile Association and timed by its officials, it became an official A. A. A. record. In 1933 alone, more than one million people saw these tractor races, and they were, in fact, repeated by popular demand for several years. With such famous drivers as Barney Oldfield, Lou Meyer, Floyd Roberts and Ab Jenkins, the tractor races were given top billing at many fairs.[49]

The effect on the farmer was gradual but cumulative. By 1935 the air tire was beginning to win real public acceptance. *Farm Implement News* estimated that from 15,000 to 20,000 farmers bought tractor tires in 1934; about 5,000 sets were original equipment on new tractors and the rest were changeovers. Of the estimated 183,000 new tractors produced in 1935, about 10 percent were equipped with rubber. In 1936 this figure rose to 20 percent and in 1937 to 45 percent. In 1935 a contest offered a tractor and other prizes for the best statement of "What I want in a tractor." Out of 15,573 entries tabulated, no less than 10,557 farmers asked for rubber tires. By 1937 a University of Nebraska survey of 674 owners of air-tired tractors in thirty-seven states revealed that 96 percent would buy their next tractor on rubber.[50]

Although air tires had been put on the Model "U," the tires and the tractor had not been designed for each other. In 1933 a new farm tractor was designed for air tires, and a few were in service before the end of the year. This new tractor, the Model "W," brought enthusiastic reports from new owners. Equipped with a four-cylinder engine of Allis-Chalmers design and manufacture, it weighed only 2,700 pounds. This tractor was officially announced

Barney Oldfield drove tractor 999 to a world record speed of 60 miles per hour.

early in 1934 as the "WC" with "Full 2-plow Power—only $675 Steel Wheels; only $825 Air Tires, f.o.b. Milwaukee." Even though they were offering the first tractor designed to have rubber tires as standard equipment the company had a real selling problem. The price was extremely reasonable and the performance remarkable, but the tractor was distinctly smaller than comparable competitive machines. Accustomed to size as a basis for judgment, many farmers could not readily believe that this tractor built of high strength steel could stand up in service as well as the heavier, mostly cast-iron tractors. But the "WC" did gain rapid acceptance; by January, 1936, the demand for it had increased so much that production was expanded from fifty to seventy tractors a day. As the spring plowing season approached, the tractor shop put on a Saturday shift to help satisfy increasing demand. The popular "WC" was for many years the bread and butter tractor of the company. In fact it became one of the most popular Allis-Chalmers tractors of all time and contributed more than any other machine to the nearly ten-fold increase in tractor division sales during the five years following the introduction of air tires.[51]

Harry Merritt felt real satisfaction in the rising production figures. With pardonable pride he said, "We regard this new development as marking the dawn of a new era in American agriculture and as the most important advancement in Tractor engineering in years." The air tire and the new four-speed tractor meant that the

Before the development of powerful tractors and later self propelled units, combines were pulled by multi-horse teams.

American farmer could plow, plant, cultivate, till and harvest in about three-fourths the time formerly required. Now he could cut in half the time he ordinarily spent in moving from field to field or in hauling on the road. He could do more work per day and work more days per year because the tractor was now adapted to more kinds of jobs. In the field, the fourth speed shortened the time of many lighter jobs. There was also a savings in fuel of up to 23 percent in ordinary field operations. But Harry Merritt was never satisfied. He estimated that if the one million farms in the United States that had tractors were all equipped with high-speed, air-tired machines, the savings to their owners would amount to 125,000 hours a year and $10 million in fuel bills.[52]

After the purchase of the La Crosse Plow Company in 1929 there was constant improvement in implement design. Investigators were sent out to farms to observe the number of operations one farmer performed in a day. They found that he often used several different tools per day and wasted much time in removing and attaching implements with clumsy mechanisms. In 1930 the company brought out the "Quick-Hitch" cultivator, the first of a series of time-saving attachments. With the "Quick-Hitch" the farmer could change implements in fifteen minutes and be on his way to the next job. In 1933 the company put out its first sponge rubber tractor seat with a back rest for greater comfort. In anticipation of the air-tired "WC", the company also developed a whole new line

of high-speed implements. During 1933, twelve new tillage tools were designed and placed in production at the La Crosse Works.[53]

While Allis-Chalmers tractor engineers in West Allis and La Crosse were developing the air tire, the "WC" tractor and a variety of improved implements, another group of engineers at La Porte, Indiana, was tackling a knotty problem—how to design and produce a small practical combine for use on the small midwestern farm. A combine is basically a threshing machine with a harvesting attachment which heads, threshes, and cleans the grain as it moves over the field. Crude combines had been devised as early as the mid-nineteenth century, and with later refinements they came to be used effectively on the great wheat farms of the far West and Northwest. These huge machines were drawn by forty to fifty horses and cut a swath thirty to forty feet wide. As use of the conventional combine crept eastward to the Kansas wheat fields, its size was reduced to twenty to thirty feet. It was successful primarily with wheat, which could simply be headed and the heads rammed through a small throat into the threshing cylinder. It could not harvest sweet clover, alfalfa, or bush beans, which inevitably clogged the small threshing cylinder. Eventually, the size was again reduced for use in the Mississippi Valley. Although

Crawler tractors eliminated the need for horses. This is a tractor powered hillside combine.

these ten and twelve foot combines were a vast improvement over the binder-thresher method of putting up grain, they still cost $1250 to $1500 and required a three-plow tractor and two or three men to operate them.[54]

Allis-Chalmers set out to develop a totally new harvester that could, as one man put it, harvest everything from bird seed to beans. Other specifications were that it had to be light enough to be pulled by a two-plow tractor, operate from its power take-off, and sell at a price low enough that a farmer with 100 acres or less of small crop seed could afford to buy it. To develop such a machine the company in 1930 purchased the rights to a small five-foot combine manufactured in California. Although cumbersome and inefficient, it was a start for experimentation and development. With the purchase of Advance-Rumely in 1931, the extensive experience of their engineers in thresher design was available in the continued experimentation and testing of the proposed combine.

The basic idea that was to make this machine different from, and better than, its predecessors was a threshing cylinder the same width as the cutter bar. This permitted the grain to be fed into the cylinder in a thin stream rather than a large quantity of it being rammed into a narrow cylinder throat. Behind the threshing cylinder a wide rack allowed the stream of straw to move toward the back of the machine, making it easier to shake the grain out of the straw. It was no longer necessary to hammer the straw to pieces. Any farmer who was feeding livestock could now save the entire

Thousands of All-Crop harvesters were sold during the 1930s and 1940s. They were adaptable for harvesting many different seed crops.

yield of straw as well as the grain. This new concept of threshing also kept the weeds and green stuff out of the grain, thus revolutionizing the harvesting of crops on the smaller farms in the United States.[55]

However, it was not nearly as simple as this sounds, for it took constant work from 1930 through 1934 to perfect the new combine. Engineers had to substitute rubber-faced iron bars for the wire brushes in the threshing cylinder because the wire brush tips broke off and killed the animals that ate the straw. Allis-Chalmers Tex-rope drive substituted for the cogs and chains made the machine lighter and more efficient. It also permitted the operator to more easily vary the speed of the cutter bar, straw rack, cleaning fan, and threshing cylinder to suit almost any kind of crop. But it took nearly two years of experimentation to find that a V-belt angle of twenty-eight degrees gave the best service. The final step was to put the combine on rubber tires to decrease weight and increase efficiency. At long last the perfect combination had been found. When the versatile WC two-plow tractor was hitched to it, the new harvester pulled so easily that there was enough reserve for the power take-off to operate the harvester in any crop.[56]

The first demonstration of the "Corn Belt Combine," as this machine was originally named, was on a farm in La Porte County, Indiana, during the harvesting season of 1934. The *Indiana Farmers's Guide* recorded that it traveled at five miles per hour and "cut and threshed wheat and oats at one operation with such ease and speed as to amaze the more than 200 spectators gathered from all parts of the country." The article concluded by saying that this machine marked "a distinctive milestone in the advancement of American agriculture, quite as much as did the advent of the reaper, more than a hundred years ago."[57]

When the "Corn-Belt Combine" was put to the test of public acceptance in 1935, it exceeded all expectations both in sales and in its ability to harvest all types of grain. By 1936 it had successfully harvested 84 different small seed and bean crops, including even rice and sunflower seeds. Eventually it was to harvest over 100 different crops and its name was changed to the All-Crop harvester. Priced at $595 each F.O.B. La Porte, 550 machines were sold in 1935. Adverse harvesting conditions throughout the country stimulated sales because it was found that the All-Crop could harvest grain when no other conventional combine could. In 1936 the sales of All-Crop harvesters jumped ten times, to 5,500. In 1937 the number was 10,500 and the following year 16,500 more replaced the binders and threshing machines. In fact, the All-Crop harvester came to be known as the "successor to the binder," for as Allis-Chalmers production increased to meet demand, binder production in the United States declined from 66,000 in 1936 to 15,000 in 1939.[58]

As with almost all new developments, there was some resistance. Malt buyers in Milwaukee for some time refused to buy barley harvested with the new combine. But the real merit of the new machine was recognized when the All-Crop harvester was awarded the Royal Silver Medal at the Royal British Agricultural Exposition at Bristol, England, in 1936. The awards committee judged this machine the most notable advancement of the year. For the first time in twelve years an American manufacturer had won this coveted award.[59]

The All-Crop harvester allowed the small midwestern farmer to compete more effectively with the large western grain grower. By combining his grain the large farmer could put it in the bin for about nine cents a bushel. By comparison, the small farmer found that the normal cost of binding, shocking and threshing during the thirties cost him about twenty cents a bushel. But, using an All-Crop harvester, the small farmer could put his grain in the bin for about ten cents a bushel, comparable to the large-scale-farmer, and pay for his machine in less than two years with the difference. Futhermore, soil conservation now took on a new meaning for the small farmer because it was obviously profitable. In the past the small midwestern farmer had to grow a fixed amount of grain to keep the cost of threshing at a profitable figure. For this reason he was unable to diversify his crops as the market and soil requirements demanded. Now for the first time he could build his soil as he built his income.[60]

Because the five-foot All-Crop harvester proved such a success, the company began work on an even smaller model. In 1938, 100 harvesters with a forty-inch cut were put on the market for the farmer with only fifty acres of grain on his farm. In 1939 about 80

The old gives way to the new. The number of work animals on farms decreased yearly as tractors were improved.

percent of the combines sold cut swaths six feet wide or less. It was true that the Allis-Chalmers harvester did not cut a very wide swath, but it did cut a mighty long one.[61]

By 1936 the tractor division was the largest one in the company and sold more than 300 products from four large plants. The West Allis Works manufactured wheel-type tractors, with the engines for the tractors and equipment built in other plants. The La Porte, Indiana, factory built the All-Crop harvester with an impressive line of threshers, big combines, clover and alfalfa hullers, and road machinery, including a complete line of power and hand-controlled blade graders, and single- and tandem-drive Speed Patrols. The plant at Springfield, Illinois, made the crawler tractors and Speed Ace hauling units. The La Crosse Works manufactured the extensive line of Allis-Chalmers farm implements, including cultivators, plows, bedders, harrows, mowers and other power machinery tools.[62]

The La Crosse, La Porte and Springfield plants were supported and supplemented during the thirties by additions of lesser importance. The Advance-Rumely holdings in Canada were reorganized as outlets for Allis-Chalmers, with the capital stock wholly owned by the American company. From 1927 until 1932 B. F. Avery & Sons of Louisville, Kentucky, had acted as the Allis-Chalmers distributors in Argentina. By the latter date the Avery Company found itself in a "straitened financial condition" and the creditors were willing to dispose of the Argentine property. Allis-Chalmers was able to acquire property valued at about $500,000 by can-

Some west coast farms required deep tillage implements pulled by track-type tractors. The Brenneis Manufacturing Company of Oxnard, California, had made these implements since 1898 and was purchased by Allis-Chalmers in 1938.

celling a note that Avery owed to the Advance-Rumely Corp. In 1931 the owners of the Birdsell Manufacturing Company of South Bend, Indiana, manufacturers of the Birdsell Clover Huller, wished to dispose of their stock and patents. General Falk arranged the purchase of this huller for $58,000 and moved the operations to the La Porte Works.[63]

On parts of the West Coast, ground conditions required the use of deep tillage equipment pulled by crawler tractors. The Brenneis Manufacturing Company of Oxnard, California, manufactured deep tillage instruments, and on July 18, 1938, the General proposed to the Executive Committee that Allis-Chalmers purchase Brenneis for about $100,000. The California company employed only twenty-three men just before the purchase but this would provide, it was thought, not only a "moderately profitable business," but also a considerable aid in the sale of the company's crawler-type tractors. Within months employment at the "California Works" of the company doubled as the Merritt spirit enlivened the West Coast plant.[64]

Although the horse has always been loved by some as an intelligent and handsome animal, the tractor division of Allis-Chalmers was dedicated to a reduction in the number of horses in the country because each one meant a potential tractor sale. The spirit of the sales department is evident in a statement attributed to Bill Roberts that he hoped to live to see the day when children would have to go to a zoo to see a horse. At the close of World War I the horse and mule population in the United States had reached its highest point, 26,723,000. At that time there were only 85,000 tractors. But from that time on, each year saw a reduction in the number of horses and a corresponding increase in the number of tractors. By 1920 draft animals had decreased by a million, and by 1928 the number dropped to 20,448,000, while tractors had increased to 782,000. A decade later the number of horses and mules had dropped another five million to 15,245,000, while the number of tractors had doubled in the same period to 1,370,000.

The simple fact is that the horse by 1918 had apparently been brought to its highest efficiency, but the tractor improved with each year. Horse power was becoming increasingly expensive because, working or not, a horse had to be fed 365 days a year. The American farm horse worked on the average for only three hours per day the year round; yet it required nearly as much food and attention when not working as when it was performing its hardest duties. When the tractor was in its shed, it required no labor or attention; the only expense was the interest on the investment and a small amount for depreciation. Farmers were obviously well aware of these facts, and by the thirties only the small farmer could afford to use horses. The struggle between the animals and tractors had become unequal by that time, and by 1938 the arguments of the Horse and Mule Association had been reduced to the rather obvious fact that tractors were inferior to animals because they produced no manure.[65]

During the mid-thirties Harry Merritt studied the farm census figures and discovered that although 4,000,000 of the 6,800,000 American farms were under 100 acres, most of the 1,200,000 tractors in the country were working farms of larger acreage. To fill this "tractor gap," Walter Strehlow, the man behind the famous "WC," designed a small tractor which was placed in production in 1938 as the Model "B," revolutionary in price, weight and adaptability, weighing 2,100 pounds and costing only $495. Compared with tractors of ten years before, it weighed and cost only one-third as much, but it would do 20 percent more work with 25 percent less fuel than any tractor of comparable power of the previous decade. It could pull a 16-inch moldboard plow at three to four miles per hour. A farmer could haul a trailer load of hay on rubber tires at about seven miles per hour. If he wanted to saw logs, he could attach a belt to the pulley wheel geared to the tractor transmission and operate a circular saw. Or, he could power shaft-driven machinery such as a mower by a take-off on the rear axle. Even Allis-Chalmers was surprised when parks and country clubs purchased many Model "B" tractors for grass cutting. Some rural mail carriers who had small acreages bought them—they delivered the mail by tractor and used it for farm work the rest of the time. This little tractor was so versatile that C. C. Gross, enterprising manager of the Columbus branch office, trained four boys and four girls each on a Model "B" to stage a full square dance routine, with music and caller, at the Ohio State Fair in 1939. Advertised as the "successor to the horse," the Model "B" was all that and a good deal more.

For the first time in agricultural history a farmer could operate a completely mechanized farm of 100 acres for an investment of $10 an acre. With the Model "B" costing $495, the next most expensive investment might be the 40-inch All-Crop harvester with power take-off, costing $345. The small farmer could then thresh all his small grains, beans, and seeds with no outside help. For plowing, he could buy an Allis-Chalmers No. 116 Moldboard Plow for $85; at twice the speed of horses, he could plow the soil deeper and pulverize it better. Finally, for $50.25 he could buy a one-row cultivator which was adaptable to all row crops. Mechanization was to spell the eventual doom of the small farmer in American agriculture, but Harry Merritt had found a new and untapped market. For the moment the small farmer was granted a reprieve by becoming temporarily more competitive with the Model "B" and its accompanying Allis-Chalmers equipment.[66]

The thirties showed dramatic growth for the Allis-Chalmers tractor division. Such revolutionary innovations as the air tire and the All-Crop harvester, with an expanded line of quality products, enabled the company to profit from the increasing farm income. Between 1932 and 1936, gross farm income increased by 50 percent, and the $6,406,000,000 of 1932 had doubled by 1941 to $13,299,000,000. The development of a superior agricultural line coincided with increasing demand for quality agricultural equipment and availability of the money to pay for it.[67]

The company's product lines other than tractors recovered from the depression slowly but steadily. But the very rapid growth in the tractor division produced an internal revolution. The once insignificant tractor tail had by 1938 come to wag the company dog. The following table graphically indicates the extent to which the tractor division became the basis for the company's returning prosperity. This was a matter of both volume and profit, for it was estimated that from 1934 through 1938 the tractor division accounted for two-thirds of the company's total net profit.[68]

Year	A-C Total Sales	Tractor & Farm Equipment Sales	% of Total	A-C Total Net Income
1931	$27,800,000	$ 7,000,000	25.2	$1,256,000
1932	14,760,000	6,240,000	42.3	d2,955,000
1933	13,280,000	5,100,000	38.4	d2,894,000
1934	20,280,000	8,950,000	44.1	d1,039,000
1935	38,780,000	20,970,000	54.0	1,985,000
1936	58,980,000	33,970,000	57.6	4,014,000
1937	87,350,000	50,950,000	58.3	7,841,000
1938	77,540,000	45,500,000	58.7	2,554,000

W. C. Johnson, vice president, Walter Geist, president, and W. A. Roberts, vice president and tractor manager, try out a new tractor.

Allis-Chalmers engineering and innovation also had a profound effect on the entire agricultural equipment industry. From a position of relative insignificance as a producer of tractors and agricultural equipment, the company shot rapidly upward to third place in this field during the middle and late thirties. It is estimated that during that decade, Allis-Chalmers had no more than one-twelfth of the salesmen in the field, but by 1937 these men were selling 13 percent of the products of the whole industry. This percentage increased during the two succeeding years so that by 1939 the company was selling more than one-fifth of the national tractor product. The fact that 1940 saw Allis-Chalmers decline to about 12 percent of the total is significant because a company regarded as an upstart by such agricultural giants as International Harvester and John Deere had forced the modernization of the tractor products of the entire industry. In the fall of 1939, International Harvester brought out four new tractor models and Deere brought out three, making the field much more competitive.[69]

The man who was largely responsible for this success was Harry C. Merritt. Recognition for his achievement and service to the company came in the form of election to the post of vice president of Allis-Chalmers in January, 1937. But recognition did not come from his own company alone: he was chosen the McCormick medalist of 1941 for "exceptional and meritorious achievement in agriculture" by the Society of Agricultural Engineers.[70]

The aim and spirit that Harry Merritt instilled in the development of Allis-Chalmers agricultural machinery is found in his statement that "this is an era of speed, and there is no reason why the farm machinery industry should sit back and allow developments of present-day engineering to pass it by. We feel that we should adopt new automotive ideas in farm machinery, especially when they not only halve the cost, but increase the output." He concluded by noting that the ultimate purpose of the company in its agricultural developments had been "the lowering of crop production costs and a reduction of the farmer's investment in farm equipment."[71]

In this instance, General Otto Falk had used the basic business technique which had made E. P. Allis such a successful businessman. Falk had given Merritt an almost free hand to run the tractor division, and he had picked the right man for the right job at the right time.

NOTES TO CHAPTER EIGHT

1   Ed Meisenheimer to Alberta J. Price, n.d.; Arthur Van Vlissingen, "50,000,000 New Dollars a Year," *Forbes*, June 1, 1938. A number of proposals to manufacture new products came to the Board and to the Executive Committee, and they did for a time endorse the manufacture of starting and lighting systems for automobiles. The committee declined an invitation by W. C. Durant to manufacture his automobile and another to produce "aeroplane" motors because "up to the present time the development of the aeroplane motor is in a more or less experimental stage and no particular type has as yet been developed which engineers engaged in this line of work feel will be the ultimate machine." But none of these proposals would help achieve the ends that Otto Falk wanted. A-C Mfg. Co. Executive Committee, March 9, 1915, March 19, 1917; A-C Mfg. Co. Board of Directors, August 6, 1914, June 3, 1915, April 5, 1917; "Allis-Chalmers: 'America's Krupp,' " *Fortune*, May, 1939, p. 55; *Power Review*, July, 1940, p. 5.

2   Bert S. Gittins, *Land of Plenty* (Chicago, 1959), pp. 50–51; Stewart H. Holbrook, *Machines of Plenty* (New York, 1955), pp. 166–168; Harold U. Faulkner, *The Decline of Laissez-Faire 1897–1917*, (New York, 1915), p. 332.; "The Development of the Tractor," *Southwest Implement Journal*, July, 1958, pp. 8–9.

3   A-C Mfg. Co. Executive Committee, August 15, December 22, 31, 1913, April 23, November 5, 1914, February 18, 1915; *Rotary Tiller*, Bulletin N. 1453, March, 1915; A. W. Van Hercke to author, June 13, 1961; Barton W. Currie, *The Tractor and Its Influence upon the Agricultural Implement Industry* (Philadelphia, 1916), p. 12.

4   A-C Mfg. Co. Executive Committee, December 22, 1913; C. R. Beck to author, July 3, 1961; Patent number 1,227,005, May 22, 1917. On September 5, 1913, the Executive Committee generally approved a proposition from C. L. Tolles of Eau Claire, Wisconsin, to sell one-half interest in his patents to the "Centipede Tractor" to the Company for $10,000 and referred it to the president, "with power." No mention is made of the outcome of this.

5   A-C Mfg. Co. Executive Committee, February 6, March 24, July 23, 1915; A-C Mfg. Co. Board of Directors, June 3, 1915; *Tractor Truck Circular Letter* No. 1, April 24, 1915. Letter No. 2, dated May 5, 1915, cautioned against the use of the word *caterpillar*, for it had been registered as early as 1910 by Benjamin Holt Manufacturing Company of Stockton, California. Allis-Chalmers adopted the term *track wheel* to describe that part of their tractor truck which corresponded to the caterpillar on the Holt tractor. C. R. Beck to author, July 3, 1961; R. M. Collette to author, November 10, 1964; *Tractor Truck*, undated brochure. For a more extensive treatment of this see Walter F. Peterson, "Twenty Years Premature: The Allis-Chalmers Tractor Truck," *Historical Messenger*, June, 1962, pp. 12–14.

6   A-C Mg. Co. Executive Committee, December 28, 1914, April 29, August 23, October 11, 1915, January 27, April 19, 1916; *Sales Bulletin*, January, 1919, p. 26; *A-C Views*, July 24, 1952, p. 5.

7   *Sales Bulletin*, February, 1919, p. 15; A. W. Van Hercke to author, June 13, 1961; *A-C Views*, July 24, 1952, p. 5; *Milwaukee Journal*, July 6, 1955. The 6–12 was not unique in principle. In the *Sentinel* of March 21 and September 19, 1915, illustrations of two other tractors show the same design. Although the tractor appearing on September 19 is even more bulky and unwieldy-looking than the 6–12, the Lion tractor appearing on March 21 was priced at only $375, which may indicate that the 6–12 was perhaps over-priced.

8   *Sales Bulletin*, December, 1918, p. 11; A. W. Van Hercke to author, June 13, 1961.

9   A-C Mfg. Co. Executive Committee, January 6, 1919; A. W. Van Hercke and W. T. Strehlow to author, June 13, 1961; *A-C Views*, July 24, 1952, p. 5; A. Frederick Collins in *Farm and Garden Tractors* (New York, 1920), pp. 156–158, deals with the Allis-Chalmers "General Purpose 6–12 Tractor" in some detail and has some good words for it even at that late date.

10 A-C Mfg. Co. Executive Committee, October 21, 1919; *Sales Bulletin*, November, 1920, p. 1; *Agrimotor*, February 15, 1921, p. 50.

11 A-C Mfg. Co. Executive Committee, January 23, 1918; *Sales Bulletin*, February 1919, p. 15, March, 1922, p. 33; *A-C Views*, July 24, 1952, p. 5.

12 *Sales Bulletin*, February, 1921, p. 5; *Agrimotor*, February 15, 1921, p. 50.

13 *Sales Bulletin*, August, 1924, p. 73; *Allis-Chalmers Tractors*, brochure of 1925.

14 Reference to tractors and the tractor department occurred at nearly every meeting of the Board of Directors and the Executive Committee from August, 1913, until February, 1924, when discussion of tractors stopped entirely. Not until July 26, 1926, did the topic again come up consistently. The Executive Committee authorized numerous deposits of money with a considerable number of banks between March 28, 1918, and May 19, 1924.

15 A-C Mfg. Co. Executive Committee, June 28, September 20, November 29, December 20, 1920, October 16, 1922.

16 A-C Mfg. Co. Executive Committee, January 17, 1921.

17 *Agrimotor*, February 15, 1921, p. 50; *Sales Bulletin*, May, 1923, p. 63.

18 *Sales Bulletin*, July, 1922, pp. 49–50, October, 1922, p. 55.

19 *Sales Bulletin*, September, 1921, pp. 15–16, November, 1921, p. 21. Illinois led all states with 23,101 tractors, Iowa had 20,270, Kansas 17,177, Minnesota 15,503, California 13,852, North Dakota 13,006, South Dakota 12,939 and Nebraska 11,108 tractors on January 1, 1920.

20 *Sales Bulletin*, February, 1924, p. 71, August, 1924, p. 73, December, 1924, pp. 77–78, March, 1925, p. 79.

21 "H. C. Merritt to be Tenth McCormick Medalist," *Agricultural Engineering*, January, 1941, pp. 30–32; "Life and Death of Harry Merritt; An Industry Leader," *Implement & Tractor*, December 4, 1943, pp. 30, 59; *President's Circular Letter*, No. 173, December 29, 1925; *Fortune*, May, 1939, p. 148; R. C. Crosby to author, June 15, 1961; W. F. Strehlow to author, June 13, 1961.

22 A. W. Van Hercke to author, June 13, 1961; *Fortune*, May, 1939, pp. 148–150.

23 *Sales Bulletin*, January, 1927, p. 91; A-C Mfg. Co. Executive Committee, January 21, 1929.

24 A. W. Van Hercke, W. T. Strehlow and John Ernst to author, June 13, 1961; R. C. Crosby to author, June 15, 1961.

25 "A Tribute to H. C. Merritt," *Farm Implement News*, August 29, 1935. *Fortune*, May, 1939, p. 150.

26 R. C. Crosby, "The Story of Persian Orange Paint," MS; Malcolm C. Maloney to author, June 19, 1961.

27 *Sales Bulletin*, April, 1926, p. 87, November, 1929, p. 109; *Seventeenth Annual Report*, March 14, 1930, p. 6.

28 *Sales Bulletin*, November-December, 1927, p. 94, May, 1929, p. 770.

29 *Sales Bulletin*, March, 1926, p. 85, March, 1927, pp. 500–501, November-December, 1927, pp. 93–94, January, 1930, p. 876; *Chicago Journal of Commerce*, March 27, 1930. In late 1929 the Executive Committee authorized Falk to buy up to 20,000 shares of Cockshutt Company stock at $25 per share so that there could be a greater interrelation between the two. The *Minutes* of the Executive Committee indicate that on November 8, 1926, Allis-Chalmers was interested in the sale of 500 tractors to Russia. This contract was apparently never completed.

30 *Power Review*, September, 1941, pp. 27–28; A. W. Van Hercke and W. T. Strehlow to author, June 13, 1961.

31 *Sales Bulletin*, December, 1925, p. 81; *Fortune*, May, 1939, pp. 148, 152.

32 A-C Mfg. Co. Board of Directors, December 2, 1927, April 5, December 7, 1928; A-C Mfg. Co. Executive Committee February 24, May 21, December 27, 1928. Before December, 1927, Falk had been investigating crawler tractors and reported "that one of the companies in this business had made an offer to the Company, on what appears to be an attractive basis." *Sales Bulletin*, April, 1928, p. 95; *Nineteenth Annual Report*, March 21, 1932, p. 4.

33 A-C Mfg. Co. Executive Committee, September 14, 1928; A-C Mfg. Co. Board of Directors, October 5, 1928; *Sales Bulletin*, March, 1929, p. 746; *Sixteenth Annual Report*, March 16, 1929, pp. 9–10; *Fortune*, May, 1929, p. 150.

34 *Sales Bulletin*, September, 1929, p. 815; A. W. Van Hercke and W. T. Strehlow to author, June 13, 1961; *La Crosse Tribune*, October 6, 1929; *The Ace Reporter: Silver Anniversary, 1929–1954, passim*; A-C Mfg. Co. Executive Committee, August 5, 1929. In addition to $275,000 in cash, Allis-Chalmers also took over an inventory estimated at about $375,000 and the good accounts and notes receivable of approximately $125,000. By March 1930 employment at the La Crosse Works had doubled, reaching 144. It then declined gradually as the depression began to be felt.

35 A-C Mfg. Co. Board of Directors, June 8, 1932; *Sales Bulletin*, October-November-December, 1931, p. 70, July to December, 1932, p. 39; *A-C Views*, July 24, 1952, p. 5.

36 A-C Mfg. Co. Executive Committee, October 15, 1928, May 13, 1929; A-C Mfg. Co. Board of Directors, October 5, 1928, January 14, 1929; Otto H. Falk, "An Act

of Faith in Modern Equipment," *American Machinist*, March 13, 1930, p. 435; *Sales Bulletin*, November, 1928, pp. 101–102, November, 1929, p. 106, November-December, 1930, p. 111; A. W. Van Hercke, W. T. Strehlow, John Ernst to author, June 13, 1961.

37 J. L. Neenan to author, July 29, 1959; Charles Allendorf to author, August 10, 1959; R. C. Crosby to author, June 15, 1961.

38 W. M. Rumely, *Personal History of Meinrad Rumely*, undated MS.; W. M. Rumely, *Rumely Company*, undated MS.; *Allis-Chalmers La Porte Works Silver Anniversary, 1931–1956, passim*.

39 A proposition to purchase Advance-Rumely was informally considered by the Board of Directors on January 24, 1930, and declined. Negotiations continued until the matter was finally settled by the Board at two meetings on December 5, 1930, and April 10, 1931. *Sales Bulletin*, April to July, 1931, p. 1049; *Chicago Journal of Commerce*, May 13, 1931; A. W. Van Hercke, W. T. Strehlow and John Ernst to author, June 13, 1961.

40 A-C Mfg. Co. Executive Committee, August 10, 1931; W. J. Shields, *Pneumatic Tires for Agricultural Tractors*, undated MS., p. 1; *Sales Bulletin*, October-November-December, 1931, p. 67.

41 W. J. Shields, *op. cit.*, pp. 4–5; J. Brownlee Davidson in *Successful Farming*, January, 1935; R. C. Crosby to author, June 15, 1961.

42 R. C. Crosby, *Rubber Invades the Farm*, undated MS., pp. 9–10; *Farm Implement News*, October 13, 1932, p. 6; *We*, July, 1947, p. 4. "A Tribute to H. C. Merritt," *Farm Implement News*, August 29, 1935, p. 40, helps to place in perspective the matter of the rubber tire. "There was nothing novel about rubber tires on tractors at that time. Industrial tractors had been equipped with high pressure truck-type tires for years. Tractors with low pressure balloon tires had been used to some extent in the citrus groves of Florida and the West. These low pressure tires, however, were not designed for farm tractors and were not satisfactory for the heavy drawbar work of general farming. A standard farm tractor tire and rim were still to be developed."

43 *A-C Milestones in Farm Mechanization*, 1953, p. 3.; R. C. Crosby, *A Decade of A-C Pioneering, 1930–1940*, undated MS, pp. 16–18; William A. McGarry, "The Farm Tractor Takes Wings," *The Magazine of Wall Street*, December 7, 1935, p. 198. The original Model "U" tractor with its airplane tires is now on permanent exhibit at the museum of the State Historical Society of Wisconsin. Alfred Lief in *The Firestone Story* (New York, 1951), p. 196, implies that Firestone took the initiative in this venture. Allis-Chalmers engineers who were involved maintain that they had considerable difficulty getting the tires for such a "hairbrained" scheme.

44 R. C. Crosby, *A Decade of A-C Pioneering*, pp. 20–23.

45 *Ibid.*, p. 24; *We*, July, 1947, p. 4.

46 Shields, pp. 2–4; "The Challenge of Low Pressure Air Tires on Farm Tractors," *Farm Implement News*, October 13, 1932, pp. 19–20. The studies by the colleges and universities largely bore out these findings. See C. W. Smith and Lloyd W. Hurlbut, *A Comparative Study of Pneumatic Tires and Steel Wheels on Farm Tractors*, Bulletin 291, the University of Nebraska College of Agriculture Experiment Station, September, 1934; H. E. Murdock, *Tests on Use of Rubber Tires and Steel Wheels on a Farm Tractor*, Bulletin No. 329, Montana State College, Boseman, Montana, April, 1937; G. W. McCuen and E. A. Silver, "Low Pressure Pneumatic Tires for Farm Machinery," *Engineering Experiment Station News*, The Ohio State University, June, 1938, pp. 22–25.

47 "A Tribute to H. C. Merritt," *Farm Implement News*, August 29, 1935, p. 40; *Sales Bulletin*, July to December, 1933, p. 37.

48 A three-page condensation of the nine-page letter from the Caterpillar Company dated January 30, 1933, was published in the *Sales Bulletin*, July to December, 1933, pp. 40–43; A. W. Van Hercke, W. F. Strehlow and John Ernst to author, June 13, 1961. It is reported that it was President Clausen of Case who made the public statement that the tractor was stuck in a bucket of water. *We*, July, 1947, p. 4.

49 *Sales Bulletin*, July to December, 1933, p. 37; R. C. Crosby to author, June 15, 1961; R. C. Crosby, *A Decade of A-C Pioneering*, undated MS, pp. 25–27; letter from Allis-Chalmers to J. Edward Schipper, American Automobile Association, Pennsylvania Avenue at 17th Street, Washington, D. C., December 19, 1945.

50 William A. McGarry, "The Farm Tractor Takes Wings," *The Magazine of Wall Street*, December 7, 1935, p. 199; *We*, July, 1947, p. 5.

51 *Sales Bulletin*, July to December, 1933, p. 39; *We*, July, 1947, p. 6; A. W. Van Hercke, W. F. Strehlow and John Ernst to author, June 13, 1961; R. C. Crosby to

author, June 15, 1961; A-C Mfg. Co. Executive Committee, January 17, March 23, 1936; General Otto H. Falk, "More Machines and More Jobs," *Factory Management and Maintenance*, May, 1936, p. 37.

52 Memorandum by Harry Merritt, 1934; Earle D. Ross in *Iowa Agriculture* (Cedar Rapids, 1951), pp. 178–179, noted that "The designing of a light, high speed, general purpose model gave the farmer a power unit of general availability. The introduction of low pressure pneumatic tires made possible increased speed, greater comfort in operation, and lowered cost of fuel and repairs. By 1939 about three-fourths of the new tractors had this equipment. Iowa led the states in the number of tractors, with 135,000 in 1939 and about 160,000 four years later—one for each 218 acres."

53 R. C. Crosby, *A Decade of A-C Pioneering, 1930–1940*, pp. 1–12, 32; *Sales Bulletin*, July to December, 1933, p. 39. The Bostrom Corporation, Milwaukee, got its start in 1935 manufacturing seat pads for Allis-Chalmers and other tractor manufacturers. Karl Bostrom, chairman of the board of Bostrom Corporation, and his father were both employed at the West Allis Works before 1935, when they set up their own manufacturing concern. Starting with farm tractor and truck cushions and seats, the firm now provides engineered seating to meet almost every need. Karl Bostrom to author, May 10, 1963.

54 R. C. Crosby, *A Decade of A-C Pioneering, 1930–1940*, pp. 37–40; Holbrook, p. 203.

55 On October 14, 1930, Otto Falk secured approval from the Board of Directors for the purchase of "a combine harvester" for $87,500 and a royalty of 5 percent. *Sales Bulletin*, November-December, 1930, p. 114; R. C. Crosby to author, June 15, 1961; R. C. Crosby, *A Decade of A-C Pioneering*, pp. 40–43.

56 R. C. Crosby, *A Decade of A-C Pioneering*, p. 43.

57 Clipping from the *Indiana Farmer's Guide*, Huntington, Indiana, n.p., n.d.; The statement by Merritt is quoted in *A-C Milestones in Farm Mechanization*, p. 4; *Sales Bulletin*, July to December, 1933, p. 38, October to December, 1934, p. 51.

58 *Sales Bulletin*, October, 1935, p. 33, July, 1936, p. 13; *Annual Review*, 1937, p. 64; *Company Facts*, 1936, n.p.; *A-C Milestones in Farm Mechanization*, p. 18; R. C. Crosby, *A Decade of A-C Pioneering*, pp. 47–49; Walter Geist in his *Newcomen Address*, December 7, 1950, p. 17, referred to the All-Crop as "the midget that does a giant's work." Production of soybean oil in the United States coincided with the All-Crop harvester. The 13,000,000 pounds produced in 1929 increased to 322,000,000 pounds in 1938. Broadus Mitchell, *Depression Decade* (New York, 1947), p. 225. The increasing production of the All-Crop resulted in expansion of the La Porte plant in late 1935 and again in 1938. A-C Mfg. Co. Executive Committee, November 18, 1935; A-C Mfg. Co. Board of Directors, September 1, 1938.

59 *Sales Bulletin*, October, 1936, p. 9; A. W. Van Hercke, W. T. Strehlow and John Ernst to author, June 13, 1961.

60 R. C. Crosby to author, June 15, 1961; R. C. Crosby, *A Decade of A-C Pioneering*, p. 47; G. E. Ryerson, "Remaking the Land," *Power to Produce* (Washington, 1960), pp. 101–103.

61 R. C. Crosby, *A Decade of A-C Pioneering*, p. 50; Broadus Mitchell, p. 222.

62 *Company Facts*, 1936, n.p.

63 A-C Mfg. Co. Executive Committee, August 10, 1931, August 2, 1932, July 18, 1938; A-C Mfg. Co. Board of Directors, December 4, 1931; *Nineteenth Annual Report*, March 21, 1932.

64 *A-C Annual Review*, 1938, p. 80.

65 *Sales Bulletin*, November, 1921, p. 21, March, 1922, p. 34. Until the early twenties the *Sales Bulletin* kept a running account of the struggle between draft animals and the tractor, but following that time they rightly assumed that the battle had been won and ignored the animals. *Agricultural Outlook Charts*, United States Department of Agriculture, October, 1952, p. 18; *Fortune*, May, 1939, p. 150.

66 *Annual Review*, 1937, p. 64; *A-C Milestones in Farm Mechanization*, pp. 10–11; Arthur Van Vlissingen, "$50,000,000 New Dollars a Year," *Forbes*, June 1, 1938; *Fortune*, May, 1939, pp. 150–151.

67 Edward C. Kirkland, *History of American Economic Life* (New York, 1951), p. 595; Arthur M. Schlesinger, Jr., *The Coming of the New Deal* (New York, 1959), p. 71.

68 The Minutes of the Executive Committee indicate the enthusiasm with which the tractor prospects were greeted during the period 1933 to 1939. There is also

frequent reference, as on July 15, 1935, to the fact that "there was no important change in the situation with respect to the other lines of the Company's products." *Fortune*, May, 1939, p. 152.

69 From 1937 through 1940 Allis-Chalmers sales appear in the following statistics:

Year	A-C Sales	Total tractor Sales	A-C's % of total tractor sales
1937	32,000	243,000	13.15%
1938	29,000	175,000	16.57%
1939	39,000	185,000	21.06%
1940	30,000	249,000	12.26%

During much the same period the sales of Model "B" and "WC" tractors were as follows:

Sales of "WC" tractor		Sales of Model "B" tractor	
1935	7,948		
1936	15,297		
1937	24,635	1937	83
1938	18,442	1938	7,584
1939	14,660	1939	17,608
1940	11,969	1940	15,169
1941	16,436	1941	14,698

All the above information from A. W. Van Hercke, June 19, 1961.

70 A-C Mfg. Co. Board of Directors, January 8, 1937; *A-C Annual Review*, January-February 1937, p. 71; *Power Review*, September, 1941, p. 27; "Harry C. Merritt 10th McCormick Medalist," *Better Farm Equipment and Methods*, January-February, 1941, pp. 4, 24. On July 1, 1941, Harry C. Merritt retired from active direction in Allis-Chalmers, but he remained in an advisory capacity. W. A. (Bill) Roberts was appointed manager of the tractor division.

71 R. C. Crosby, *A Decade of A-C Pioneering*, p. 47.

Max Wellington Babb
President 1932–1942

CHAPTER NINE

DEPRESSION
AND SURVIVAL

"THE NEW YEAR seems to hold considerable promise," wrote General Otto Falk on December 28, 1928, in his New Year's message to the employees of Allis-Chalmers. He felt that the expected improvement in general business conditions should be reflected in many of the company's products. He concluded with a statement of faith: "We have confidence that the Federal Government Administration will do its part in co-operation with industry to improve general business conditions." For Allis-Chalmers the New Year did, statistically, hold real promise. The company's agricultural sales in 1928 had doubled from $5 million to $10 million. Total sales increased from $36 million to $45 million, and net income rose from $2.9 to $4.3 million.[1]

Riding a wave of confidence and optimism, the New York Stock Exchange on September 3, 1929, set a record with price averages reaching their highest to that time. The next morning the *New York Times* reported the Allis-Chalmers transactions for that record day: with a dividend rate of 7 percent, Allis-Chalmers stock closed at 316½, up five points. On September 20 the stock was split, four shares of the new stock for each share of the old. What General Falk along with most other businessmen had not foreseen was that October, 1929, would mark the beginning of the Great Depression. On October 29, the disastrous sixteen million share day on the Exchange, Allis-Chalmers stock opened at 46½, dropped to 37½ and closed at 41. With the stock off five points, 15,500 shares were sold.[2]

This, unhappily, was only the beginning. The dollar volume of business fell from 15 percent above normal in July, 1929, to 49 percent below normal in July, 1932. Save for a slight rise in early 1931, the descent was a fairly straight line between the two points. No previous depression had ever deflated the dollar volume of business by more than 35 percent, but this drop amounted to 56 percent. National income, which had been $87.2 billion in 1929, fell to $41.7 billion in 1932. Unemployment increased in inverse proportion to the decline in national income: four million in 1930, eight million in 1931, twelve million in 1932. Nearly one out of every four workers was then seeking a job. As a producer of durable goods, Allis-Chalmers slipped more slowly into the depression than did many other firms.[3]

In late 1929 and early 1930 no one knew how bad the depression would be or how long it would last. Unlike Detroit and the automobile industry, Milwaukee and Allis-Chalmers were at first little effected. The *Milwaukee Sentinel* reported on January 1, 1930, that "The receding tendency of general business had been considerably less sharp in Milwaukee than elsewhere." Thousands of job-hunters came to Milwaukee after a federal survey indicated that production was higher there than elsewhere.[4]

On October 31, 1929, the week of the disaster on the stock exchange, 8,671 employees were at work at the Allis-Chalmers plants. Because demand for the durable goods produced by the company was not immediately effected, the number of employees continued to rise until a high point was reached on February 28, 1930. At that time there were 5,764 workers at West Allis, 900 at Norwood, 532 at Pittsburgh, 538 at Springfield, 132 at La Crosse and 1,522 in the general office, a total of 9,388 persons. But bookings declined from $6,128,000 in March to $1,529,000 in December. By mid-July Otto Falk reported "a noticeable decrease in active sales negotiations" to the Executive Committee. During July alone 942 men were laid off, 712 of these at West Allis. It was not, however, until December 1, 1930, that the General used the word *depression*. At this point Milwaukee's officials admitted that the city was facing "the worst business depression in its history."[5]

Whatever private concerns the General expressed to the Executive Committee, like other business and government leaders he remained publicly optimistic. In his New Year's message for 1931 he said, "Public confidence is gradually being restored and it is the belief of most business people that the year 1931 will bring better conditions than we have experienced during 1930, which has presented many perplexing problems in both business and domestic life." One full year later he was still able to write, "We enter the New Year with faith that our resourceful Country will soon work out of this serious depression." But company sales and income figures from the high of 1929 to the low point of 1933 clearly indicate that the worst was yet to come.

Year	A-C Total Sales	A-C Total Net Income
1929	$45,300,000	$4,331,000
1930	41,470,000	3,605,000
1931	27,800,000	1,256,000
1932	14,760,000	d2,955,000
1933	13,280,000	d2,894,000

The thrifty citizens and diversified industries of Milwaukee could not cushion the collapse indefinitely. By 1933 it was clear that Milwaukee had actually been effected more directly and severely than almost any other industrial community in the nation. The number of wage earners employed had dropped to 66,010 in 1933, a decline of 57 percent from the 1929 total of 117,658, and less than 50 percent of the general property taxes levied could be collected.[6]

In January, 1933, at the depth of the depression, General Otto H. Falk took stock of conditions. He was not impressed with the arguments of the technocrats and others that replacement of men by machines had been an important cause for unemployment in pre-depression years or that it had been a factor in bringing on the industrial collapse. "Machines have provided more jobs than they ever have taken away," he said. "Besides, they have brought many comforts and luxuries the average man never would have known had it not been for mass production made possible by these machines." Instead, he blamed the business collapse mainly on the speculation that had preceded the crash. "We can't all become rich by speculation and that's about what we tried to do, from the big stock market plunger down to the modest wage earner and the farmer who played the grain market. In such a buying splurge we can create no new wealth, but only fictitious values. Well, we had a grand spree, but now we must sober up."

By January, 1933, General Falk was convinced that "no magic shots in the arm" would help. Very simply, the national economy had to "get down to bed rock" and then gradually begin to work its way up again. He was sure in his own mind that the unemployment

These roller gates which control the flow of the Mississippi River, installed in 1937, are the largest in the United States. Each is 129.5 feet long and 29.5 feet high.

problem was never quite as grave as the politicians made it seem, for in a country the size of the United States there would always be from two to two and one-half million people who were "incapable of doing any kind of work" as well as a sizable number of "bums and floaters who don't want to work." Furthermore, by employing only the heads of families, "we would have the unemployment pretty well cleared up."[7]

The General had more to offer than observations; he also had some positive suggestions which he felt could not fail to help the country get back on its feet. As early as December 5, 1929, in a report on the machinery industry to the Business Survey Conference, he recommended that "this conference go on record in asking that this Congress pass a tariff bill without further unneces-

Allis-Chalmers engineers developed the automatic step-type feeder voltage regulator in the 1930s for regulating the voltage on power distribution lines regardless of the power load. Voltage regulation under load was also built into many power transformers.

sary delay so as to relieve the uncertainty of business on this subject. It is my candid opinion that this would be more helpful than any other one thing in stabilizing business." In 1933 he said, "I don't think the ship is going to sink; I am sure it won't if . . . the public will insist on an immediate and commensurate deflation of the government and if the legislative bodies will inspire confidence by letting the people know nothing is to be done to add to taxes, things will soon be on the upgrade." The General's economic philosophy was solidly based on nineteenth-century principles. But this was a twentieth-century economic catastrophe which had to be met with a totally new plan of action on a different set of principles.[8]

President Herbert Hoover on December 5, 1929, addressed the Business Survey Conference called by the United States Chamber of Commerce and asked businessmen to continue as usual. He urged industrialists not to cut wages or abandon construction programs, and he requested labor leaders not to rock the boat. Allis-Chalmers complied with these presidential requests. Wages and salaries were maintained at 1929 levels long beyond those in many other industries, the first reduction not occurring until December 1, 1930, a full year after the depression had begun. Because "there appeared to be no definite prospect of early substantial improvement," the Executive Committee reduced monthly salaries between $150 and $199 by 5 percent and monthly salaries over $200 by 10 percent. This reduction also applied to employees working by the hour whose full time earnings fell within those limits. With ample reserves at hand, the General could follow President Hoover's request to continue construction programs. In spite of declining sales, Falk authorized the extension of the erecting shop and the construction of a No. 7 building to run parallel with the No. 6 Shop for the electrical department. Begun in May of 1930, this was the most substantial building in the West Allis complex and was equipped with the first 100-ton crane. This crane would allow the company to build the largest transformers that could be handled by the railroads. Again in the fall of 1932, when the National Committee on Industrial Rehabilitation requested companies to repair, reconstruct and replace structures and equipment, Allis-Chalmers was prepared. A sum of two hundred thousand dollars was promptly appropriated as "part of the National Program to get industry back in gear."[9]

Sitting behind his massive leather-topped desk at West Allis, General Falk saw this depression as an opportunity "to exert the maximum selling effort to influence the users of our lines of machinery in the rehabilitation of their old equipment and replacement of obsolete and worn out machinery and to make a reasonable investment in inventories of materials and supplies." But the company's essential function, the engineering of capital goods equipment, made it especially vulnerable in a depression. Although the General constantly talked and wrote about salesmanship, the company representatives had always found it extremely difficult to sell the "big stuff." The idea of building a new cement plant, opening a mine, building a dam or spending several hundred thousand dollars on a new power installation had to originate with the buyer. In a profound depression such buyers were extremely rare.[10]

As the number of new orders declined and the company, outside the tractor division, began increasingly to subsist on replacement parts, Allis-Chalmers, in common with most big corporations, was faced with heavy overhead costs. It proved difficult to reduce administrative expenses very much and virtually impossible to reduce the fixed charges on the plant and machinery. Allis-Chalmers was faced with a real dilemma. Between 1929 and 1932 the average price of capital goods as a whole fell only 20 percent; yet production in the capital-goods industries fell 76 percent. Working against these fixed expenses, the General had to cut expenses elsewhere.[11]

As early as July of 1930, President Falk began his constant searching for places to cut costs. Examining individual expense reports, he found excessive costs for entertainment, hotel accommodations, and long distance telephone calls. He asked "that every member of the organization try to realize his responsibility to the company and to regulate his expenditures in the same manner that he would if the business were his own and if he were spending his own money. I shall expect your whole-hearted cooperation." In line with other economy measures the fee paid to directors for attending board meetings was reduced from fifty to twenty-five dollars, and the fee to members of the Executive Committee was reduced from one hundred to fifty dollars. Actual savings from these measures were minimal and other areas of economy had to be found.[12]

Although research and development was the "very foundation of a technical business," this was one area where expenditures could be curtailed during a depression. Increasingly, research was limited to the development of new products which promised early commercial value. For example, a new method of boiler water treatment was developed, and the material was marketed under the trade name Akon. Also in 1932, the transformer department brought out the step-type feeder voltage regulator for automatic regulation of the voltage of power lines. It made Allis-Chalmers a leader in this type of equipment, a position it has maintained since. The tractor division, capturing a larger share of the market and producing a greater proportion of both income and profits for the company, consistently received substantial sums for research and development. The pattern of expenditures for industrial research can be seen in the following figures:

1931	$1,169,844.95	1934	479,089.68	1937	2,053,581.37
1932	654,056.33	1935	1,041,367.72		
1933	605,158.95	1936	1,434,517.17		

Although research and development funds were cut substantially at the depths of the depression, they were increased beyond the levels of the previous decade as soon as recovery was in sight. By 1941 expenditures for research and development amounted to $3,470,446.[13]

The orders booked by the company steadily and relentlessly declined for three consecutive years. Orders totaling $6,128,000 were entered on the books during March of 1930. One year later only $2,707,000 in orders were booked, but this total was far better than the $970,000 in January of 1932. Orders continued to decline until the low point was hit in January of 1933 when only $402,000 in orders trickled in to the stricken company. By January, 1933, vacations with pay had been eliminated and additional salary reductions had been effected including one-half month vacations without pay; net salaries were, however, held at the minimum of $125 per month. "In all of these reductions the employees accepted the situation with fine spirit and good will, and their degree of cooperation has not been lessened," reported Otto Falk at the close of 1932. According to sales engineer John L. Neenan, his salary at the depths of the depression had been cut 52 percent. With the indication of better times in 1933–1934, the first of a series of salary increases began on May 1, 1934. But the company was forced to reduce salaries once again during the recession of 1937–1938. This time, however, only those of $225 per month or more were affected. Not until the beginning of World War II did salaries reach pre-depression levels.[14]

The largest installation of diesel engine driven generators ever was in Vernon, California; these five generators were each rated 7,500 kva at 167 rpms.

Although the wage rate was reduced, the real wage and salary reduction stemmed from a reduction of hours of labor. General Falk did his best to keep his regular employees at work regardless of cost and profit. Although Allis-Chalmers was by no means alone in this, the extent to which it was done and the spirit of the company officers is worthy of note. During 1930 and 1931 the number of employees declined only slightly from 6,861 to 6,473. But as conditions grew worse the number was finally reduced to 4,170 on December 31, 1932. This was the core of the firm's skilled workers and experienced office personnel. If the company was to operate effectively in the future, these employees had to be kept together. According to Irving Reynolds, "from 1930 to 1933 salesmen would come in with jobs they had taken sometimes not even at factory cost. As a temporary expedient during the depression, they'd take them to General Falk and he'd say, 'How much labor? O.K., give the boys a job. Nothing for profits.' "[15]

The General felt responsible not only to the employees but to the community as well. He said in 1933, "Where would the city of West Allis be without the A-C plant? We have got to weather the storm, both for our own sakes and the community's." As the depression deepened and more banks began to go under, Falk and the Executive Committee were willing to write off the company interest in the Monongahela National Bank of Pittsburgh because the company plant in that area was relatively small and the Allis-Chalmers interest in the bank "comparatively remote." But for the West Allis banks the situation was different. The officers felt that it

This is the motor room of a 76-inch continuous hot strip mill of the Inland Steel Company, Indiana Harbor. Six 3,500 horsepower direct current motors power finishing stands at speeds from 175 to 350 rpm.

was important for the employees and the company to have a strong bank in West Allis and bent their efforts especially toward saving the First National Bank of West Allis although, unhappily, the attempt was unsuccessful.[16]

General Falk broke many precedents to hold the company together during this difficult period. The long-standing practice of not installing its own flour mill equipment was abandoned for the Milwaukee Western Malt Company and others. Installation at least gave more work to the men in the shops. Through means such as this the company did its best to provide maximum employment. Because there were not enough jobs for all company men the General in 1931 discontinued his pride and joy, the Graduate Training Program, which had been maintained since 1904. Its return in 1935 was considered one of the best signs that the company was emerging from the depression.[17]

In the face of great difficulties, the morale of the men still at work was remarkably high. Moreover, concern for less fortunate fellows which had produced the Allis Mutual Aid Society came to the fore on the part of those who still had jobs. General Falk and William Watson, general works manager, appointed a Relief Committee which investigated and passed on applications. In January, 1931, the employees contributed $4,447 to the fund, matched by the company with $4,600. That Christmas some 200 baskets of food,

A unique system of home heating using special heat exchangers was developed by Allis-Chalmers in the 1930s. This model burned a fine grade of coal, while others used gas or oil.

clothing, and toys for children were distributed to needy Allis-Chalmers families. This practice of sharing continued throughout the depression. Because the company was deluged with appeals from numerous cities and many local charities, it adopted the policy of contributing only to funds in those cities where it had plants. During this critical period the company contributed $11,000 per year to the Milwaukee Community Fund, with lesser amounts to the fund drives in the other cities where it maintained plants.[18]

In view of the downward economic spiral which continued for three agonizing years, the company understandably sustained losses for three successive years. In 1932, faced with a loss of $2,955,042.87, the General felt it wise to establish a $2,000,000 special reserve fund out of the surplus to help ride out the storm. The following year closed with company losses of $2,893,905.30. With some improvement in sales during 1934, particularly in the agricultural area, losses were reduced to $1,039,405.95. The General, who had had the foresight to build up the company surplus during the balmy days of the twenties, was able to guide the company through the worst of the storm with a comfortable margin. Not including the $2 million reserve fund, the earned surplus of the company in 1934, after sustaining a $1,039,405 loss, still stood at $7,791,325.24. Although conditions were substantially better in 1935 with a net income of $1,985,136.86, the preceding years had left their mark on the company in a number of ways.[19]

Otto H. Falk had become president of Allis-Chalmers during the difficult period of 1913–1914, and his leadership and the prosperous war years had made Allis-Chalmers a leading manufacturer of heavy machinery during the twenties. Once again he faced the prospect of depression. But this time he chose to share the responsibilities of company leadership. Over the years Max Wellington Babb had become the alter ego of Otto Falk, even though the two men were a study in contrasts. Otto Falk was the extrovert who enjoyed publicity; Max Babb was less outspoken and did not seek publicity. The General was willing to follow a hunch in developing company policy while Max Babb was a legal technician who worked primarily with logic. Nonetheless, Otto Falk chose Max Babb to succeed him as president of Allis-Chalmers in 1932, when he elevated himself to chairman of the board.

Max Wellington Babb was born in Mount Pleasant, Iowa, in 1874. He graduated from Iowa Wesleyan College in 1895 and from the University of Michigan Law School in 1897. After practicing in his father's law firm from 1897 to 1904, he moved to Chicago to become an attorney for the Allis-Chalmers Company. In 1913 he was named vice president of the Allis-Chalmers Manufacturing Company and subsequently became a director and a member of the Executive Committee. By the early years of the twentieth century lawyers had become an integral part of big business in the United States. Vanderbilt had his Chauncey M. Depew, Huntington had his Charles H. Tweed, Harriman had his Robert S. Lovett, and General Otto Falk had his Max W. Babb.[20]

But the General had not given over full direction of the company to the new president. The Board of Directors on June 3, 1932, amended the constitution to read "That the Chairman shall be the medium of communication to the Board on all matters presented for their consideration and have general charge of the affairs of the Company." General Falk made it perfectly clear that "this is not meant that I am relinquishing my active connection with the affairs of the Company, but hereafter I desire to give more attention to matters involving the Company's general policies and less to details of operation." The new chairman concluded his statement in words not entirely clear or precise, "It is therefore suggested that questions pertaining to the detail operations of the Company be taken up with the President and other matters with me as formerly." Old company hands knew that Otto Falk continued to run Allis-Chalmers. Titles changed but the *locus* of power remained the same.[21]

The depression had a profound effect on every manufacturing department, but the impact was never quite the same. No other department of Allis-Chalmers was affected so directly or so seriously as the sawmill department. A tradition of leadership in sawmill equipment extended back to the days of George M. Hinkley and the E. P. Allis Company. Through the prosperous twenties the average sales of this department had exceeded half a million dollars a year. Since these sales bore a rather close relationship to construction in the United States, the precipitous decline in production by the sawmill department reflects the total collapse of the building industry by 1932.

1929	$441,505
1930	280,001
1931	79,376
1932	17,246

Although sawmilling continued as a company department for another two decades, this department never recovered from the depression and was eventually dropped.

When the Federal Housing Administration was established as part of the New Deal recovery program, the lumber industry was "solidly behind the move" because it offered hope for recovery. The *Sales Bulletin* in 1934 entered into this spirit. "After the Federal Housing Act gets under way, lumber demands will naturally increase. In turn it should also create a demand in repairs for the sawmills, as well as new and up-to-date equipment." Company publications implied that this had already taken place. *Company Facts* in 1936 observed that Allis-Chalmers was the only builder of sawmills that could also provide its own power plant and electrical equipment and had designed and equipped many of the largest sawmills in use. But this was, in a sense, a statement of past greatness. No new developments had taken place to revolutionize the industry and create a demand for new mills. Recovery from the depression meant that departmental sales from 1936 to 1943 leveled off at about $125,000 per year, but since there were no new mills this amount represented repairs and replacement of individual machines.[22]

During World War II sales did increase somewhat, largely because of the new "Streambarker" in 1942. This machine removed the bark from paper mill logs by powerful hydraulic action, and eliminated the work of many men. But this was no more than a temporary reprieve. On May 15, 1950, Allis-Chalmers sold its sawmill equipment, patterns, drawings, and good will to the Prescott Company of Menominee, Michigan, who later moved to Portland, Oregon. After three-quarters of a century one of the company's oldest product lines was closed out.[23]

A number of factors contributed to the decline and eventual termination of the sawmill department. One of these was the growth of the company in other directions which left this department a very small part of the larger whole. For a small department many of the business procedures that are necessary in a large organization are unneeded red tape and serve to increase the cost of the department. There is also pattern storage, pattern maintenance, and the charge for office space, until the cost of maintaining

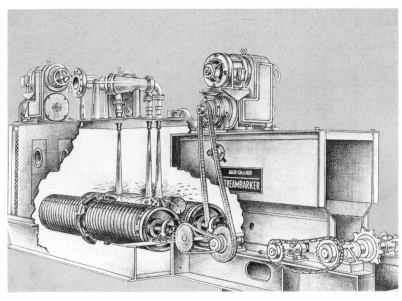

To reduce time spent in paper mills removing bark from logs, Allis-Chalmers introduced the "Streambarker," which stripped the log by hydraulic action. This development helped ease the manpower shortage caused by World War II.

a small department finally becomes too great for the company to bear. Another factor in the department's decline was that the company found that selling sawmill equipment required specialists in the field as well as experts in the shops. The Graduate Training Program had developed college-bred salesmen who generally did not care to rough it with the sawmill men in the woods. Because these salesmen had a quota to meet, they sold the saleable lines in the accessible areas, but the new generation of salesmen simply did not have the practical experience or the ability of their predecessors to sell sawmills. In an earlier period Fred Schlitz and others could draw and price an entire mill on the spot. On more than one occasion, Schlitz had been known to get wrapping paper from a grocery store, clear off a desk and draw up the plans then and there. By the mid-thirties no man in the department could do that.

A third factor was lack of leadership. A man with real ambition had little interest in the sawmill department because it had no future. With eight different managers in the last sixteen years of its existence, the department obviously had neither continuity nor leadership. When Robert K. Prince came to the department in 1943 he found letters from district salesmen that had gone unanswered for six months. The department was, as one man put it, "allowed to go to hell internally."

The basic system of milling flour had changed very little since William Dixon Gray developed the roller mill, but milling machines were improved. For example, these twenty-two double roller mills were formed into four batteries, each driven by one motor, rather than being driven by individual belts.

When the company began to take a hard look at the department in the mid-forties, it found that as the quality of the personnel and products had decreased, costs had risen so that the company was no longer competitive in this area. As early as 1947 it seriously considered closing out the department. To make a case for such action the sawmill department priced itself out of the market. This policy achieved its purpose, and the sawmill department was terminated in 1950 with the sale to the Prescott Company.[24]

By contrast flour milling, another old and respected department, held up very well during the depression. Except for the sudden increase in sales with the acquisition of the Nordyke line in the mid-twenties, sales for that decade averaged around half a million dollars a year. Compared with other product lines, flour milling sales of $297,382 in 1933 were very respectable. Late in 1932 the *Sales Bulletin* speculated on why "The milling field has not felt the fangs of the depression to any extent." The writer suggested that because of the depression "our people have gone back to the more simple, nutritious and inexpensive foods, in consequence of which we see a bigger demand for bread; that means more flour is used." New machines were purchased "at a good rate."[25]

Another reason for the unusual stability of this department was that flour milling machinery was increasingly being adapted for special purposes. Installations were made for fuller's earth, soya bean processing, cork grinding and salt milling. Quite another source of large sales came from the brewers and distillers who moved rapidly to quench the American thirst after the repeal of prohibition. When Hiram Walker & Sons, Inc., built their $13 million plant at Peoria, Illinois, in 1934, Allis-Chalmers secured the contract for the milling machinery. The twelve three-pair high roller mills built in West Allis could grind 1,000 bushels of grain per hour in this giant distillery, at that time the largest in the world. The Hiram Walker sale was the largest single flour mill order during the depression, and it provided a maximum number of jobs for Allis-Chalmers employees because it included both manufacture and installation. The impact of the repeal of prohibition should not be exaggerated, however, for most distilleries merely wanted their grain ground up, and any kind of grinding machine would do the job.[26]

Because no improved grinding technique was devised to make milling machines obsolete, departmental sales leveled off after the mid-thirties at an amount in excess of $600,000 per year. Since the Allis-Chalmers machines were so well made and lasted so long, too many of the company's orders were merely for replacement parts. Although milling machinery was no longer one of the main product lines, Allis-Chalmers nonetheless was proud to be the only company in the world that produced all the equipment for the production of flour—from tractors, cultivators and harvesters to roller mills and sifters, and including the engines which generated the power to operate the mills. As the largest producer of milling

(Top) *The end of prohibition brought many orders for new machinery from breweries and distilleries. These twelve mills for grinding grain were installed in the Peoria, Illinois, plant of Hiram Walker & Sons.*

(Bottom) *Overhead lineshafts and individual belts were required in machine shops before the advent of the Texrope drive and individual motor drives.*

machinery in the United States during the 1930s, Allis-Chalmers was manufacturing from 75 to 90 percent of the total production.[27]

The Multiple V-Belt Drive, a new form of power transmission and a by-product of the flour milling department, became very profitable for the company and was a source of considerable support during the depression. During the early twenties the age old problem of power transmission continued to trouble textile manufacturers because belts slipped and chains jerked, tearing delicate threads of the cloth and tying up production. These problems were brought to the engineer in charge of flour mill transmission, a young man named Walter Geist. After working unsuccessfully for over two years with rope drives, Geist satisfactorily solved the problem of power transmission. He discovered that the difficulty came from ropes or belts which were designed to ride the bottom of the grooves, and the solution to the problem lay in the principle of wedge contact. His experiments in developing a satisfactory textile mill drive then centered on the concept of a V-shaped belt and sheaves. Geist patented a design for using them in multiples, similar to the multigrooved pulleys of the English rope drive system.[28]

The potential advantages of the Multiple V-Belt Drive were of great significance to the manufacturer. In factory planning it meant a valuable saving of space and the elimination of shafting with all its handicaps, such as shutting out light, endangering operators, and taking up excessive space. The location of a machine could now be determined by the pattern of manufacturing rather than by the source of power. The potential efficiency of transmission was

Texrope Drives could operate on very short centers without belt slippage, as shown by this oil well rig.

(Top) Long belts from overhead line shafts were especially bad in textile mills because of oil drip and the breakage of threads upon startup.

(Bottom) Texrope Drives eliminated these problems. Motors could be connected to individual machines by short center drives, and the rubber belts acted as a shock absorbing medium between the motor and the machine, eliminating broken threads.

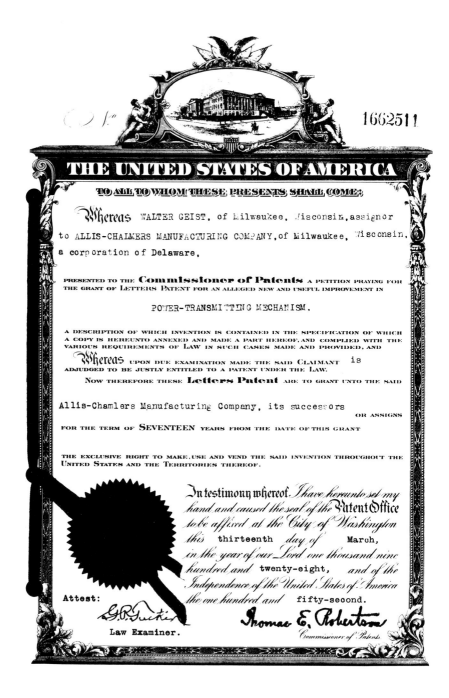

Of the thousands of patents issued to Allis-Chalmers and its engineers, none was more important than that for the Multiple V-Belt, or Texrope Drive. Issued to Walter Geist in 1928, it gave the company the exclusive right of manufacture and sale for seventeen years. Allis-Chalmers soon negotiated license arrangements with other companies.

very high, running between 96 and 99 percent on well-engineered drives. Furthermore, belt slippage was prevented because each belt, either when starting or when under an increased load, could momentarily sink deeper into its groove as well as extend longitudinally. The reduced tension made possible by this drive allowed for longer life in the motor bearings.

The Multiple V-Belt Drive was also much safer for employees, since it was compact and could be easily and inexpensively placed behind a guard. Moreover, the drive was silent. Just when the effects of noise on employee efficiency were beginning to be appreciated, real strides forward in noise elimination were possible.

The potential advantage of this drive for the Allis-Chalmers Manufacturing Company cannot be underestimated. In replacing older and less desirable forms of power transmission, the Multiple V-Belt Drive stimulated the sale of a variety of company products, particularly that of electric motors and all types of auxiliary equipment for the production, transmission and use of power. Sales figures indicate an inter-relation of all company electrical products during the rapid electrical expansion of the late twenties.[29]

The Texrope Drive, the Allis-Chalmers trade name for the Multiple V-Belt Drive, was officially placed on the market in March, 1925. Company letters indicate that there was no immediate thought of using it in industries other than textiles. By the end of the year, however, about one thousand of the drives had been built and had proven their superiority in refrigeration, fans and blowers, machine tools such as boring machines, drill presses and planers, paper-making, and flour milling, as well as in a number of less significant applications. Despite these successes Otto Falk did not announce until April, 1926, that "The Allis Texrope Drive has developed from an experiment to a commercial proposition of respectable proportions."[30]

The Texrope department developed rapidly. At the end of its first five-year period, with rapid acceptance and wide application, over 75,000 Texrope Drive installations aggregated considerably more than one million horsepower. Its growth also placed the Texrope department in a new perspective within the company. Having developed originally as a part of the flour mill department, it soon outgrew its subordinate position and, having had full departmental status briefly, was raised to divisional status in August of 1929 as the Texrope division of the Allis-Chalmers Manufacturing Company.[31]

A United States Government Patent covering the Texrope Drive was applied for in 1925 and granted on March 13, 1928, to the Allis-Chalmers Manufacturing Company as assignee of Walter Geist. It was company policy not to maintain absolute control over ideas as simple and attractive as the Texrope Drive; others were sure to develop the same thing or something perhaps even better regardless of patent restrictions. Under the circumstances it was better to share the patent and earn all possible royalties. By 1933 eleven leading manufacturers of power transmission machinery were licensed by Allis-Chalmers; by 1944, fourteen. These manufacturers also engaged in cooperative research to improve the drive.[32]

Allis-Chalmers kept well in the forefront in developing the sheave, the grooved wheel over which the V-Belt ran. Constant improvements had been made in the original cast iron sheave until in 1931 a complete line of pressed steel sheaves under the trade name Texteel was introduced and received wide acceptance. By 1934 the Duro-Brace Texteel sheave was introduced. Stronger, better in appearance, more suitable for stocking purposes and designed to accommodate larger bores, this new sheave represented a notable advance in the product. In 1938 the Texrope Vari-Pitch Speed Changer was placed on the market, first appearing in Boston at the Electrical Manufacturer's Trade Exposition. On this unit the speed was controlled electrically through a pilot motor and indicated on a large tachometer.[33]

In the beginning the Texrope Drive was planned for a maximum capacity of 15 horsepower, but very soon the company ventured into larger capacities. By late 1934 the largest Texrope had a cross section of 1½ by 1 inch, and the largest drive was a 2,000 horse-

Allis-Chalmers was the first to develop the close-coupled pumping unit, where the pump impeller was keyed to an extension of the motor shaft. This installation shows a group of close-coupled pumps driven by explosion-proof motors.

power for diesel engines used in a wind tunnel for the National Advisory Committee for Aeronautics. Very simply, the introduction of the Texrope short center V-Belt Drive revolutionized mechanical transmission practice and was generally recognized as one of the great steps forward in the history of power transmission.[34]

Sales of Texrope Drive equipment, not including the royalties from licensed manufacturers, had risen to $2,834,763 by 1929. Sales dropped to $619,108 in 1932 but soon began to rise again, and by 1940 sales totaled $3,680,677. The development and rapid growth of this new company product was in large part the work of Walter Geist. Born on December 1, 1894, on Milwaukee's south side, he was one of seven children of a Norwegian metal pattern-maker and machinist. When he was fifteen years old, family circumstances forced him to leave South Division High School to join Allis-Chalmers as a messenger boy in the blueprint department at a wage of 10 cents an hour. Soon he entered the milling department as a tracer and was rapidly promoted to draftsman. At thirty-one he was made engineer in charge of transmission and began the experimentation that led to the development of the Multiple V-Belt Drive. His patented development combined with his basic ability and inherent drive brought rapid advancement, and in 1928 he was made assistant manager of the milling department.

The ability of Walter Geist to develop a new product and to sell it effectively to the public was recognized in 1933 when he was named general sales representative for the general machinery division, and again in 1939 when he was elected a vice president of that division. Early in 1942 he was named executive vice president, and in May of that year was elected president; he served in that capacity until his death on January 29, 1951. The V-Belt Drive produced both profits and leadership in the field for Allis-Chalmers.[35]

"The electrical business during the year 1931 had been characterized by replacements and necessary additions to existing plants rather than new and larger developments." This statement from the *Sales Bulletin* at the close of 1931 meant that the depression had seriously hurt sales of the electrical departments of the company. Switchgear sales had slipped to a disastrous low of $87,471 in 1932 from $1,172,067 in 1929. Five years later, however, they would exceed this predepression high point by an additional half million dollars. The Pittsburgh plant, which made most of the smaller transformers, had a predepression record production of $3.5 million which had declined to $882,009 in 1933. From that point on, transformer sales increased rapidly with the revival of business until sales of $6,378,335 were reached in 1937. Sales of electrical equipment from the West Allis Works in 1929 had been $7.5 million, a figure which stood in strong contrast to the $1.5 million in 1933. But the heavy electrical equipment manufactured at this plant rebounded rapidly to nearly $8 million in 1937.

However, the small motors and similar equipment produced at the Norwood Works showed the least resiliency. Having achieved slightly over $4 million in sales in 1929, sales of its small, mass-produced equipment dropped to $639,929 in 1932. The best post-depression year before the war was 1937, with sales of $3,264,272. After the acquisition of the Bullock Company in 1904 the electrical departments of the company had been consistent money-makers, and after the depression this characteristic reasserted itself.[36]

In spite of the depression General Falk had implicit faith in the future of the American economy in general and in Allis-Chalmers in particular. In the twenties he had built up reserves which not only tided the company over the difficult financial period of the early 1930s but also permitted the acquisition of additional lines at the most advantageous terms possible. Pursuing the policy of acquisition in 1931, Allis-Chalmers bought the principal assets of the American Brown-Boveri Company, Inc., of Camden, New Jersey, and purchased the capital stock of the Condit Electrical Manufacturing Corporation of Boston, Massachusetts.

In 1925 the Brown-Boveri Company of Switzerland had established itself in the United States by incorporating the American Brown-Boveri Electric Corporation, with headquarters at the former New York Shipbuilding Corporation of Camden, New Jersey. American Brown-Boveri then acquired several electrical manufacturing companies, the most important of which was the Condit Electrical Manufacturing Company of Boston. This latter company was among the first to offer oil circuit breakers for commercial sales, the first to develop trip-free devices, removable tanks and the downward break. By 1926 Condit also had widened its lines to include oil circuit breakers of the largest indoor and outdoor sizes. In 1938, seven years after control passed to the Milwaukee Company, the name was changed to the Boston Works of the Allis-Chalmers Manufacturing Company.[37]

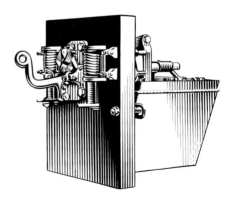

One of the main products of the Condit Electrical Manufacturing Company, which Allis-Chalmers acquired in 1931, was oil circuit breakers.

American Brown-Boveri brought to Allis-Chalmers a line of mercury arc power rectifiers, generator voltage regulators, transformers, and heavy-duty electrical railway locomotive motor and control equipment. The mercury arc rectifier economically converted alternating current to direct current for industrial and municipal purposes and had rapidly achieved wide acceptance. Through this acquisition, Allis-Chalmers also became the sole licensee in the United States for the manufacture and sale of turbo-blowers and turbo-compressors built to designs of Brown-Boveri and Company. The turbo-blowers furnished the large quantities of low-pressure air required for blowing blast furnaces, copper converters, and sewage aeration. The Camden plant was closed following the purchase in 1931, and the manufacture of rectifiers, voltage regulators and turbo-blowers was shifted to the West Allis plant works and the ABB transformers to the Pittsburgh plant.[38]

Despite the addition of new product lines and new facilities, the company did not drop its research program in the electrical field even in time of depression. The engineers took this opportunity to standardize synchronous motors and generators, as well as direct current and induction motors. Conscious of the fact that Allis-Chalmers transformers which supplied current to transmission lines must not fail in service, the company in 1932 installed a costly

During the 1930s expanding aluminum and chemical plants needed large amounts of direct current power. Mercury arc rectifiers built by Allis-Chalmers provided much of this power, as illustrated by these twenty grid-controlled units at the Alcoa plant of the Aluminum Company of America.

two million volt testing laboratory at the West Allis Works. Here it was possible to simulate lightning, the most common cause of transformer failure, and to study its effects on transformers and other equipment for long distance power transmission. A new circuit breaker testing laboratory, consisting of an outdoor concrete test chamber with complete observation and control equipment for the testing of switchgear and interruption devices, was also begun in 1932 at West Allis. By 1936 research had already resulted in a number of important improvements in circuit breaker design. The testing facilities supplemented the research facilities and testing plants of A. Reyrolle & Co. in England, and Brown-Boveri Co. in Switzerland, and all three pooled their engineering data. During the same period insulation research on high voltage generator coils materially improved that product. Other studies of low-loss steel and non-magnetic materials made it possible to design turbo-alternators which used much less material than before, and used this at lower costs.[39]

Because Allis-Chalmers was able to continue its research and to consolidate its position in the electrical field during the depression, the Norwood Works, Boston Works, Pittsburgh Works and the electrical department of the West Allis Works were all operating at capacity by 1939. Allis-Chalmers had become the third largest electrical manufacturer in the United States after General Electric and Westinghouse. These three were the major competitors in the

This three million volt impulse generator was installed in the West Allis plant in the 1930s. It was used for testing transformers and other electrical equipment under lightning conditions.

heavy electrical equipment field, with General Electric and Westinghouse dividing the bulk of the sales and Allis-Chalmers running well behind. In the manufacture of "package goods" such as small motors and control equipment, however, the company competed with over one hundred other firms. Despite this competition the electrical department consistently provided half or more of the industrial business of the company.[40]

If Allis-Chalmers suffered from the depression it also profited from some of the New Deal legislation designed to combat the economic slump. With the inauguration of Franklin D. Roosevelt on March 4, 1933, the pattern of American society began to undergo a series of changes. During the famous "hundred days" and shortly thereafter, an enormous amount of legislation was passed by Congress to move the economy forward. Because of, or in spite of, some of this legislation, the economy did begin to revive. But not all departments of Allis-Chalmers were directly affected or affected in the same degree. The attempts at recovery sponsored by the New Deal seemed to have little effect on sawmilling, flour milling or the Texrope division. Nor did the electrical department benefit directly since the major government agency in this area, the Rural Electrification Administration, had no appreciable effect on departmental sales. Some areas within the company, however, did profit from the attempts at recovery. The tractor division, for instance, which by the late thirties accounted for over half the

This bank of eighteen transformers was one of the world's largest at its installation in 1932 for the Public Service Electric & Gas Company of Newark, New Jersey.

Boulder (now Hoover) Dam of the United States Reclamation Service was the largest hydroelectric project in the world in 1937. Allis-Chalmers designed and built the project's metal-clad switchgear units for controlling the generated power; these were also the world's largest.

Allis-Chalmers originally received orders for the Boulder Dam project of four 115,000 horsepower turbines, two of which are shown here on the erecting floor at West Allis. By 1940 the company had installed or had orders for seven of the project's eleven turbines.

company sales, was supported in large part by the prosperity created by agricultural subsidies.

Governmental assistance to the financial health of the company varied from the indirect type in the agricultural program to the rather direct form in the Public Works Administration. Allis-Chalmers began hiring again just before Roosevelt took office, and the upward trend in employment continued until November 30, 1937, when a recession began and the number of employees decreased. At that time West Allis had 9,681 in its shops, Norwood 772, Pittsburgh 1,176, and Springfield 466. LaPorte employed 778 men on November 30, La Crosse 1,219, Condit 706, and there were 3,065 people in the general office. The total of all Allis-Chalmers employees was 17,863.[41]

The Public Works Administration, organized on June 16, 1933 with an appropriation of $3,300,000,000, was designed to stimulate heavy industry by fostering public works projects that required huge quantities of material. PWA allotments brought Hoover (then Boulder) Dam to completion two and one-half years ahead of schedule. The first Allis-Chalmers order of four hydraulic turbines for this dam was of enormous benefit to the hydraulic turbine department and to the company as a whole. This contract for approximately $1,100,000, the largest for the department until then, was most opportune. Because major hydroelectric developments require a great deal of time—often several years for engineering and construction—this department was still enjoying the benefit of orders which had been booked before the depression, when other departments were in financial trouble. As late as 1931 this department had sales of $1.5 million. On the other hand, it eventually felt the depression at a later date and to a greater degree. In 1934 when many departments were recovering, the hydraulic turbine department had sales of only $71,079. Even though sales for 1935 were over $1.3 million, the intense competitive bidding for such a large sale might indicate little or even no profit.

The size and power of the turbines for Hoover Dam added to the many world's records already held by Allis-Chalmers. Each turbine had a three-piece forged steel shaft 38 inches in diameter, reaching 75 feet from the turbine to the top of the generator. The water from Lake Mead was admitted through pipes to a spiral turbine casing of cast steel, weighing about 200 tons, the inlet of which was ten and one-half feet in diameter. For transportation the casing was made in several sections and then joined together on the site. Each turbine discharged close to 3,000 cubic feet per second when developing the maximum power, over 1,250,000 gallons per minute. The first four turbines weighed 3,000 tons and filled about 150 carloads. By 1940 Allis-Chalmers had installed or had on order 7 of the 11 turbines in the Hoover Dam power plant. The turbines, and especially the pressure regulators were so eminently suitable that a substantial premium above the prices of other bidders was conceded in the contract.[42]

These turbines were not the company's only contracts for the Hoover Dam project. In early 1934 Allis-Chalmers was awarded the contract for the oil circuit breakers for the generators. These 4,000-ampere, 23,000-volt, 2,500,000 kva units which the company designed and manufactured were larger than any previously built. By June of 1935 the first three of twelve ordered had been shipped. Each had passed all tests, impulse, puncture, heat runs and interrupting capacity, conducted at the new West Allis laboratory, thus enabling engineers to place full momentary current of 120,000 amperes on the units.[43]

Allis-Chalmers also contributed to other PWA sponsored power generating projects, of which the Tennessee Valley Authority was the most famous. Once again the hydraulic turbine department secured its share of the contracts. For the Pickwick Landing Station on the Tennessee River, Allis-Chalmers built the two largest adjustable vane propeller turbines ever made. This contract also brought orders for two complete governor systems for the turbines. Under maximum head of 60 feet, the turbines could develop approximately 63,000 horsepower at full gate opening, but the governors were designed to limit the output to 55,000 horsepower to correspond with the capacity of the generators. These

Allis-Chalmers also provided equipment for the many dams and power plants of the Tennessee Valley Authority. Shown here is one of two adjustable blade propeller turbines with a capacity of 63,000 horsepower that were installed at Pickwick Landing.

governor systems included Allis-Chalmers actuator type governors, motor-driven oil pumps, pressure and receiving tanks, oil piping and switchboards. Fifty-three miles above Pickwick Landing, the company placed four vertical shaft, Francis-type turbines in plate steel spiral casings, each rated at 35,000 horsepower at 100 rpm under a 92 foot head.[44]

Since Allis-Chalmers built both hydraulic and electrical equipment, the company could contract to supply apparatus for complete hydroelectric installations, insuring the purchaser undivided manufacturer responsibility. Moreover, Allis-Chalmers was the only company in the world building all three types of hydraulic turbines—Francis, propeller, and impulse. As the government began to develop major power projects, however, the normal stream of business from private utilities diminished. In short, the federal government had become the most important customer of the Allis-Chalmers hydraulic department.[45]

Municipal governments also proved to be excellent customers. A growing number of cities in the late 1920s started building sewage disposal plants for their increasing populations, but construction of many of these was suspended following the crash for lack of funds. Construction and completion of sewage systems became one of the favored enterprises of the PWA. Because most of these were

When the Milwaukee area sewage treatment plant was expanded, Allis-Chalmers furnished the high volume centrifugal compressors needed to aerate the sewage. A few years later the company also supplied the axial compressor in the foreground.

"activated sludge" plants, Allis-Chalmers was able to supply centrifugal (turbo) blower equipment for many. An excellent example of this type of system subsidized by a PWA loan was in Milwaukee. On April 12, 1934, the company received a contract from the Sewage Commission of Milwaukee for two steam turbine driven blowers, with condensing equipment and accessories for aeration of sewage.[46]

Another PWA project in Milwaukee was the water purification plant, constructed on made land on the shore of Lake Michigan at an approximate cost of $5,250,000. The modern Gothic architecture of the superstructure conceals a wide variety of Allis-Chalmers products. The several different types of pumps are controlled by a flush type control board. These, with 54-inch and 60-inch butterfly valves complete with piping, were all produced in the company's plants. Once again, all units installed substantially exceeded the guarantees and operated free from mechanical noise and vibration.[47]

The excellent plant facilities as well as the core of trained personnel kept intact through the depression years made it possible for the company to profit from special jobbing contracts during the difficult period from 1933 to 1936. Rolled steel base

When Milwaukee built a water filtration plant to control the quality of their water supply from Lake Michigan, much of the major pumping and control equipment was supplied by Allis-Chalmers. This included centrifugal pumps, the synchronous motors to drive them, control boards, and valves.

plates for the San Francisco-Oakland Bridge were surfaced on the forty-foot boring mill and edged on the twelve-foot planer. The steel saddles for the caps on the cable towers were also machined at West Allis. Because work was slack in the engine department, the company re-entered the general forging field, making rudder stocks for the United States Navy. A similar Navy order was received for 128 rough machined steel castings for destroyer turbine casings. All of these and other jobbing contracts, mostly from the government, helped keep men at work in the mid-thirties.[48]

The 1920s had been the heyday of the crushing, cement and mining department, with sales in 1926 of $8 million. The depression actually affected this department more than others, for by 1933 sales had fallen to only one-tenth of those seven years before. In 1932 the volume of metallic ore mined was only 30 percent of the 1929 level. But sales of mining machinery showed a marked improvement in 1934, largely from the government gold and silver purchase policies which made it profitable both here and abroad to reopen mines once considered uneconomical to operate. Allis-Chalmers was prepared to secure this business because no other single manufacturer could offer such a complete line of mining and metallurgical machinery. This advantage, however, amounting to a government subsidy, was insufficient to keep mining afloat as a separate department. In mines other than gold and silver operators found themselves with a surplus of equipment, and even the

New equipment increased the efficiency of many cement plants. An example is this huge dry grinding Compeb mill at the Missouri Portland Cement Company plant in Independence.

demand for repair parts was drastically reduced by the lower level of operations. By 1941 sales had declined to such a point that the mining department was discontinued and its functions dispersed to other areas within the company.[49]

The crushing and cement departments had fared much better during the depression years. The extensive building of roads and dams under the PWA and other government agencies helped bring sales up to half their high point reached in the 1920s. Allis-Chalmers pulverators prepared limestone sand for the Norris Dam and limestone for fertilizer in the soil conservation program. Allis-Chalmers provided all the vibrating screens for the immense amount of materials used in the Grand Coulee Dam. The fact that the concrete was being poured at a rate equivalent to handling 2,500 tons of gravel per hour over a 21-hour operating period per day adds significance to the fact that these screens consistently handled more than the estimated tonnage with complete satisfaction.[50]

In 1895 cement production had been less than one million barrels per year. Increasing demand brought production to 170 million barrels by 1935, and government building continued to increase this demand until the "pump priming" was stopped in 1937. By that year Allis-Chalmers had sold 438 of its Compeb Mills throughout the world. The greatest single unit cement grinding mill of the thirties was built by the company for the Universal Atlas Portland Cement Co., Leeds, Alabama. The drive required 1500 horsepower, and the mill was mounted on two of the largest roller bearings ever made—each bearing almost five feet in diameter and capable of supporting 175 tons.[51]

Despite the relative decline of crushing and cement in the thirties as compared with the previous decade, in 1937 it still ranked second among the industrial departments. Roughly one-third of its business was in maintenance and replacement parts, another third for new equipment abroad and the remaining third for new domestic orders.[52]

Perhaps no department in the company enjoyed greater stability in annual sales during the 1920s than the steam turbine department, yearly averaging about $2 million. By contrast, it was one of the more erratic during the decade of the 1930s. Having declined in 1932 and again in 1934 to $670,000, sales shot upwards to about $4 million in 1938, only to decline to $1.5 million the following year. The most significant installation in design and efficiency for the entire decade was the Port Washington Plant of The Milwaukee Electric Railway and Light Company, a tandem compound unit considered in 1936 the latest word in turbo-generator design. The unit was rated at 80,000 kw with steam turbine capacity of 135,000 horsepower, and Allis-Chalmers had also supplied the steam condenser, the circulating pumps, and other equipment. Because of its high efficiency in operation, the Port

Washington plant was called America's premier station.

The most significant development in steam turbines came from the company's re-entry into the marine propulsion field. Following World War I the company lost interest in this area largely because of the decline of the American merchant marine. Following 1933 the United States Maritime Commission began increasing subsidies of the shipbuilding industry, and the Navy began a building program under the PWA. In 1935 Allis-Chalmers took its first marine turbine orders in many years, a departure from traditional practice which proved to be most profitable for the company. By late November, 1940, the steam turbine department had on order 120 units. Of this total, 934,000 kw was the aggregate capacity of marine propulsion turbines, as compared with 549,000 kw for the total capacity of land generating units. Orders in both areas by 1940 came primarily from the National Defense Program.[53]

Allis-Chalmers had survived the depression. In fact, once economic recovery began sales soared to a point almost twice as high as that reached in the boom year of 1929. But the course had been rocky, as can be seen from a study of the sales billed from 1929 through 1940.

Year	Power and Industrial Equipment Division	Tractor Division	Total
1929	$34,618,401	$10,683,955	$45,302,356
1930	29,043,249	12,432,700	41,475,949
1931	20,399,552	7,401,087	27,800,639
1932	8,510,580	6,253,484	14,764,064
1933	8,134,009	5,152,759	13,286,768
1934	11,399,506	8,887,642	20,287,148
1935	17,860,803	20,926,204	38,787,007
1936	25,252,056	33,729,307	58,981,363
1937	36,376,762	50,976,854	87,353,616
1938	31,641,536	45,901,902	77,543,438
1939	27,321,649	47,017,902	74,339,551
1940	37,696,312	49,400,654	87,096,966

The industries group had rebounded to the 1929 sales level, and with the support of the enormously profitable tractor group, the team of Falk and Babb had led the company through the wilderness of the depression to a more prosperous land by the eve of World War II.[54]

Otto Falk was a child of the nineteenth century. Born into a prosperous family, his conservative schooling had been reinforced by military service. As late as mid-January, 1933, it was questionable whether Falk as a conservative businessman who believed in conservative economics would be willing to accept the terms of the New Deal under which prosperity might return to Allis-Chalmers. At that time he expressed great concern over "the extent of (government) regulation of industry." He pointed out that it had "become so ridiculous that the head of a firm like this that operates in many states has to have a staff of attorneys at his elbow to be sure he isn't violating one of the myriad regulations." Taking at face value Franklin D. Roosevelt's campaign promise of economy in government, he was convinced that the federal government could cut its expenses enough to balance the budget, "and I hope the new administration will have the courage to do it."[55]

Whatever the background and training of Otto Falk, as president of Allis-Chalmers he had a simple philosophy: whatever was good for Allis-Chalmers was good for the country and good enough for him. Just as he had made shells in 1915 over the protest of the German-Americans and just as he had gladly sold thousands of tractors to Communist Russia, even though he abhorred the political philosophy of the New Deal administration, he was prepared to work with it after March 4, 1933. As federal agencies were established to prime the American economic pump through grants and subsidies of all kinds, they received the support of General Falk.[56]

The Agricultural Adjustment Act of 1933 was an emergency measure to increase farm income and reduce farm surpluses. This act received the General's approval in February, 1934, when he wrote, "In the south where the government's cotton subsidy is operating, the purchasing power of the farmer has stimulated business to its normal level. The soundness of the President's plan is further substantiated by the fact that cotton spinning activities have maintained their recovery to the extent of 56 percent increase over 1932." The benefits of the Civil Works Administration were acknowledged in the same article. "The CWA is giving employment to over 4,000,000. As soon as Congress convenes, the President plans to ask for further appropriations to be spent on additional civil works projects. Thus with more than four million of pay envelopes being distributed every week where before there were none, a new consumer market has been created which should defy even the most chronic pessimist."[57]

Falk's philosophy, based on opportunity for the company, was also reinforced by the promptings of the company's financial adviser, Charles Hayden. In October, 1933, Hayden of the New York firm of Hayden-Stone & Co., called for a "gigantic credit expansion program" by the government. "If this country could risk $25,000,000,000 to help France whip Germany, it can risk $25,000,000,000 to help itself out of this depression," he declared. Hayden also had good words for the National Recovery Administration. "The new administration and the NRA have done much in restoring confidence and in obtaining tangible results in putting people back to work. But we must set our teeth and make up our minds to a long and difficult pull back to prosperity."[58]

Allis-Chalmers subscribed to the NRA. On July 1, 1933, Leo W. Grothaus was appointed assistant to the president to direct the company's participation in the program. This action was pursuant to a resolution by the board a month before authorizing the officers of the company to accept the NRA codes as rapidly as they were formulated, and "that in the meantime the Company adopt, as of August 1st, the minimum rates and maximum hours, as specified in the President's Blanket Code." Although at its height the NRA enrolled more than 2.25 million firms involving more than 22 million persons, the internal operation of Allis-Chalmers was little affected except by regulations concerning wages and hours. The inept leadership of General Hugh Johnson made the provisions of the poorly drawn law even less effective. Almost all concerned, New Dealers as well as industrialists, were relieved when the Supreme Court on May 27, 1935, invalidated the law, and the Blue Eagle was grounded. The greatest virtues of the NRA were that it was a psychological stimulus during the spring and summer of 1933, and it had improved working conditions for some.[59]

While the economic recovery of the company during the New Deal can be attributed mostly to direct or indirect federal support, the profits thus accrued were by no means clear cut. What the government gave with one hand it frequently took away with the other. During the eighteen-year period from 1913 to 1931 the Allis-Chalmers Manufacturing Company paid $19,730,758.28 in federal, state, and other taxes, the average being somewhat over

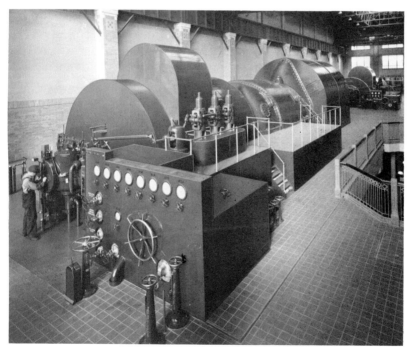

America's premier power station in the late 1930s and early 1940s was the Port Washington Power Plant of The Milwaukee Electric Railway and Light Company. Among Allis-Chalmers equipment supplied to the plant were the 80,000 kilowatt steam turbine generator units, the steam condensers and the condenser water circulating pumps and motors.

$1 million per year. In 1936 alone the taxes paid by the company, including federal income tax, surtax on undistributed profits, capital stock tax, excess profits tax, state income tax, and realty and personal property taxes, amounted to $2,950,000, or the equivalent of $1.67 per share on the outstanding common stock. In the more prosperous year of 1937, these taxes increased to $5,221,220.88, equivalent to $2.94 a share on the 1,773,341 shares of common stock. These taxes were equivalent to $322 per employee.[60]

In the 1938–1939 recession taxes paid declined to just over $3 million per year. In 1940, however, sales rose to approximately those of 1937 and taxes to $5,160,825.75, equivalent to $343 per employee and $2.91 per share of common stock. A breakdown of this $5 million provides a picture of where the company tax dollars went.

State and Local Taxes	
(Including State Income Taxes)	$1,414,771.05
Social Security Taxes:	
State and Federal Unemployment	
Compensation	802,405.35
Federal Old Age Insurance	
Contribution	265,325.05
Total:	$1,067,730.40
Federal Taxes:	
(Excluding Social Security)	
Income Taxes Including Excess	
Profits Tax	2,518,000.00
Capital Stock Tax	160,324.30
Total:	2,678,324.30
Total of all taxes:	5,160,825.75

No word of complaint came from Falk or Babb because the company had profited from federal programs during the thirties, and in 1940 it began to book orders at an unprecedented rate under the National Defense Program. These orders were reminiscent of the war a generation before, for they included $10 million for marine turbines for fifteen destroyers, shafting for twenty destroyers, machining on ninety 16-inch gun slides and fifty-five 12-inch gun slides. By February 1 of 1941 the backlog of unfilled orders came to $68 million. Allis-Chalmers had become, either directly or indirectly, a constant beneficiary of government programs.[61]

One-fifth of the taxes paid in 1940 went for the support of state and federal unemployment compensation as well as Federal Old Age Insurance. For government-sponsored social legislation, the company began contributing under the Wisconsin Unemployment Reserves and Compensation Act on July 1, 1934, and employees were covered by the Federal Social Security Act shortly thereafter. Allis-Chalmers had developed an informal arrangement for the aging employee. Generally speaking, a man was allowed to continue work as long as he was able. A study in 1930 showed that about 4.5 percent of all shop employees at the West Allis Works were between 60 and 69 years old, 1.5 percent were 70 or older, and the oldest employee was 84. But the company had no provision for the elderly employee who had given long and faithful service to the company and was no longer employable. This matter was the main subject at the board meeting on October 20, 1933. Charles Hayden suggested that for such employees the officers of the company should "fix a rate of compensation to be paid on retirement." Although not a satisfactory program by later standards, it did indicate a recognition of responsibility.[62]

While the 1930s are remembered most for government social welfare programs, it must be recalled that the Allis-Chalmers Company had been concerned for the welfare of its employees since the days of E. P. Allis. That the spirit of the company went beyond compliance with state and federal legislation can be seen in the Group Insurance Plan which Allis-Chalmers started in May, 1930, during the first stages of the depression. The company plan at first provided for up to 10,000 employees, insurance in the aggregate of about $20 million. The insurance contract, underwritten by the Metropolitan Life Insurance Company of New York, was the largest group insurance contract written in Wisconsin until that time. Under this plan each employee who had been with the company for three months or more was entitled to insurance in the amount of $2,000 payable in full to the beneficiary in the event of death, or payable in monthly installments for sixty months in the event of permanent disability. In addition, each insured employee was entitled to visiting nurse service. For this coverage the insured paid a premium of $1.20 per month and the company paid the balance of the total premium.[63]

At the end of the first decade of operation, the beneficiaries of 564 employees had received death benefits under the Group Life Insurance Plan in the amount of $1,329,704. In addition, benefits totaling $107,037.56 had been paid to insured employees in conformity with the total and permanent disability provision. As of December 31, 1939, 13,328 employees were insured for an aggregate amount of $28,843,892.[64]

The *Power Review* and the *Sales Bulletin* can be considered official voices within the company. They spoke with authority on products and sometimes carried the personal views of General Falk and of the management. A brief sociological essay which appeared in the *Power Review* of July, 1938, would indicate that Allis-Chalmers recognized the problems of the urban, industrial society more than many other firms, and that the company's accommodation to the philosophy and practices of the New Deal was far greater than that of some other industries and industrialists. The essay pointed out that the transition from a domestic to a highly efficient industrial system of production had been accompanied by "numerous perplexing problems involving politics, economics,

A composite of representative products of the Allis-Chalmers Manufacturing Company in the 1930s and 1940s.

and sociology." Stating that whereas Americans had once enjoyed a relative domestic security, "we now have industrial insecurity, would be bludgeoning the obvious." But, the essay continued, few would wish to return to a previous era.

J. C. Freeman of Allis-Chalmers felt it hardly necessary to point out that one of the greatest evils accompanying industrial advancement had been the increase of unemployment.

> Since 1929, a feeling of doubt and anxiety has gripped the heart of the American worker. We have experienced, in order of events, a depression, a fitful season of prosperity, and an economic lapse, which we have termed a recession. Prior to these events, when work was available and wages high, we felt no need for social insurance. But when factory chimneys stood gaunt against the noonday sun; when lathe and riveting machines had been stilled; when unemployment stalked across the land, the bread lines were in evidence everywhere; then did we as a nation begin to realize the value of proper safeguards against want and distress.

Other industrial societies such as Germany, France, England, and Austria had been compelled to face the problem, and the writer noted that they had "met it, with relative success, by social legislation. So likewise, did the United States."[65]

The company management recognized the changes in the political and social climate in the United States. But it was one thing to recognize the impact of the depression on labor and quite another to develop satisfactory policy or to learn how to cope with a dynamic and fluid labor situation. Since from 1907, the only labor difficulties had been relatively minor occurrences in 1916 General Falk and Max Babb were no better equipped by previous experience than most other company executives to deal with the labor problems which were to become perhaps the most critical of all for the company officers in 1939, 1941 and 1946.

The NRA, to which Allis-Chalmers had subscribed in 1933, was conceived as an experiment in industrial self-government under mild federal supervision. It involved price-fixing and regulation of competition to stimulate recovery and to eliminate ruinous competition during the depression. However, since the NRA had suspended the anti-trust regulations, industry agreed to concede to labor a similar right to consolidate. This guarantee of collective bargaining was in section 7a of the NRA act; the whole of section 7 concerned labor rights and labor conditions under the codes, but section 7a was crucial. It declared that every code must assure "That employees shall have the right to organize and bargain collectively through representatives of their own choosing, and shall be free from the interference, restraint, or coercion of employers . . . in the designation of such representatives." In response to this invitation to organize, membership in all unions, in and out of the American Federation of Labor, grew rapidly from the 1933 low of about 2,973,000 members.

To umpire relations between management and labor President Roosevelt created the National Labor Board in August, 1933. This was succeeded in June, 1934, by the National Labor Relations Board with quasi-judicial functions. Two months after the NRA was declared unconstitutional in 1935, Senator Robert F. Wagner of New York salvaged and fortified the collective bargaining provision of section 7a through the National Labor Relations Act of July 5, 1935. This act declared that "employees shall have the right to self-organization, to form, join, or assist labor organizations, to bargain collectively through representatives of their own choosing, and to engage in concerted activities, for the purpose of collective bargaining or other mutual aid or protection." The three new members of the National Labor Relations Board were also empowered to prohibit "unfair practices" of employers.[66]

Between the summer of 1935 and early 1937 the Committee for Industrial Organization was formed within the American Federation of Labor. Because it proposed to organize workers by industry rather than by craft, it was expelled from the AF of L and emerged as a rival federation of unions, the Congress of Industrial Organizations (CIO). Under the protection of the new law, organized labor grew. By 1939 union strength had been recruited to some 8,500,000; about 4,000,000 in the AF of L and roughly the same number in the newly organized CIO. In six years nearly 6,000,000 workers had joined the union ranks.

According to the stated theory, collective bargaining was designed to quiet and perhaps eliminate labor disputes, attempting to maintain the peaceful relations of the 1920s and the early years of the depression by placing bargaining on a sound basis. This assertion, however, did not prove true, for a period of unparalleled turbulence now ensued in which Allis-Chalmers encountered more than its share of problems.

From the days of Edward P. Allis, the company had not discriminated against labor organizations within the shops, and many employees had been affiliated with a variety of craft unions. But until the NRA, no significant collective bargaining relations had developed. Under section 7a the company sponsored the organization of the West Allis Works Council, originally composed of twenty-five members, five representing management and twenty representing the employees. The five management representatives were soon dropped from the Council; this new Council, made up entirely of representatives elected by the employees, bargained with the company on wages, hours and conditions of employment. The request of several AF of L craft unions for company recognition prompted Allis-Chalmers in June, 1934, to re-evaluate and revise its policy towards unions. It decided to recognize *any union* as bargaining agent for its employee members, and not to discriminate against any employee because of union membership or activity.

As prosperity began to return, labor agitation in Milwaukee and throughout the country continued to increase. In spite of this, the Executive Committee mistakenly believed that "by far the larger portion of employees in the shops are satisfied with existing conditions, and wish to be let alone in their employment." Neither this attitude nor the policy statement of June prepared the management for a serious threat of a strike by six labor unions in October. This strike was averted because labor had no recognized leadership to direct it. The situation was changed in 1935 when Harold Christoffel, a militant young union leader, secured a Federal Union charter from the American Federation of Labor for organizing all company employees not eligible for AF of L craft union membership.

The impact of increasing union pressure on the company was obvious by April of 1936. After negotiations with the various collective bargaining representatives, the company instituted a bonus and vacation-with-pay program for hourly employees that cost an estimated $350,000. The bonus plan stipulated that on December 23, 1936, all hourly employees would be paid a bonus of 3 percent of actual earnings between March 1 and December 12. The vacation-with-pay program was designed to reward hourly workers for "prompt, efficient, and continuous service." One day of vacation with pay was to be granted for each year of continuous service to a maximum of five years. On November 13, Falk and Babb brought to the board a proposal to increase the bonus to 5 percent to allay the continuing labor unrest. But before the year ended labor made additional requests for higher wages and other adjustments.

By mid-1936 Harold Christoffel had persuaded most of the employees of the West Allis Works to join the AF of L Federal Union. In 1937 he promoted a change of union affiliation to the newly organized UAW-CIO. Under a charter issued by the International UAW to Local 248, Christoffel was named president, and the Works Council was subsequently dissolved.

The organizational drive of the CIO began in Milwaukee in the summer of 1936. By 1937 it had won the support of the workers at the Allis-Chalmers Manufacturing Company, the Bucyrus-Erie Company, the Harnischfeger Corporation, the Heil Company, the Harley-Davidson Motor Company, the Briggs and Stratton Corporation, and the A. O. Smith Corporation. The success of the CIO in gaining a contract in March, 1937, from the United States Steel Corporation, a corporation as traditionally hostile to unions as it was powerful, was not lost on Milwaukee labor; by July 7, CIO unions were able to form the Milwaukee County Industrial Union Council.

After early 1938, when the NLRB certified Local 248 as the exclusive collective bargaining representative for the employees in production and maintenance, the history of labor relations at the

West Allis Works was strife-ridden and confused. These difficulties culminated in a thirty-day strike in 1939, over the issue of
the union demand for a closed shop. Following its traditional
policy, the company resisted this move, and in the strike settlement did not grant the closed shop clause. This settlement only
foreshadowed the following year when a more serious seventy-six
day strike held up production of $45 million worth of military
equipment.[67]

As prosperity began to return in the late 1930s and the company
showed a profit in spite of its labor difficulties, it again contributed
substantially to such causes as Milwaukee Hospital, gave $25,000
to Marquette University in 1939, and raised its contribution to the
Community Fund to $25,000 in 1940. However, the new image of
the "employer" which arose during the depression and the company's continuing labor difficulties with its West Allis employees
produced an unfortunate and untrue picture of the company in the
public mind.[68]

Despite the difficulties faced by the company throughout the
thirties, by 1940 it was challenged only by Krupp of Essen as the
company which produced the greatest assortment of tools to wrest
power from the earth. The significance of the company was found
in the diversity of its products rather than in its volume. It had
shown a remarkable ingenuity and adaptability. Although its past
had been with the "big stuff," such developments as the Texrope
Drive proved that it could devise equipment for and capitalize on
the trend toward small individual power units. It had also led the
way in relegating the monstrous and unwieldy tractor of the
twenties to agricultural museums, and the tractor division had
assumed an important place within the company.

On May 21, 1940, when Nazi columns were pressing deep into
France, General Otto H. Falk died at the age of seventy-four in
Milwaukee. The Board of Directors of the Allis-Chalmers Manufacturing Company, which he had led for twenty-eight years,
passed a resolution "That the business world has been deprived of
a useful and constructive member, and the country has lost an
upright and patriotic citizen." These words are similar to those
mentioned in the press at the death of Edward P. Allis. These two
men had created and shaped a great Milwaukee industry. Neither
was an engineer but both led a vast engineering firm which
manufactured some of the heaviest equipment made. E. P. Allis
had built his firm on the genius of Hinkley, Gray and Reynolds;
Falk continued these product lines and used the same basic
technique in choosing Harry Merritt to develop the tractor division. Public service had been important in the lives of both men,
and the greatest bequest of each was the manufacturing firm they
left to become part of the fabric of Milwaukee and the nation.[69]

June, 1940, was the end of an era. Falk's men, Max Babb and
Harry Merritt, would die within two years, and a new type of

leadership would be in charge of the company. The company itself would change; in fact it was already changing under the impact of war. In 1940 the General was proud that sales were projected toward the $100 million mark. At the close of a decade they would surpass the half a billion mark. A whole list of new products came to bear the Allis-Chalmers name as the steam engines, sawmills and flour mills of the Allis era were supplanted by new and different product lines. Change took place in society as well. The purpose and pattern of government were transformed under the impact of depression and war, and relations with labor assumed violent and sometimes sinister proportions.

The key to the success of Edward Allis and Otto Falk had been their faith in the future, confidence in their own abilities and the willingness to adapt themselves to changing patterns of industry, society and government. But the world that they had known was no longer to exist once World War II was concluded. June, 1940, was the end of an era and the beginning of another for both the world and for Allis-Chalmers.

NOTES TO CHAPTER NINE

1   *Sales Bulletin*, December, 1928, p. 706.
2   New York Times, September 4, October 30, 1929. Over the period January 10, 1930, to October 2, 1931, Otto Falk, with authorization from the Board of Directors, in sporadic attempts to bolster the stock value, picked up on the open market at least 73,000 shares of Allis-Chalmers stock at prices ranging from 43 down to 27. These shares were offered to the officers and employees of the company under the additional compensation plan. A-C Mfg. Co. Board of Directors, January 10, March 7, 1930, May 8, October 2, 1931.
3   Thomas C. Cochran, *The American Business System* (Cambridge, Massachusetts, 1957), p. 111; Arthur M. Schlesinger, Jr., *The Crisis of the Old Order* (New York, 1957), p. 248; Dixon Wecter, *The Age of the Great Depression* (New York, 1948), p. 66.
4   *Milwaukee Sentinel*, January 1, 1930; Bayrd Still, *Milwaukee, The History of A City* (Madison, 1948), pp. 478–479.
5   A-C Mfg. Co. Executive Committee, November 18, 1929, March 10, July 14, 1930; Board of Directors, February through December, 1930; Still, p. 479.
6   *Sales Bulletin*, December, 1930, p. 1008, October, November, December, 1931, p. 1; *Fortune*, May, 1939, p. 152; Still, p. 479.
7   *Milwaukee Sentinel*, January 15, 1933.
8   *Sales Bulletin*, December, 1929, p. 864; *Milwaukee Sentinel*, January 15, 1933.
9   *Sales Bulletin*, December, 1929, p. 863; A-C Mfg. Co. Executive Committee, December 1, 1930; *Sales Bulletin*, May, 1930, p. 932; *Chairman's Circular Letter* No. 233, October 31, 1932.
10 *Chairman's Circular Letter* No. 233, October 31, 1932.
11 A-C Mfg. Co. Board of Directors, July 1, 1932; Frederick Lewis Allen, *The Lords of Creation* (New York, 1935), p. 407, quoting computations of Frederick Mills.
12 A-C Mfg. Co. Board of Directors, March 11, 1932. Director O. C. Fuller wrote "I fully approve of this. I have failed to attend several meetings, knowing I was not needed for a quorum, to save the company the fee and expense of my attendance."

13 *Annual Reports*, 1931–1941; M. C. Maloney, *Complete History of Liquid Conditioning Products, General Products Division, Allis-Chalmers Mfg. Co.*, MS, April, 1964, pp. 1–6.

14 A-C Mfg. Co. Executive Committee, March 10, 1930, through September 20, 1938, *passim; Circular Letters* from No. 237, January 17, 1933, to No. 311, August 18, 1939, *passim.*

15 Irving H. Reynolds to Alberta J. Price, August 31, 1945; J. L. Neenan to author, July 29, 1959; Charles Allendorf to author, August 10, 1959; J. J. Kern to author, June 15, 1961. The *Minutes* of the Executive Committee for March 8, 1933, offers the precise number of employees at each of the plants at the low point of December 31, 1932, as compared to the month before and the month following.

	Nov. 30 1932	Dec. 31 1932	Jan. 31 1933
West Allis	1830	1723	1864
Norwood	241	263	253
Pittsburgh	187	173	195
Springfield	188	200	181
LaPorte	207	213	197
LaCrosse	20	19	23
Condit	112	116	100
Gen. Office	1496	1463	1467
	4281	4170	4280

16 *Milwaukee Sentinel*, January 15, 1933; A-C Mfg. Co. Executive Committee, January 16, 25, June 1, 1933, February 5, 1934.

17 *Sales Bulletin*, July to December, 1932, p. 859, October, 1935, p. 7.

18 A-C Mfg. Co. Board of Directors, January 18, October 2, December 4, 1932, October 7, November 4, 1932, May 5, October 20, 1933, April 6, 1934; *Sales Bulletin*, September–October, 1930, p. 994, January, February, March, 1931, p. 1034, October–November–December, 1931, p. 7.

19 *Annual Reports* for the years 1934, 1935, 1936; A-C Mfg. Co. Executive Committee, March 8, 1933.

20 *Sales Bulletin*, April, 1930, p. 918, January to June, 1932, pp. 1, 3; William Miller, "The Business Elite in Business Bureaucracies," pp. 287–288, in William Miller, ed., *Men in Business* (New York, 1962).

21 A-C Mfg. Co. Board of Directors, May 6, June 3, 1932; *Sales Bulletin*, January to June, 1932, p. 1; Charles Allendorf to author, August 10, 1959; J. L. Neenan to author, July 29, 1959.

22 *Sales Billed* (Corporate) Industrial Equipment Division; *Sales Bulletin*, October to December, 1934, p. 70; *Company Facts*, 1936, n.p.; Jack Haase and M. Jensen to author, August 5, 1959.

23 Jack Haase and M. Jensen to author, August 5, 1959; Ernest Shaw to author, August 3, 1959. The company did retain the rights to the Streambarker but the increasing availability of labor following the war reduced the desirability of this machine, and almost none of them were sold.

24 Ernest Shaw to author, August 3, 1959; Jack Haase and M. Jensen to author, August 5, 1959; R. K. Prince to author, August 6, 1959; Earl Lashway to author, August, 7, 1959.

25 *Sales Billed* (Corporate) Industrial Equipment Division; *Sales Bulletin*, July to December, 1932, p. 859.

26 *Sales Bulletin*, January to June, 1933, p. 34, July to September, 1934, p. 53; *Twentieth Annual Report*, March 15, 1933, p. 9.; J. L. Neenan to author, July 29, 1959.

27 *Company Facts*, 1936, n.p.; *Fortune*, May, 1939, p. 56; J. L. Neenan to author, July 29, 1959. An indication of the kind of extended service rendered by Allis-Chalmers milling machines can be seen in the policy of 1960 not to furnish repair parts for machines made prior to 1910. But even then, as a special favor, it did supply a part for an 1884 machine.

28 Texrope memorandum, 9–6–51, p. 2. On August 3, 1959, Ernest Shaw told the author that "Walter Geist damn near burned up the office experimenting with rope drives." *Sales Bulletin*, March, 1925, p. 240; *From the Shadoof to the Dominant Drive*, A-C Brochure, 1944, p. 32. Researchers in the automobile industry at this time were working with a "V-belt" to run a single-grooved sheave driving the radiator fan. Both groups of experimenters discovered the principle of wedge contact at roughly the same time. However, it might be more

precise to say that they "rediscovered" this principle, which had been used in English rope drives. Based on this concept the V-belt became a standard automobile accessory.

29 Unmarked MS. pp. 1–3. C. A. McCormack remarked to Alberta Johnson Price on July 19, 1945, "The wonderful thing about Texrope is that they wear out—replacements are necessary frequently."

30 *Sales Bulletin*, March, 1925, p. 239, August, 1925, p. 253, October to December, 1934, p. 905; *Thirteenth Annual Report*, April 9, 1926.

31 *Sales Bulletin*, December, 1925, p. 271, August, 1929, p. 685.

32 Texrope Drive is United States Patent #1,662,511. *Sales Bulletin*, March, 1928, p. 495; Kinball Wyman to author, August 11, 1959; *Texrope Circular Letter* No. 95, November 3, 1933.

33 *Sales Bulletin*, October to December, 1934, p. 906; *Texrope Circular Letter* No. 146, March 8, 1938.

34 *Sales Bulletin*, October to December, 1934, p. 905.

35 *Sales Billed* (Corporate) General Products Division; Executives' Biographies for Publicity Files. The Texrope Drive was very successful, and Walter Geist was amply rewarded, but it is interesting that during the fifteen year period 1925–1940 there is only one mention of *Texrope* in the *Minutes* of the Executive Committee (February 15, 1932) and one in the *Minutes* of the Board of Directors (October 5, 1928). Compare this with the constant mention of tractors for an indication of the main interest of General Falk.

36 *Sales Bulletin*, October, November, December, 1931, p. 47. The writer quoted also added, "The public utilities and industry as a whole have proceeded cautiously, with few major developments undertaken, and this has not been conducive to development of apparatus of extraordinary sizes or characteristics." Also, *Sales Billed* (Corporate) General Products Division, and Power Equipment Division.

37 W. S. Edsall, *History of the Boston Works*, MS, 11–4–40, pp. 2–4; *Sales Bulletin*, July to September, 1934, pp. 6–7. The American Brown-Boveri Electric Corporation was purchased for 85,000 shares of Allis-Chalmers common stock at $40 per share as authorized by A-C Mfg. Co. Board of Directors, June 5, 1931. On July 31, 1931, there were 280 employees at the Condit plant; this increased to 318 by August 31, but declined as the depression deepened. A plan for the complete liquidation of the Condit Corporation was approved by the Board on November 13, 1936. Founded in 1899, Condit Electrical Manufacturing Company brought to Allis-Chalmers some 250,000 square feet of manufacturing space with up-to-date equipment, covering about 15 acres in the Hyde Park District of Boston. Sears B. Condit, after selling the Condit Electric Manufacturing Company to Brown-Boveri, continued as president until 1936 when the Condit Electrical Manufacturing Corporation was dissolved and it became the Condit Works of the Allis-Chalmers Manufacturing Company. G. A. Burnham was then appointed assistant manager of the Electrical Department, in charge of the Condit Works. In 1938 the name was changed to the Boston Works of the Allis-Chalmers Manufacturing Company, and W. S. Edsall was appointed assistant manager of the Electrical Department in charge of the Boston Works.

38 *Sales Bulletin*, August-September, 1931, pp. 3, 23, 11, July to December, 1932, p. 63, July to December, 1933, p. 63; "Blowers and Compressors," *Company Facts*, 1936, n.p.

39 *Sales Bulletin*, July to December, 1932, p. 19; *Twentieth Annual Report*, March 15, 1933, p. 8; *Company Facts*, 1936, n.p.

40 *Annual Review*, 1940, p. 25. The importance of a steady supply of balanced orders is seen in this statement. "It is gratifying to build record-breaking machines since it is a symbol of advancement, engineering and shop skill. Nevertheless, a steady stream of standard machines of the smaller and medium sizes is equally essential, not only to complete the lines of types and ratings but, even more important, to maintain office and shop organizations balanced and a steady level of employment. This past year's business has filled these requirements admirably." "Electrical Equipment," *Company Facts*, 1936, n.p.; *Fortune*, May, 1939, p. 56.

41 A-C Mfg. Co. Executive Committee, December 20, 1937.

42 *Sales Bulletin*, July to December, 1932, p. 807, January to June, 1933, p. 809, July to December, 1933, p. 817; *Power Review*, December, 1940, p. 14. On January 16, 1933, Dr. William Monroe White, appeared before the Executive Committee and presented the specifications for the hydraulic turbines for the

Hoover Dam. The specifications "contained a number of conditions which were considered unusually drastic and unfavorable to the contractor." The conditions specified that no payments would be made until the machinery was ready to ship, in the event of failure to perform guaranteed efficiency the machinery was to be replaced at the contractor's expense, and a provision that if the Congress did not appropriate the necessary funds the government would be relieved from meeting the payments as they became due. After careful consideration the committee authorized Dr. White to submit bids for the work "in accordance with the Government's form of contract and specifications on a basis of what might be considered a fair price."

43 *Sales Bulletin*, July to September, 1934, p. 6, January to June, 1935, p. 29.

44 *Annual Review*, 1937, pp. 39–40; *Power Review*, December, 1940, p. 14.

45 "Hydraulic Turbines," *Company Facts*, 1936, n.p.; *Fortune*, May, 1939, p. 57.

46 *Sales Bulletin*, January to June, 1934, p. 55.

47 *Power Review*, May, 1939, pp. 22–26.

48 *Sales Bulletin*, July to December, 1933, pp. 35–36, October to December, 1934, p. 49, January, 1936, p. 111, November, 1936, p. 8. In view of the completion of more recent bridges such as the Mackinac Straits Bridge the comment in the last *Sales Bulletin* noted is of interest. "On November 12 (1936), this engineering epic was completed—an 8¼ mile transbay highway, the erstwhile subject for dreamers. Probably it will not be surpassed for a thousand years because over no other great body of water would there be traffic enough to justify more than the $77,600,000 cost of the Bay Bridge."

49 *Sales Bulletin*, October to December, 1934, p. 16. *Annual Review*, 1936, p. 15; "Mining Machinery," *Company Facts*, 1936, n.p.; J. N. Friedbacher to author, August 5, 1959.

50 *Sales Bulletin*, January to June, 1935, p. 26; "Crushing and Cement Machinery," *Company Facts*, 1936, n.p.; *Industrial Review*, July, 1937, p. 4.

51 "Crushing and Cement Machinery," *Company Facts*, 1936, n.p.; *Annual Review*, 1937, p. 9; Memorandum to George Smith, April 14, 1947. In 1947 some claimed that the Leeds, Alabama, mill was likely to remain the largest in the world because new developments had made these giants impractical. In 1962 this prediction was invalidated when the company shipped the sections of a new and larger cement plant by barge on the Great Lakes to the Atlantic Cement Co. in Ravena, New York. The two kilns, each 580 feet long, were capable of producing five million barrels of cement a year. *Milwaukee Journal*, October 6, 1961.

52 *Annual Review*, 1940, p. 11; *Fortune*, May, 1939, p. 56.

53 *Sales Bulletin*, July to December, 1932, p. 67, July to December, 1933, p. 68; *Annual Review*, 1937, p. 53, 1938, p. 71; "Steam Turbines," *Company Facts*, 1936, n.p., *Annual Review*, 1940, p. 91; *Power Review*, 1940, p. 23.

54 *Sales Billed* (Consolidated) 1913–1956.

55 *Milwaukee Sentinel*, January 15, 1933. Otto Falk had a high regard for the Coolidge administration as noted in his holiday greeting of December 20, 1936. "While the factors determining business for the next twelve months are varied in their significance, I am nevertheless convinced that with the sound business-like administration of our National Government and the conservative optimism of the Public, we have nothing to fear for the year 1927."

56 On January 30, 1930, Albert M. Marsh and James J. Bickler of the Allis-Chalmers flour mill department left Milwaukee to fulfill an 18-month to 3-year contract with the MacDonald Engineering Company to build flour mills and grain storage for the Soviet Republic and to teach the Russian engineers the American system of milling. The *Sales Bulletin* of January, 1930, p. 873, noted, "Information coming to us indicates that the Soviet Government is contemplating the erection of several large flour milling plants in various sections of the U.S.S.R. and we feel that in loaning Mr. Bickler and Mr. Marsh to the MacDonald Engineering Company for this work, that it will go a long way towards getting our machinery equipment into the plants to be erected." Allis-Chalmers, however, did not get these contracts.

57 General Otto H. Falk, "Those Who Can Take It," *How-to-Sell*, February, 1934, n.p. In the same article Falk endorsed the processing tax of the AAA.

58 *Milwaukee Sentinel*, October 21, 1933. Hayden had come to Milwaukee for the Allis-Chalmers board meeting.

59 A-C Mfg. Co. Board of Directors, June 1, September 7, 1933, December 17, 1934; *Circular Letter* No. 245, June 29, 1933; *Sales Bulletin*, July to December, 1933, p. 63. In the *Twenty-First Annual Report*, March 20, 1934, p. 9, Falk and

Babb held that the immediate impact of the NRA had not been profitable for the company since while costs and selling prices had increased under the codes, they had not been reflected "to any important extent in contracts shipped and billed to customers."

60 *Twenty-Fourth Annual Report*, March 10, 1937, n.p.; *Twenty-Fifth Annual Report*, March 12, 1938, n.p. The *Minutes* of the Executive Committee for August 2, 1932, carry a breakdown of profits and expenditures made by the company during the period January 1, 1913, to December 31, 1931:

1.	Profits earned		$46,307,933.46
	Mfg. Profit	$46,307,933.46	
	Other Income	10,897,789.85	
	Less debenture int., discount	3,760,649.60	
		7,137,140.25	
			53,445,073.71
2.	Preferred dividends declared		15,570,149.25
3.	Common Dividends declared		19,384,610.85
4.	Taxes, Federal		10,732,095.88
5.	Taxes, Wisconsin		7,844,673.11
6.	Taxes, All other		1,153,989.29
7.	Capital Expenditures		25,208,359.73
8.	Wages & Salaries		180,223,979.00
9.	Materials Purchased		240,810,141.25
10.	Maintenance & Repairs		24,379,413.20
11.	Depreciation		14,107,540.29
12.	Standard Development		9,363,135.08
13.	Employees Accident Compensation		950,005.16

61 *Annual Report for 1938*, March 15, 1939; *Annual Report for 1939*, March 15, 1940; *Annual Report for 1940*, March 20, 1941; A-C Mfg. Co. Executive Committee, June 25, October 21, 1940, February 20, 1941.

62 A-C Mfg. Co. Executive Committee, June 25, 1934; *Power Review*, July, 1938, p. 33; *Sales Bulletin*, May, 1930, p. 931, November, 1936, p. 23, *Annual Review*, January-February, 1937, p. 77; A-C Mfg. Co. Board of Directors, October 20, 1933.

63 A-C Mfg. Co. Board of Directors, April 4, 1930; *Sales Bulletin*, May, 1930, p. 931. It was estimated that this insurance would initially cost the company approximately $40,000 per year. Participation by the employees was very nearly 100 percent.

64 *Annual Report for 1939*, March 15, 1940.

65 *Power Review*, July, 1938, p. 33.

66 Broadus Mitchell, *Depression Decade* (New York, 1947) pp. 254, 277–278. The first of a series of federal measures designed to strengthen the bargaining position of organized labor was the Norris-LaGuardia Act of 1932, which stated that as a matter of public policy employees should be free to unionize. Section 7a of the NRA code made this mandatory. The code also stipulated that "no employee and no one seeking employment shall be required as a condition of employment to join any company union or to refrain from joining . . . a labor organization of his own choosing." Writing nearly a generation later, Eric F. Goldman, *Rendezvous With Destiny* (New York, 1952), p. 365, candidly called the Wagner Act "probably the most bluntly anti-corporation legislation the United States has ever known."

67 No attempt has been made to write a detailed history of Allis-Chalmers labor relations during the period 1933–1940; instead, the company's labor relations are placed in the perspective of the decade as a whole. For a detailed but undocumented study of the relations between the company and organized labor during this period, see *Resume of Pertinent Phases of Collective Bargaining History at West Allis Works*, MS by William McGowan, n.d. Labor matters enter the *Minutes* of the Executive Committee on August 20, 1934, and continue in detail through the remainder of the decade. For the impact of the CIO on another Milwaukee firm, the Bucyrus-Erie Company during the same period, see Harold F. Williamson and Kenneth H. Myers, *Designed for Digging* (Evanston, 1955) pp. 237–244.

68 A-C Mfg. Co. Executive Committee, April 17, 1939, May 3, September 23, 1940.

69 *Milwaukee Journal*, June 21, 1940; A-C Mfg. Co., Board of Directors, June 6, 1940. On March 6, 1939, Otto Falk owned 20,817 shares of Allis-Chalmers stock.

Walter Geist
President 1942–1951

CHAPTER TEN

EXPANSION AND RECONSTRUCTION

WAR AND DEPRESSION have been the historic watersheds of the twentieth century. The first of these wars, World War I, was important to the Allis-Chalmers Manufacturing Company because it brought financial stability. Had it not been for the capacity production and substantial profits generated by the wartime economy, even the managerial genius of an Otto Falk might not have been able to carry a solvent company into the good years of the twenties. Because of the General's excellent management, Allis-Chalmers had both the financial reserves and the range of products needed to carry it through the depression and reach sales of more than $80 million in the late 1930s. But apart from the growth of the tractor division to prominence by 1937, the character of the company had remained largely the same.

The death of General Falk in June of 1940 terminated an era in the history of the company as had the death of Edward P. Allis half a century earlier. Although Max W. Babb had been president since May 6, 1932, Otto Falk had remained as chairman of the board until a few months before his death. Under Falk the management of the company had been largely controlled by one man; although there were capable men in the company, none of them had been trained in overall management. As a consequence, Babb had no trained, reliable staff. Because of ill-health Babb resigned the office of president on January 5, 1942, and was elected chairman of the board, in which office he served until his death early in 1943. William C. Buchanan, a member of the Board of Directors, served as interim president of Allis-Chalmers from January 5 to April 16, 1942. Born in Johnstown, Pennsylvania, on May 8, 1888, Buchanan had steadily advanced in a succession of steel mills in New York, Connecticut, Pennsylvania and Ohio. In 1929 he became vice president and general manager of the Keystone Steel and Wire Company and in 1935 was elected president of the Globe Steel Tubes Company of Milwaukee. Shortly thereafter he joined the board of Allis-Chalmers.[1]

William C. Buchanan
President 1942

Neither Buchanan nor the board thought of his election to the presidency as anything but temporary. The board was thus given time to appraise the present and to plan for the future. Their eventual choice for president was Walter Geist. A self-made man with little formal education, Geist had joined the company at the age of fifteen as an errand boy. Primarily an inventor and engineer, later a salesman, he had not been a member of either the Board of Directors or the Executive Committee before he was named president on May 7, 1942.

The instrument which catapulted Allis-Chalmers into a new era was the industrial dynamism created by World War II. As a manufacturer of some 1,600 different products, most of which were essential to the nation's industry, Allis-Chalmers was inevitably drawn almost completely into the vortex of the war effort. The company had the plants, the reservoir of skilled personnel, and a tradition of workmanship that was called on to the full as the United States rapidly converted into an "arsenal of democracy." Under the circumstances, Walter Geist found himself captive to this pattern of change, unable to mold the company as he chose. The forces of change can be seen in statistics. Whereas billings in 1940 nearly equalled the 1937 high of $87 million, they jumped to $122 million in 1941; $196,040,342 in 1942; $286,368,271 in 1943; $364,716,653 in 1944; and $282,219,214 in 1945. Employment increased from 1932 to 1941 by 404.2 percent. During the period from June, 1940 to May, 1942, when Allis-Chalmers was quite without executive leadership, the employment figures rose from 22,168 to 31,120, with more than half of this increase employed in the West Allis Works. The number of workers rose to 35,974 in 1943, remained roughly the same with 36,832 the following year, and declined to 28,515 in 1945 as the war effort began to slacken.[2]

As the United States moved into total war in an effort to crush the Axis powers, company sales increased dramatically. Significantly, direct war orders largely approximated the difference between the sales billed and the pre-war company production. In 1942 war orders accounted for $107,300,000 of the $195,600,000 in sales. The remaining $88,300,000 represented roughly the sales figure of

1937. In 1943 sales mounted to $295,600,000, of which $226,500,000 represented war orders, and the following year the sales totalled $378,900,000, of which $305,300,000 were war orders. The difference between war orders and total sales in 1943 and 1944 was $69,100,000 and $73,600,000 respectively. These last figures approximate fairly closely the total sales for 1938 and 1939.[3]

The milling department was one of several little effected by the booming war economy. The nations supply of bread was a day-to-day contribution to the American war effort made possible through Allis-Chalmers installations in nearly 90 percent of America's flour mills. The high point of milling sales, $1,175,495 in 1944, was, however, less than in the comparable period in World War I and only slightly higher than sales in 1939. Sawmilling remained in a dismal decline. The best war year for this department, amounting to $361,684 in 1945, was less than one-third the sales of 1920. The crushing, cement and mining departments fared better than flour milling and sawmilling. As the producer of at least three-fourths of the cement manufacturing equipment and as a respected manufacturer of crushing and mining machines, Allis-Chalmers had sales of close to $7,000,000 in 1942 and 1943; this figure was, however, still short of its sales records set in the twenties. Hydraulic turbine sales approached $3,000,000 in 1943, only slightly above the average for the best years of the twenties and thirties. Clearly, the total war effort had no very noticeable effect on these departments. The

President Geist signing autographs for women war workers. At the left is vice president William Johnson.

Texrope division fared somewhat better. Peak sales of $5,026,940 in 1944 nearly doubled the sales of this division in the best years of the two previous decades.[4]

The vast bulk of sales during the war was concentrated in a few regular lines and in some special projects. Allis-Chalmers had contributed primary power units to the American economy in peace and in war since the first years of the century. The largest single order it had filled prior to World War II was for destroyer steam turbines during World War I. From 1940 through 1945 the company continued to produce steam turbines, significant in both number and size, for land power installations. These were allocated to war plants, to American allies and to large power stations. But the principal source of orders was the United States Navy. Turbines for destroyers, baby flat top escort carriers, and Victory ships poured from the West Allis Works. The impact of these orders on the steam turbine department and on the company can be seen in sales billed during the war years:

1939	$ 1,614,245
1940	4,279,377
1941	7,344,788
1942	19,799,677
1943	30,181,830
1944	17,518,706
1945	17,076,419[5]

Even before the outbreak of World War II, Allis-Chalmers was an important supplier of defense equipment, especially for the Navy. In 1940, President Franklin D. Roosevelt toured the West Allis plant with Wisconsin governor Julius Heil, company president Max Babb, and vice president Walter Geist.

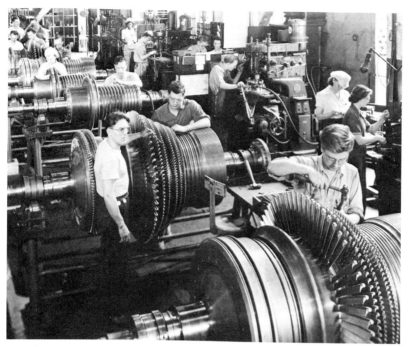

As it had in World War I, Allis-Chalmers again produced propulsion turbines
for destroyers, as well as turbine driven auxiliary generators. This view shows
the blading of turbine spindles.

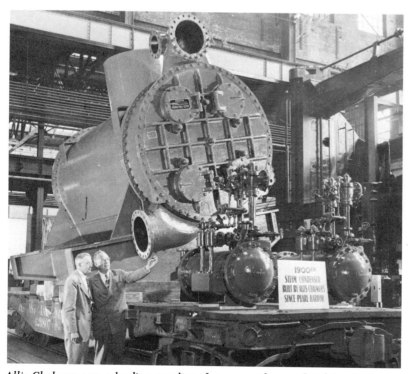

Allis-Chalmers was a leading supplier of steam condensers for ships, producing
more than one a day.

*During World War II, the
company built propulsion
steam turbines for
destroyers. Here the casing
is opened for inspection.*

*Many submarines were
equipped with Allis-Chalmers
propulsion and control
equipment. This is the incline
test of a propulsion motor.*

*Many new refineries were built during the war years to supply high octane gasoline
for airplanes. These refineries used the Houdry process, for which Allis-Chalmers
supplied gas turbine driven compressor units.*

As the United States moved from preparedness to total war, the impact on the electrical department was tremendous. Although the sales of the Boston plant which made circuit breakers and the Pittsburgh plant which manufactured transformers only doubled during the war years, the products of other plants were in far greater demand. At the Norwood Works the production of motors during the war averaged 34,198 per year, and the value of its production nearly tripled. At West Allis, where the larger motors and generators were manufactured, sales quadrupled from $8,413,293 in 1940 to $32,730,583 in 1942. A new building, later known as the Hawley plant, was erected at West Allis under a Navy facilities contract for the manufacture of propulsion control equipment. Upon completion its production increased from $9,468,905 in 1943 to $13,163,625 the following year. Hardly a vessel put to sea during the war was not equipped entirely or in part with electrical equipment from Allis-Chalmers. Although propulsion equipment, electrical controls, motors for winches and the like are not innately dramatic, they were essential to the war effort and helped publicize the company name.[6]

The tractor division, which had been developing rapidly during the thirties—its annual sales were eight times greater in 1941 than they had been in 1934—continued to grow during the war years. In 1937 and again in 1940 tractor sales were approximately

Allis-Chalmers manufactured pumps for the pipelines which carried petroleum products from western oil fields to the middle eastern states.

(Top) A wide assortment of electrical control units for ships from submarines to aircraft carriers was built in the Hawley plant.
(Bottom) Allis-Chalmers crawler tractors, many with bulldozers, played an important part in construction work, such as the clearing of land for air strips.

$50,000,000. Under the pressures of war, they rose rapidly until in 1944, with sales of $195,101,326, the pre-war peak was practically quadrupled. The main emphasis in the thirties had been on the farm tractor, which had been streamlined and made more efficient. During the war, thousands of these tractors equipped with Allis-Chalmers implements contributed to make the United States the world's principal source of food. The track-type tractors, improved during the 1930s, now found a new market as the production of farm tractors was restricted by late 1942 to 23 percent of 1940 production. The heavy-duty crawler type tractor cleared roads and leveled airfields, filled bomb craters and hauled ammunition; it became the muscle of the construction battalions. In 1943 the engineering department completed the two largest high speed military prime movers used in the war. The eighteen-ton M4 traveled thirty miles per hour, pulled the 90 and 155 millimeter anti-aircraft guns and carried the crew and the initial rounds of ammunition. The thirty-eight ton M6 pulled the 255 millimeter gun and was equipped with high speed transmission, front-connected towing winch, and electrical air brake controls for the gun. A two-star general in the Pacific Theatre once said, "In this war there are two items of equipment which should be memorialized in marble—the tractor and the jeep." Allis-Chalmers was a leading manufacturer of the tractor.[7]

Experience of more than four decades in the manufacture of turbines led to negotiations in 1941 with the Army Air Force for the manufacture of superchargers to enable American planes to fly at altitudes up to 40,000 feet. A preliminary plan drawn up on April 17, 1941, called for 4,000 units per year, to be constructed in a building costing $5,000,000. But the tempo of the war rapidly

The two largest prime movers used by the Army during the war were built by Allis-Chalmers. The M4 weighed 18 tons; the M6, 38 tons.

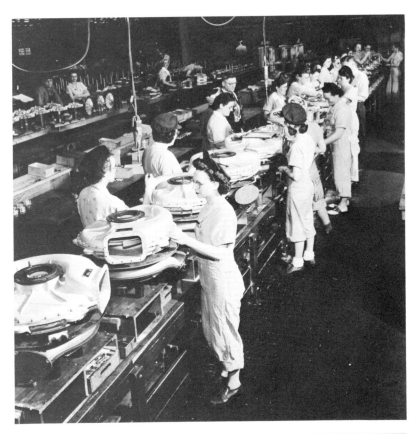

One of the great production stories of the war was the work done by women in the nation's factories. At Allis-Chalmers, they built airplane superchargers, which increased possible flight altitudes.

increased orders and soon demanded a far larger plant. When the model factory, air conditioned and prepared for blackouts, was dedicated on August 27, 1942, it covered 985,495 square feet and cost $25,045,065. In this excellent facility Allis-Chalmers production personnel, directing a work force which was 78 percent women, brought supercharger production up to 4,500 per month at a decrease in cost of 50 percent over the original $3,100 per unit. On land, on the sea and in the air, Allis-Chalmers served the war effort. [8]

The wide range of company experience and the quality of its personnel led to the involvement of Allis-Chalmers in interesting and highly significant research projects. Robert C. Allen, chief engineer of the steam turbine department, and Dr. John Theodore Rettaliata, manager of the research and gas turbine development section of the same department, were key men in the Navy H-1 jet propulsion program which made experimental models available by the end of the war. In the complex development of the atomic bomb, Allis-Chalmers produced more equipment by weight for the "Manhattan District Project" than any other company. A wide variety of products flowed from the company shops at West Allis to the Manhattan Project; part of the Hawley plant was built especially for this work. The company's part in the production of the bomb can be divided into four sections: first, the electromagnetic process and equipment for separating the different kinds of ura-

The production of the atomic bomb was one of the best kept secrets of the war. This view shows the manufacture of electromagnets, wound with silver, for separation of different types of uranium.

(Top) Labor-saving tractors, such as this one developed especially for small farms, were an important company contribution to Allied food production.
(Bottom) Anti-aircraft guns were produced at the La Porte plant.

nium; second, the equipment especially designed and built for gaseous diffusion, or atom sorting; third, the large power plant equipment for extraction of plutonium from uranium; and fourth, the betatron and then the synchrotron, a 300-million electron volt atom smasher for atomic research. Research and production at Allis-Chalmers was not only pointed at the immediate needs of the war but also at possible uses for the future.[9]

The contribution in materials by Allis-Chalmers to the war effort was not without its price in labor difficulties. In fact, the basic cause for a decade of continuous labor strife from 1937 to 1947 became increasingly clear: it lay in Communist domination and control of Local 248 UAW-CIO which represented the production and maintenance employees. Although the company bargained in good faith in the hope for a sound labor relations program, the union leadership deliberately prevented genuine accord. Through 1940 the company made every effort to work with the union, giving in to more and more demands to ensure continuous production for the preparedness program.[10]

On January 21, 1941, the uneasy relationship between Allis-Chalmers and Local 248 was broken by a strike. The union announced that an overwhelming majority of the employees had so voted by secret ballot in accordance with the requirements of the Wisconsin Employment Relations Act. Production at the West Allis Works came to a standstill for seventy-six days. Some employees, however, charged that the strike vote was not taken by a proper secret ballot. Company counsel obtained permission from the Board for a supervised examination of the ballots by John Tyrrell and Clark Sellers, nationally known examiners of ques-

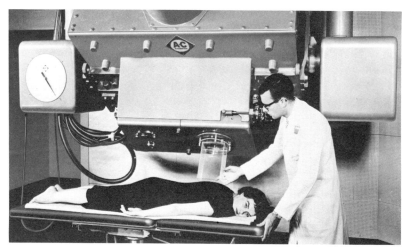

The Betatron, a high power X-ray unit, was developed during World War II. It was ordered by industries, research units, and hospitals.

tioned documents. These experts conclusively proved that a gross fraud involving at least 40 percent of the ballots had been perpetrated by the officers of Local 248.[11]

The Allis-Chalmers management was divided over the reasons for both previous labor difficulties and for the strike of 1941. One group, mostly those in production, held that the problems were only a matter of union aims. Others held that the strike went far beyond this and amounted to obstructionism, relating directly to overall Communist Party policy. The position of the latter group was vindicated in 1947 when Louis Budenz, former editor of the *Daily Worker*, revealed that in November of 1940 he had been sent to Milwaukee as an emissary of the Party to meet with Eugene Dennis, Harold Christoffel, and other local labor functionaries. Their orders were to strike the West Allis Works in order to hamper the defense program and prevent aid to Great Britain. According to Budenz, Christoffel agreed to bring about a strike by whatever means necessary.[12]

As the strike wore on, Secretary of the Navy Frank Knox telegraphed appeals to the company to open the West Allis Works and to the employees to return to work. A back-to-work movement developed among the employees that was violently opposed by the leadership of Local 248. The company, after serious rioting developed, felt compelled to keep the plant closed to prevent further bloodshed and destruction of property. At this juncture Circuit Court Judge Charles L. Aarons upheld the right of the Wisconsin Employment Relations Board to conduct a new strike vote. His decision concluded:

> The workers are the direct beneficiaries of the Union organization. The Union officers are trustees to carry out the purposes of the Union, which is to promote the welfare of the workers. Breaching their obligation as trustees, as found by the Board, they have deliberately deceived the beneficiaries, and they here seek to prevent these deceived workers from having an untrammelled opportunity to express their true opinions. . . . That is a proposition that must necessarily shock the conscience of all right thinking men. Certainly no court, of law or equity, can stand idly by in the face of such a contention.

Two days before the scheduled date of the strike vote under supervision of the WERB, settlement was effected by officials of the Federal Government.[13]

While the exposure of the vote fraud was a serious blow to the leadership of Local 248, Christoffel maintained control of the union through coercion and intimidation and rode out the storm of protest. The leaders of Local 248 did not, however, escape public castigation by Walter Reuther and other top union officials at the International UAW Convention in Buffalo in July, 1941. At that time Reuther charged that Local 248 was "dominated by political racketeers of the Communist stripe" and characterized the election of the Local 248 delegates to the convention as "the worst kind of strong-armed political racketeering."[14]

Immediately following the German invasion of Russia on June 23, 1941, Local 248 changed its tactics. To provide all-out aid to Russia and Great Britain through Lend-Lease and to fill American military needs, the union made every effort to support the production of war goods which had become the principal source of orders for Allis-Chalmers. Instead of impeding production the leadership of 248 turned to harassment of company management. As union demands and abuses of power increased, the management realized that eventually a confrontation would be necessary to maintain shop control.

Negotiations for a new agreement were in progress on VJ Day. With the expiration of the War Labor Board and the relaxation of wartime controls, the company tightened the grievance procedure to prevent continued union abuses. It also announced that it would once again adhere to its traditional policy that "no employee's job at Allis-Chalmers shall depend on membership in the Union." The company, however, offered to match the national wage increase of eighteen and one-half cents per hour and to submit the contract proposal to a vote of the employees in a supervised secret ballot. This offer was rejected by the leadership of 248. In the spring of 1946 the United Electrical Radio and Machine Workers of America at the Boston Works, Pittsburgh Works and Norwood Works, and the Farm Equipment and Metal Workers of America at the La Porte and Springfield Works went out on strike; on April 29, the workers at the West Allis Works struck. The strikes at the outside works were all settled by early fall of 1946, leaving Local 248 and the Allis-Chalmers management to fight the issue to a finish at the West Allis Works. This strike was not settled but was officially terminated on March 23, 1947, after it was broken by a back-to-work movement which spelled an end to the ten-year reign of Harold Christoffel.[15]

There had been no unified opinion or plan of action by company officials during 1940 and 1941. However, by the time of the 1946 strike, the question had been resolved, and President Walter Geist provided the leadership for company success in such a confrontation. While management was fully aware of the implications of the strike and the forces behind it, many employees and the general public were not. Despite the fact that Russia had only recently been our "great and gallant ally" and that this was a period when management protests of Communist control of unions were usually regarded as "red baiting" and "Union busting," Walter Geist decided that it was time that the company reveal the facts concerning Local 248. During the late summer of 1946 a brochure detailing Communist domination of Local 248 was distributed to company employees. By early fall the *Milwaukee Sentinel* began daily publication of a series of articles by "John Sentinel" which exposed Communist control of Local 248 and of the Milwaukee County CIO Council. The *Milwaukee Journal* also took up the fight with a similar series of articles and in its editorials accused certain union officials of being secret Communists.

These startling revelations precipitated a back-to-work movement which resulted in serious violence; other CIO unions were called upon for picket line support to save the bargaining agency at the West Allis Works. While more and more men returned to work, as many as 1,000 uniformed officers were required to keep order at the gates of the West Allis Works when shifts changed. An independent union composed of disillusioned 248 members petitioned the Wisconsin Employment Relations Board for an election which was held in January, 1947. Because neither Local 248 nor the independent union obtained a majority, Local 248 lost the right of employee representation.[16]

With the majority of the employees back at work, Local 248 officials announced on March 23, 1947, that the strike was being called off, but that the fight would be continued within the plant. The company then moved immediately to discharge Harold Christoffel and ninety other members and officials of Local 248. After an extensive investigation, the National Labor Relations Board upheld the company in these discharges.

The final phases of the strike coincided with hearings for the amendment of the National Labor Relations Act before the House Committee on Education and Labor. The result was the Taft-Hartley Act, which was passed over President Harry S. Truman's veto in late June, 1947. Testimony by Allis-Chalmers officials concerning their experience with a Communist-controlled union deeply effected the hearings and prompted the inclusion of certain provisions in the bill, including the requirement that unions file non-Communist affidavits as a condition of NLRB certification. One of the critical questions before the UAW International convention at Atlantic City in October was whether the UAW should comply with the non-Communist affidavit requirement of the Taft-Hartley Act. Although Harold Christoffel bitterly opposed adoption of the resolution for compliance, the delegates voted their acceptance. This action forced the resignation of Local 248 officials, who would have been compelled to sign affidavits to obtain NLRB certification. Walter Reuther, who had gained control of the Executive Board of the CIO, thereupon appointed Pat Greathouse as administrator of Local 248. Christoffel and other Local 248 officials were then brought to trial by the union and formally expelled from the UAW. Christoffel's difficulties did not end here. While being questioned under oath by the House Committee on Education and Labor in 1947, he had denied Communist affiliations. He was subsequently indicted for perjury, convicted by a jury in Federal District Court in Washington, D. C., and after two appeals to the United States Supreme Court, began serving a two- to six-year term in the Federal Penitentiary in 1953.[17]

Company officials offered the new union administrator full cooperation, and the NLRB finally certified Local 248 as bargaining agent for the production and maintenance units. Walter Reuther stated the official position of the UAW-CIO on the long period of labor difficulties at West Allis in a Milwaukee speech in December, 1947:

> The Allis-Chalmers situation . . . was not something of which the labor movement could be proud. It was a black spot on the whole CIO and Milwaukee and Wisconsin labor in particular . . . We did not lose the Allis-Chalmers strike because we were fighting a tough management . . . At Allis-Chalmers the Union lost its membership, collective bargaining rights and its contract despite the fact that the workers favored UAW-CIO. We lost these because there were people in position of leadership who put their loyalties outside of their Union, outside of the rank and file and outside of their country.[18]

A difficult time in the history of Allis-Chalmers had been brought to a satisfactory conclusion. Unfortunately the name Allis-Chalmers had been synonymous with turbulent labor relations for nearly a decade. The misunderstandings that had developed between the front office and the plant would soon disappear, but it would take longer for a satisfactory understanding to emerge between the company and the communities in which it had its plants. The eleven-month strike left a battle scar in which the company can even take some pride, in its choice of battle with a Communist-dominated union rather than appeasement.

On June 30, 1950, Allis-Chalmers entered into a five-year agreement with Local 248, based on the recently-signed General Motors contract. The agreement provided for pensions, cost-of-living wage adjustments, annual productivity wage increases, and an expanded vacation-with-pay plan. It also granted a modified union shop effective upon a secret vote of a majority of the employees. In 1955 renegotiation of this contract for three years took place without particular incident.

By 1958 the question of a master contract covering all company plants had become an issue. The company held that its widespread operations in a variety of fields and in many localities were not compatible with a master contract. This division of opinion on an issue of trade union policy culminated in a strike which began on February 2, 1959, and lasted for eleven weeks. In 1962 there was a four-day strike, and in 1964 a one-week strike before negotiations were settled. It should be noted, however, that the issues involved in 1959, 1962, and 1964 were all normal collision-type issues between management and union which were resolved without continuing ill will or major disruption of production. In contrast with the unhappy decade up to 1947, R. L. Day & Co., investment counsellors, reported as early as August, 1952, that the Allis-Chalmers labor position was "one of the soundest of any major corporation in the nation."[19]

During much of 1946 and early 1947, while other companies had been converting to peace-time production and tooling up for the post-war boom, Allis-Chalmers had been strike-bound for eleven months. Sales of $282,219,214 had been counted for 1945, but one year later total sales equalled roughly one-third that amount, $93,840,030. Certainly the 1946 sales were higher than the $87,000,000 pre-war record. But these had been achieved in large part through the product of company plants other than the strike-ridden West Allis Works. Allis-Chalmers had beaten the Communist-dominated union, but the strike cost the company market losses that were never regained.

No sooner was the strike settled than the company launched into full production on all fronts; by the end of 1947 sales had rapidly mounted to $211,949,890 as Allis-Chalmers, at long last, began to enjoy post-war prosperity. Sales continued to rise until two years later they hit $351,097,878. This, however, was no more than a response to the general national prosperity as the company continued the same organization, products and techniques.[20]

Walter Geist recognized the magnitude of the job confronting him. Since it was obvious that he could not run a company with sales over a quarter billion dollars the same way that Otto Falk had run it when sales were only fifty million dollars, he appointed two vice presidents: W. A. (Bill) Roberts to head the tractor division and William C. Johnson to head the general machinery division. Johnson, in turn, appointed Joseph L. Singleton to take charge of all general machinery sales. However, these two appointments and the election of a vice president for company-wide research were all the real major changes that Geist made. In his day Otto Falk had contended against excessive departmentalization, but during the forties the divisions and departments once again achieved far too much autonomy.

Working from the profits of the late forties, Geist gradually began to purchase new plants and to slowly expand the company's product lines. In November, 1948, he approved the purchase of a World War II ordnance works at Gadsden, Alabama from the Landsdowne Steel & Iron Company. Now called the Gadsden Works, this plant manufactured regulators. Two foreign works also came into the company system in 1950, expanding it worldwide. One was a small works at St. Thomas, Ontario, while the first manufacturing venture overseas was the purchase of a plant at Essendine, England. In response to English needs the Essendine Works specialized in tractors, loaders, harvesters, vibrating and gyratory screens, centrifugal pumps and compactors.[21]

The presidency of a major company takes its toll on the men thrust into leadership positions. The enormous pressures of World War II were immediately followed by those of the eleven-month strike. The determination of Walter Geist, more than anything else, held the company to the task of beating the Communist-dominated union. But even without such extraordinary pressures, the man-

W. A. Roberts
President 1951-1955

agement of a large and widespread company without adequate staff organization can prove too much for a relatively young man. Walter Geist was only fifty-six years old when he died on January 29, 1951.

Because the presidency of Allis-Chalmers had traditionally been held by a general machinery man, it was something of a surprise when the board chose Bill Roberts over W. C. Johnson for the post of president. The decision was based on Roberts's age (fifty-three compared with Johnson's forty-eight), his greater breadth of experience, and his leadership of the most profitable end of the business. Johnson was re-elected to the post of executive vice president to provide support for the new president, but he died less than six months later. Bill Roberts, a different kind of president, was good for Allis-Chalmers. He was a political maverick who often introduced himself "I'm Bill Roberts. I'm a Democrat and I'm from Missouri." He was also a self-made man who rose to win the Horatio Alger Award from the American Schools and Colleges Association in 1952.[22]

In the spirit of the salesman which he had been for so many years, Roberts disliked the routine of desk work. He preferred, like Otto Falk, to spend hours in the product department offices and to brush elbows with the man in the shop. Whereas Falk had called periodic meetings of department heads, Roberts instituted the president's weekly conference on Monday mornings with department and division heads. These meetings kept him abreast of all company activities. Instead of always holding the monthly board meetings at the same secluded director's room, he convened them at the various plants so that he and the directors could gain a thorough view of company operations. All of this was in keeping with his personality and also with his statement issued upon taking office, "The only thing we're thinking about doing is doing a better job in the future than in the past."[23]

In exploring huge, old, reliable Allis-Chalmers, Roberts found that it did not really exist as a corporate entity. It was a loose union of many good but small companies, with each product line practically autonomous in engineering, manufacturing, and sales. Communications between departments and divisions was often minimal and sometimes nonexistent. Although this ungainly and outdated system worked to a degree, it certainly limited flexibility and stifled the ambition of many able men. In a fiercely competitive economy, Allis-Chalmers could not realize the potential inherent in a strong, integrated company. Roberts decided that change must and would take place.

In many companies reorganization would begin with a rash of hiring and firing and the drafting of a new organizational chart. But Roberts did not follow the usual pattern. He fired no executives and filled not a single executive post from the outside. He could not reorganize an organizational chart because he did not believe in one. On the surface this seemed much like the philosophy of Otto Falk. But Allis-Chalmers was quite a different company by the early 1950s. During the four short years that Bill Roberts was in the presidential office, company sales increased nearly $150,000,000 to a total of $492,948,963 in 1954. The corporate name remained the same but the fifty million dollar company of the 1920s had become a half billion dollar company in the 1950s.[24]

While Roberts was improving communications between the two major divisions and between the departments through the Monday morning meetings and his own personal diplomacy, he also spread the managerial load. When he became president in 1951, Allis-Chalmers was probably the only large company in the country suffering from a shortage of vice presidents. Roberts gradually remedied this problem by increasing the number of vice presidents from six to fifteen, with a series of deliberate and purposeful appointments. Robert S. Stevenson, who had joined the company as a tractor salesman in 1933 and worked his way to the head of the tractor division by 1951, was named to the post of executive vice president the following year. Stevenson, one of the company's youngest officers at forty-five, became a true assistant to the president. Reporting to Roberts and Stevenson were two "over-all" vice presidents: Joseph L. Singleton, a graduate engineer, was vice president of the general machinery division and Willis G. Scholl, who began as a tractor salesman in 1936, was vice president of the tractor division. All of these vice presidential appointees, together with staff officers, began to assume precise responsibilities as they improved communications through uninhibited discussion at the Monday morning meetings.[25]

This was a period of amazing growth in sales and staff, when the company vision expanded to bring in a variety of new products and new plants. During his four years in office Roberts persuaded the board to purchase new plants at the rate of two a year. The Canadian Allis-Chalmers, Ltd., at Lachine, Montreal, Quebec, was repurchased, having been sold in the 1930s. The Lachine Works manufactured many of the products found in the general machinery line. The facilities of the Victor Electric Products Company of Cincinnati were purchased in 1952 to provide additional

space for the manufacture of electrical products as Norwood Works
No. 2. In a burst of enthusiasm and optimism, Roberts added plants
at Eling, England, and Terre Haute, Indiana. These plants and
facilities largely provided duplication of existing output, but he
moved to fill out the company lines, particularly in the profitable
tractor division. The La Plant-Choate Manufacturing Company of
Cedar Rapids, Iowa, which manufactured motor scrapers and
wagons, was added in 1952. In 1955 the Gleaner Harvester Company of Independence, Missouri, was purchased to produce giant
self-propelled combines. The Baker Manufacturing Company of
Springfield and Beardstown, Illinois, was also added the same
year, and its complete line of bulldozers and auxiliary equipment
was incorporated into the Springfield Works. Boyd Oberlink, who
had been elected a vice president by the Board and appointed
assistant to the president by Roberts, was responsible for sur-
veying plants and negotiating their purchase.[26]

The acquisition of the La Plant-Choate and Baker companies
was part of a major struggle developing in the heavy tractor and
construction equipment field. In August of 1953 General Motors
had acquired the Euclid Road Machinery Company of Cleveland.
This acquisition by a major competitor was probably the principal
reason why Allis-Chalmers purchased the Buda Manufacturing
Company of Harvey, Illinois in November. General Motors had
been the sole supplier of diesel engines for the Allis-Chalmers
heavy tractors. But Buda, now the Harvey Works, made diesels too,

*The Ontos, armed with six rocket launchers, filled the postwar need of the Marine
Corps for a heavily armed, lightweight assault vehicle. It was built in La Porte.*

and Allis-Chalmers was thus assured of self-sufficiency in that area. Buda had more to offer than that, however. It also produced materials-handling equipment, gas engines, and natural gas engines which helped the company to round out its line in the general machinery field.[27]

During the war Allis-Chalmers had built the Hawley plant at West Allis to manufacture materials for the Manhattan Project. Although work in this field was discontinued after the war, Bill Roberts felt very strongly that the company must not just plan ahead in limited fashion but must project itself well into the future. If it was to be a leading producer of power it had to be involved in atomic energy for peacetime use. Westinghouse, moreover, had entered the field, and Allis-Chalmers had to move to maintain its position. Toward the end of 1954 the company began to work on the commercial applications of atomic energy when in October, the Atomic Energy Commission disclosed that Allis-Chalmers had been chosen to design and build the power generating unit for an atomic power test plant at Lemont, Illinois. This was only the first of many similar contracts. Once again, continued company growth came largely because of an ability to assess and meet new power needs.[28]

With his eyes to the future, Roberts began spending over $5 million a year on research. Under the guidance of Dr. Harry K. Ihrig, Allis-Chalmers research not only promoted company profits but also explored new commercial possibilities. The research

In 1954, the Atomic Energy Commission selected Allis-Chalmers to design and build the power generating units for Argonne, the first atomic power plant near Chicago.

division tackled a number of problems directly related to customers' needs. For example, company engineers developed a process for concentrating taconite iron ore into manageable pellet forms. This was not only a valuable contribution to the industry but, as a result, Allis-Chalmers was selected to build giant machines for the crushing and processing of taconite ore.[29]

Bill Roberts had taken the company in hand, initiated a necessary reorganization of its administration, and reoriented its direction. An inveterate optimist, he once said "My glass is always half full, not half empty." This dynamic man who literally wore himself out in the company's service also provided a new spirit before he died of a heart attack on April 12, 1955. A most perceptive appraisal of Roberts' role in the history of the company was given by a vice president who said, "A man like Roberts was needed at that point—he was a strong man and the company was in a period of transition."[30]

The death of Bill Roberts at the age of fifty-seven took everyone by surprise, but the appointment of his successor was a surprise to no one. Roberts had been an excellent judge of men. He chose carefully, demanded the highest performance and rewarded those who had imagination, ability and energy. Almost from the day in 1933 when Robert S. Stevenson joined the tractor division as a salesman, Roberts's keen eye had followed the career of this young minister's son who had been born in Seattle, Washington, on November 10, 1906. After attending high schools in Spokane,

Allis-Chalmers developed the Grate-Kiln method of iron ore production, which made economical use of lower grade ore. The ore was pulverized and sorted, then formed into pellets and treated with heat to resist breakage during transport to steel mills.

Robert S. Stevenson
President 1955-1965

Washington, and Portland, Oregon, Stevenson studied at Whitworth College in Spokane and was graduated from Washington State College in 1929. He joined Allis-Chalmers four years later as a tractor salesman in the Kansas City branch.[31]

The swift expansion of the tractor-division to national prominence as the third ranking producer provided unlimited opportunity for able men within the division, and Stevenson proved that he could succeed in any assignment handed him by Roberts. In 1935 he was appointed assistant manager of the Omaha branch, and the next year he was transferred to company headquarters in Milwaukee. With the constantly accelerated production of the tractor division in peace and in war, Stevenson moved upward through the division until he was named general sales manager in 1950.

Upon Roberts's election to the presidency of Allis-Chalmers, Stevenson succeeded him as vice president in charge of the tractor division. Later that year he was elected to company's Board of Directors. On the basis of his performance in working with the division's field organization and dealers, and for his role in the introduction of new products, Stevenson was appointed executive vice president in July, 1952. He soon became a member of the Executive Committee of the Board of Directors.[32]

As an articulate and thoughtful corporate executive, Stevenson was called upon more and more often to represent the company in the governmental and public sectors. He appeared before the Congressional Joint Committee on Atomic Energy in July of 1953, advocating an amendment to the Atomic Energy Act of 1946. Later that year Governor Walter Kohler appointed him to the Milwaukee Expressway Commission. Stevenson's agricultural and business experience also gained him memberships on the executive committee of the Farm Equipment Institute, the Farm Equipment Advisory Committee and the Construction Machinery Advisory Committee for the National Production Authority.

When Stevenson became president, he knew that his company was unique. Externally, Allis-Chalmers appeared to be a half-

These direct current, twin drive blooming mill motors, powered by the flywheel motor-generator at the left, were used in the steel industry.

These power transformers are provided with automatic voltage control under load and are cooled by forced-air coolers, thermostatically controlled.

This City of Chicago sewage treatment plant is equipped with Allis-Chalmers pumps and motors.

Large volumes of high pressure air are needed to test modern jet plane prototypes. This Tennessee test center features Allis-Chalmers motor-driven compressors that can supply 600,000 cubic feet of air per minute, with a compression ratio of 2,500 to 1.

billion dollar corporation. But internally the company was made up of about thirty distinct, fairly small businesses. In construction equipment it ranked behind such tough competitors as Caterpillar Tractor and International Harvester. In farm equipment it ranked behind Harvester and Deere & Company. In materials handling it was smaller than Clark Equipment, Yale, Towne and Hyster; in electrical equipment it ran third to General Electric and Westinghouse. Still, no other company in the country could claim its diversity.

Because of the uniqueness of its spread of products, Allis-Chalmers was sometimes thought to be a barometer of business prosperity. But this did not take into consideration the internal condition of the company. In 1954, it netted over 5 percent on sales and 10 percent on stockholders' equity, and it paid two dollar annual dividends on its stock which sold as high as 40 in 1955. But the company had hit a half billion dollar sales plateau which would continue for a decade. The company had not taken advantage of opportunities, it had not been sufficiently reorganized—although Roberts had made a good beginning—and some thought of it as the conservative old gentleman on 70th Street.[33]

The problems Allis-Chalmers faced were not those that could be overcome by a few shifts in personnel and a change in advertising. Progress was only possible through complete reappraisal of the company and its products in relation to the future of American society. But reappraisal and reorganization was to mean a decline in earnings—a dividend rate that was to be cut four times following 1957—and a decline in value of its stock. Things would worsen before they improved because intelligent changes in an established corporation required careful planning and study and a good deal of time.

Only six weeks after assuming the presidency, Stevenson took his first important step in revising operations. He split the two major divisions—tractor and general machinery—into five operating divisions, each with its own profit and loss statement: farm equipment, construction machinery, industrial equipment, power equipment, and general products; the Buda division (later to become the engine-material handling division) remained separate.[34]

Many changes had taken place over a full century, and more were soon forthcoming. The E. P. Allis Company had built its reputation on sawmilling and flour milling equipment and steam engines. The last steam engine, however, had been built in 1930, the sawmill department had been discontinued in 1950, and flour milling offered little future, with the last big mill built about 1960. No regular sales force was maintained for these products. Although the flour milling market was declining, some of the same equipment, such as roller mills and sifters, came to be used more and

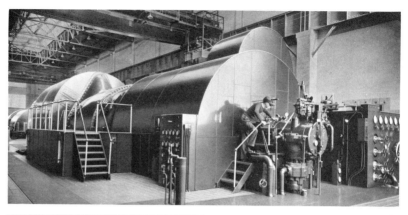

*(Top) In the years 1936-1940, the Port Washington plant of the Wisconsin Electric
Power Company was the world's most efficient. It was equipped with
Allis-Chalmers steam turbine generator units, condensers, and pumps.
(Center) The close-coupled compound steam turbine generator was an
Allis-Chalmers development, as was the center line at floor level unit (bottom).*

more for other purposes. An overall re-evaluation of approach was developing under Stevenson, away from the old orientation toward products to a new approach of market evaluation.

This new approach can be seen in contrast to the "Four Powers" of the old Allis-Chalmers Company—gas, steam, water and electricity. The last and largest gas engine was completed in 1930. Steam turbines had been a principal product since about 1903, and the company had built some of the largest and most famous. Even as late as 1960, Allis-Chalmers brought out a totally new turbine line. But a close study of the market and of profits and losses indicated that the $40 to $50 million in steam turbine orders each year was unprofitable. The utilities were using fewer but larger turbine-generators. By the early sixties the total United States market was down to less than forty units per year, and competitors were trimming prices to get these orders. With only about 10 percent of this market, Allis-Chalmers could not support the space devoted to steam turbines at the West Allis Works. Accordingly, in December, 1962, the company bowed out of this field except for servicing all units it had built, leaving it largely to General Electric and Westinghouse. As Stevenson said, "It was tough abandoning a business we'd been in for 50 years and it was tough taking a $40 million write-off." This was one of Stevenson's difficult decisions that was reversed by subsequent management who later re-entered the steam turbine field under a totally new concept.[35]

Condensers increased in size with steam turbines. An example is this 115,000 square foot single-pass unit on the erecting floor of the West Allis plant.

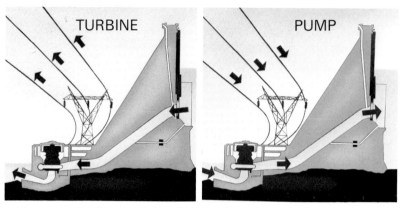

An important development during this period was the pumped storage system for electric power production. During times of maximum power needs, the unit operates as an hydraulic turbine. When less power is required, it pumps water back into the storage reservoir.

The production of aluminum requires vast amounts of low voltage direct current. Much of this is supplied by mercury arc rectifiers such as the line shown here. Later installations were supplied with solid state rectifiers.

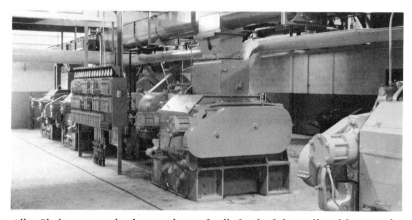

Allis-Chalmers was a leading producer of mills for the flaking of breakfast cereals.

Hydraulic turbines and steam turbines had become part of the company product lines at about the same time. While the latter was losing money for the company by the late 1950s, the former was still profitable. To obtain facilities near the source of heavy steel plate, the company purchased the S. Morgan Smith Company of York, Pennsylvania in 1959 and moved the production of all Allis-Chalmers hydraulic turbines there. The York plant produced hydraulic turbines, pumps, valves, hoists and water control gates. The hydraulic division became one of the leaders in the pump-storage field and, among others, produced twelve units for the Tuscarora project of the New York State Power Authority, then the world's ninth largest hydroelectric project. The company also completed a project that could pump 5,000 cubic feet per second and supply enough power in one hour to last a major city for a full day. There seemed to be a particularly bright future in reversible pump-turbine units. These act as hydraulic turbines during the day when power demand is at a peak, by generating electricity with falling water. At night, when power requirements are low, the generator becomes a motor to drive the turbine as a pump, thus returning the water to storage.[36]

Allis-Chalmers had first entered the electrical field with the acquisition of the Bullock Electric Company of Norwood, Ohio, in 1904 and had been a supplier of motors and power equipment for utility companies ever since. Continually expanding this line through the purchase of works in Boston and Pittsburgh, it has also manufactured all major components for power lines and substations including transmission and distribution transformers and regulators. In 1963 the company purchased the Schwager-Wood Company Inc. of Portland, Oregon, to add high voltage switches to its regular line of switches and switchgear.[37]

The entire electrical industry sustained a sobering blow in 1959 when the federal government brought suit against more than fifty individuals representing twenty-nine companies—substantially the entire industry—for price-fixing activities in violation of anti-trust laws. Four Allis-Chalmers men were charged and the company was indicted in eight of the twenty separate cases. The company and its employees were found guilty, the four men paid a total of $8,500 in fines and the company paid fines of $127,500. The company was also sued by a number of its customers. Stevenson maintained that the men had participated in the meetings and price arrangements without the knowledge or consent of company officers. Official statements by the company had prohibited price fixing and stipulated compliance with the laws during the Falk administration, but had not formally done so in more recent years, although it was assumed that such arrangements were forbidden. To remedy this situation, President Stevenson issued a statement of policy on February 8, 1960, forbidding company employees to become involved in any price, production or marketing arrangements.[38]

(Top) This C-Stellerator unit, built by Allis-Chalmers and RCA, was used in nuclear fusion research at Princeton University.
(Center) Allis-Chalmers also contributed to research on direct current transmission of high voltage electric power with this test facility for the Bonneville Power System. It consisted of rectifier columns, insulating transformers, rectifying transformers, voltage dividers and regulators.
(Bottom) Installation of hydraulic turbine driven generators in one of the plants of the Tennessee Valley Authority.

For most of the mid-twentieth century, the tractor division and its offshoots had provided the bulk of the sales and a substantial portion of the profits for the company. To broaden this line Allis-Chalmers in 1959 bought the Tractomotive Corporation of Deerfield, Illinois from whom it had purchased tractor auxiliary equipment for many years. Production here concentrated on tractor loaders, tractor dozers, and backhoes.[39]

In attempting to secure a wider market, Stevenson sponsored a deliberate thrust into international sales. After spending the early part of 1953 in Europe, he became convinced that there was an ever-expanding market there for American products. On September 11, 1957, he announced the formation of Allis-Chalmers International as a major operating division. The new division was to be responsible for all manufacturing, engineering and sales operations and activities outside the United States and Canada. Of the dozen plants added to the system in the decade following this move, nearly half were abroad. The purchase of Thomas C. Pollard Pty. Ltd. of Newcastle, Australia, provided for the manufacture of construction and processing machinery and centrifugal pumps in the Pacific. The following year Allis-Chalmers acquired Industrial Dufemex, S. A. of Mexico City. In 1963 these operations were expanded and transferred to a new plant in San Luis Potosi. Mexican production has centered on the manufacture of lift trucks. When an ample market for crawler tractors and allied equipment opened in Italy, Stevenson arranged the purchase of Vendor, S. P. A. in Cusano in 1959. The next year a controlling interest in Etablissements de Constructions Mecaniques de Vendeuvre, S. A. was secured. With headquarters in Paris, its manufacturing plant in Dieppe, and a warehouse erected in 1964 at Vendeuvre, France, a new base for the manufacture of engine-generator sets and motor graders was established in the area of the Common Market. In 1963 the company's Canadian subsidiary, Allis-Chalmers Rumely, Ltd. opened a manufacturing facility in Guelph, Ont., to build lift trucks and tractor loaders.[40]

In the United States Allis-Chalmers acquired the Valley Iron Works Corporation of Appleton, Wis., in 1959, and operated it as a wholly-owned subsidiary manufacturing papermaking machinery. In 1965 the subsidiary became the Appleton plant of the company.

Overseas sales have been heaviest in processing machinery, farm equipment and construction, but Allis-Chalmers products can be found across the globe. It has been said that "the sun never sets on the Allis-Chalmers trademark." Plants in seven countries help to provide approximately 15 percent of the total sales of the company. Stevenson's concept of Allis-Chalmers International can be seen in his statement, "We are confident that this approach to our many possibilities throughout the world will provide a means for us to enlarge our stake in the world's power and equipment development and provide new opportunities for Allis-Chalmers people everywhere."

Although Stevenson had expanded the company's venture into atomic energy, by 1965 this field appeared less promising. The construction of nuclear power plants was fraught with cost and developmental problems. With little or no prospect of profit from these activities, the company announced its long-anticipated withdrawal from the field in March, 1966. It continued, however, to produce integral components for nuclear power projects; with the re-entry of Allis-Chalmers into the steam turbine business early in 1970, even greater participation in the nuclear power field became possible.[41]

To insure the place of Allis-Chalmers in the world economy, Stevenson expanded research, development and engineering expenditures by more than 40 percent. One of the most dramatic fruits of the research program was the development of fuel cells which could convert chemical energy directly into electrical energy. Even in their most primitive form, a group of such cells could generate enough electricity to power a vehicle. For demonstration purposes, a group of such cells was installed in an Allis-Chalmers farm tractor, which became the world's first vehicle ever powered by fuel cells. This tractor is now in the Smithsonian Institution.[42]

Of more practical concern however, was the application of fuel cells to power the instruments on board the space capsules launched in the rapidly developing space program under the National Aeronautics and Space Administration. Accordingly, the company sought and won a succession of contracts with both the military and NASA. But after being awarded the major contract it needed to make its fuel cell effort visible, the company suffered the same fate dealt other aerospace firms when NASA cut back its programs and cancelled the crucial Allis-Chalmers contract. Later the company announced the discontinuation of its advanced elec-

This installation near La Crosse, Wisconsin, was the last complete nuclear power plant built by Allis-Chalmers.

The company received a number of contracts from NASA for the development of fuel cell systems for space projects. Here President Stevenson observes one being tested.

The first vehicle powered by fuel cells was this Allis-Chalmers tractor, now at the Smithsonian Institution. Other test vehicles included fork lift trucks.

trochemical products department and sold the fuel cell activity to another firm.[43]

While management was looking into the future, projecting the company's manufacturing and sales abroad and looking at its products in the light of changing markets, it did not neglect the problems of costs and efficiency of production. To support this internal appraisal, outside consultants were called in; their studies led to the plugging of profit drains and increased efficiency through a concentration of manufacturing facilities. In all, four plants, including the rather new Terre Haute transformer works, were promptly shut down, eliminating a floor space of 882,000

This Allis-Chalmers lawn and garden tractor, driven by Milwaukee Braves pitcher Warren Spahn, is representative of equipment built by the Allis-Chalmers subsidiary in Port Washington, Wisconsin.

Tractor shovels and wheel loaders were products of the construction division.

Other construction equipment included crawler tractors and earth movers, shown here on a highway project.

(Top) The agriculture division built harvesters for grains of many types. This is a hillside combine of the early 1960s. (Center) One of the popular farm tractors of the era was the D-21, shown here pulling a multiple gang plow. (Bottom) Many types of lift trucks were built for indoor and outdoor use, such as this unit shown loading lumber.

Willis G. Scholl
President 1965-1968

square feet. Attention was then focused on the old, sprawling West Allis Works. Studies indicated that Edwin Reynolds had planned very well indeed; although the original works was sixty years old, the plant could not only be saved but could still function as the main company works. Some of the top floors were vacated to cut down costly vertical movement in the manufacturing process, while scores of unneeded machines were torn out and sold. The task of modernizing the venerable West Allis facility, with the goal of making it one of the most efficient of its type, proved a continuing program which extended well into the 1970s.

The reappraisal and expansion of products and plants, combined with research and a new international sales thrust, were accompanied by a continuing reorganization of staff. The brief period when Roberts and Stevenson had worked in tandem had been happy, and the concept was not lost on Stevenson when he became president of Allis-Chalmers. Willis G. Scholl proved to be the man with whom Stevenson could best work, and his career had been similar. After joining Allis-Chalmers in 1936 as a tractor salesman at Columbus, Ohio, he rose to the post of manager of the Toledo branch in 1943. In 1947 Scholl was appointed eastern territory manager for the tractor division and in 1951 was elected a vice president. Because of the breadth of Scholl's experience, Stevenson chose him in 1956 for the post of executive vice president. Soon they were, as Stevenson put it, "assigning things back and forth, working out a doubling arrangement." In August, 1959, Stevenson told his Monday morning meeting that thereafter Scholl would act as the chief operating officer, which would allow Stevenson to devote more time to the corporate planning so essential to the development of the company.[44]

More than two years were spent in a thorough study of the entire organization of the company and its products, using the best talent within the company combined with that of outside consultants. Nearly all executives and district or branch managers were interviewed during visits to all of the plants. "We needed an organization to focus on engineering, manufacturing and sales," Stevenson said. To accomplish this he set up in November, 1963, several of the basic operating divisions plus Defense Products and A-C International, each of which constituted a separate profit center. The divisions were reorganized in terms of the markets they served, thus doing away with the old manufacturing-oriented system under which Allis-Chalmers traditionally had operated.[45]

Before the reorganization one sales department had served all of the capital goods products, which meant that the salesmen were unable to specialize or concentrate their efforts. Under the new system each division manager had his own sales department and could direct his salesman's efforts. The company's experience and talents were thus concentrated on the markets they were best equipped to serve. Each division was streamlined to serve its customers better, faster and with less red tape. The restructured divisions became more cohesive units, with more direct authority, shorter lines of communication and less paper work. Among the reorganizations and readjustments in the company's century-long history, Willis Scholl held that the reorganization of November, 1963, "would rate a place right near the top."

The reappraisal and reorientation of the company under Stevenson's leadership began to produce some significant results. On February 23, 1965, Allis-Chalmers announced that 1964 sales had reached an all-time record of $629,067,412 (later restated at $631.2), or 15 percent higher than those of 1963, and that net earnings were 85.4 percent higher than those of the previous year. In January of 1965 the elections of Willis G. Scholl as president and Robert S. Stevenson as chairman of the board were announced. By 1966, sales had risen to $860.2 million and net income was $26.1 million or $2.67 per common share.[46]

The first vibrating screens were developed in the 1930s and their use in mining, crushed stone and other industries spread rapidly. These screens are processing iron ore.

At this point Allis-Chalmers seemed well on the way to re-gaining the success it had enjoyed under General Otto Falk earlier in the century. Certainly, no one could have predicted the rapid series of events which occurred in the closing years of the 1960s. As we shall see, these events posed the gravest threat ever to the management of Allis-Chalmers and threw a roadblock squarely athwart any further progress for the time being.

NOTES TO CHAPTER TEN

1   *Annual Report*, 1941, p. 9, 1942, p. 17.
2   *Annual Report*, 1942, p. 7, 1943, p. 12, 1944, p. 3, 1945, p. 13.
3   Standard Form of Contractors' Report for Renegotiation, RB Form 1, 1942, 1943, 1944; *Annual Report*, 1937, p. 10, 1938, 1939.
4   Department Product Line Statements, 1944, 1945; A-C Mfg. Co. Board of Directors, 1942, 1943, 1944.
5   A-C Mfg. Co. Executive Committee, Jan. 7, May 27, 1918; A-C Mfg. Co. Board of Directors, 1939–45.
6   A-C Mfg. Co. Board of Directors Reports, 1939–40, 1942–45.
7   A-C Mfg. Co. Board of Directors Reports, 1934, 1937, 1940, 1941, 1944; *Annual Report*, 1942, p. 1, 1943, p. 61, 1944, p. 62.
8   A-C Mfg. Co. Board of Directors Reports, 1942.
9   Stephane Groueff, *Manhattan Project*, ms.
10 *Milwaukee Journal*, March 14, 1947.
11 *Milwaukee Journal*, March 25, 1941; *Milwaukee Sentinel*, March 26, 1941.
12 *Milwaukee Journal*, March 14, 1947; *Milwaukee Sentinel*, March 14, 1947.
13 *Milwaukee Journal*, April 3, 7, 1941.
14 *Buffalo Courier Express*, August 10, 1941.
15 Walter Geist to West Allis Employees, March 5, October 17, 1946, March 24, 1947; W. C. Van Cleaf to West Allis Employees, June 19, 1946; *Annual Report*, 1946, p. 3.
16 *Annual Report*, 1946, p. 3.
17 *Annual Report*, 1947.
18 *Milwaukee Journal*, March 4, 1948.
19 R. L. Day and Co. to A-C Mfg. Co., statement, August, 1952.
20 *Annual Report*, 1945–47.
21 *Annual Report*, 1947, p. 10; 1950, pp. 1–2.
22 *Annual Report*, 1950, p. 13.
23 *Annual Report*, 1954, p. 22.
24 *Annual Report*, 1954.
25 *Annual Report*, 1952, p. 13.
26 *Annual Report*, 1950, p. 2, 1951, p. 28, 1952, p. 24, 1953, p. 3, 1954, pp. 3–4.
27 *Annual Report*, 1953, p. 13.
28 *Annual Report*, 1954, p. 4.
29 *Engineering in Action*, 1957, p. 13; *Annual Report*, 1958, p. 13.
30 *Annual Report*, 1955.
31 *Annual Report*, 1955, p. 7.
32 *Annual Report*, 1952, p. 13.
33 *Annual Report*, 1954, p. 2.
34 *Annual Report*, 1955, p. 4.
35 *The Story of One Year at Allis-Chalmers*, p. 23; Allis-Chalmers Product Line Statements, 1950–60; *Annual Report*, 1962, p. 4.
36 *Annual Report*, 1958, p. 3; "Qualified Contractor," March, 1963, statement.
37 *Annual Report*, 1963, p. 12.
38 *Annual Report*, 1960, p. 2.
39 *Annual Report*, 1959, p. 8.
40 *Engineering in Action*, 1957, p. 26; *Annual Report*, 1957, p. 9, 1960, p. 1, 1963, pp. 16–19.
41 *Annual Report*, 1969, inside cover.
42 *Annual Report*, 1959, pp. 2–3; *Engineering in Action*, 1960, p. 3.
43 *Annual Report*, 1962, p. 14, 1968, p. 12.
44 *Annual Report*, 1955, p. 7.
45 *Annual Report*, 1963, p. 6.
46 *Annual Report*, 1964, pp. 26–27, 1966, p. 26.

EPILOGUE

by

C. Edward Weber, Ph. D.

David C. Scott
President 1968–

FIGHTING TAKEOVER, TURNAROUND STRATEGY

The Allis-Chalmers Manufacturing Company appeared to be emerging into the light of a new day in the early years of the 1960s. The respected and venerable firm, having suffered economic stagnation in the previous decade, was being energetically revitalized to meet the demands of the changing industrial world. Sales and earnings were beginning to climb as a result of the thorough reappraisal and reorganization initiated by Robert S. Stevenson and continued under the leadership of Willis G. Scholl. By 1966, sales had risen to $895.8 million and net income to $26.1 million, and prospects for continued growth were promising.

This very success, however, made the company a target for the mergers and takeovers which were forming the giant conglomerates of the 1960s. Allis-Chalmers was soon engaged in a protracted struggle to prevent an unwanted takeover. While the company initially sought a favorable merger as a defense against takeover, it later determined to fight for its independence as a separate corporate entity.

The takeover struggle, coming just as Allis-Chalmers was beginning to prosper under this continuing reorganization, thwarted any further progress until it could be resolved. Only then, its independence and corporate unity assured, could the Allis-Chalmers Manufacturing Company continue the strategies which eventually turned the company around in 1971.

The experience of Allis-Chalmers in the 1960s was part of a quickened pace of mergers and acquisitions which marked the decade and dramatically altered the organization of American industry. Speculation intensified during the sixties as investors bought stock in anticipation of profits due to mergers. This movement gave rise to conglomerates, formed by the merger of firms which did not compete in the sales of their products and which did not sell to each other in significant amounts.

Firms growing in income and sales as a result of conglomeration became favorites of investors seeking a quick appreciation in the value of their stock. Trading in a conglomerate's stock pushed its price up, confirming investor expectations. The higher stock prices gave the conglomerates command over additional capital to exchange their stock for that of "weaker" (lower price of stock relative to earnings) companies. Often the conglomerate borrowed short term to buy the stock of another company, in expectation of repaying its loan from the financial benefits flowing from the acquisition. These benefits depended too often only upon the continued speculative rise in the value of the conglomerate's shares.

A frantic search took place during the 1960s for firms with prospects for improved earnings which had not yet been reflected in the prices of their shares. When a business was identified for acquisition, negotiations commenced, in most cases for the mutual benefit of stockholders of both companies.

However, the acquiring company sometimes sought to take another company over without the approval of its directors. Forced takeovers were undertaken with surprise and often without opportunity for deliberative evaluation by the stockholders of the company being acquired. Stockholders were under pressure to decide quickly on offers to sell their stock, in an atmosphere often charged with recriminations and strong feelings. The quality of management and employee morale suffered because of uncertainty about the company's future, and the potential for strategic management errors increased both during and after the forced takeover.

Once a company was targeted as "weaker" and "ripe" for takeover, the public discussion of this view increased anticipation of takeover. Investors then bid for the stock in hopes that they would be able to sell it at a profit to another trader or, ultimately, to the raiding company. With the momentum created by rising expectations and prices, the targeted company's directors and management were swept toward merger to fulfill stockholder expectations for a quick increase in the value of their holdings. Or to resist takeover, the directors might thwart these hopeful expectations and cause a decline in the value of the stock, which risked the ire of stockholders and speculators alike.

Allis-Chalmers became a target for negotiated merger or forced takeover because of the turnaround in its performance begun under Robert Stevenson and Willis Scholl in the early 1960s. Its attractiveness was enhanced by its prestige as a blue chip, old-line firm with a name that stood for quality and had good prospects for future growth.

Rumors that a tender offer would be made for Allis-Chalmers were reported widely in the press and financial community in 1967. The possibility formed that the company was to be sold to the highest bidder, submerged within another corporate entity, and divided into its several manufacturing and product components.

Allis-Chalmers was on the block, facing extinction as an independent company.

Such rumors and expectations about the company's potential heightened trading in its stock. While the market closed between $24 and $26 per share during the last two weeks of July, 1967, the stock was bid up each day until it reached a high of $44 on August 17. Allis-Chalmers was reported to be the second most actively traded stock on August 3, and it became the most actively traded on August 7 when the New York Stock Exchange placed a 100 percent margin requirement on it to dampen speculation. Yet the Allis-Chalmers stock continued to command widespread interest and be among the most actively traded on the New York Exchange.

One of the companies mentioned in connection with the possible takeover of Allis-Chalmers was Ling-Temco-Vought, Inc. LTV had increased its assets and earnings from previous acquisitions and mergers, and the market anticipated continued growth. The price of LTV stock had multiplied under chairman and chief executive James Ling, who was widely admired as a leading contender for the title of "takeover champion."

On August 10, 1967, Ling informed Allis-Chalmers president Willis G. Scholl that LTV had authorized a tender offer to acquire for cash any and all shares up to 4,750,000 of the common stock of the Allis-Chalmers Manufacturing Company. The price offered was $24 per share. An offer was made for any and all shares up to 143,000 of the 4.20 percent convertible preferred stock of Allis-Chalmers at a price of $150.00 per share. The offer was subject to a resolution by the A-C Board of Directors that the company would not oppose the offer or authorize any corporate action to enjoin or otherwise obstruct it. Ling said that the tender offer would not be made unless such a resolution was adopted, and he gave the Board until 3 p.m. on August 16.

As well as sending a telegram to Scholl, LTV announced this tender offer in Dallas the same day; the announcement was carried widely in the national press that evening and the next morning. There followed many interpretive reports in the press describing the financial success of LTV and Ling and analyzing the attractiveness of Allis-Chalmers to companies such as LTV and its vulnerability to takeover.

Reports in the financial press provided an important source of information to stockholders and potential investors, and the favorable reports on LTV provided Ling a significant tactical advantage in the proposed tender. Without question the tone of reports on LTV's proposed offer was one of excitement and approval. The potential of overwhelming support of Allis-Chalmers stockholders for the takeover of the company could have easily developed as these rumors, expectations and financial "analyses" were reported. A clear indication of stockholder reactions was the rapid rise in the price of Allis-Chalmers stock during August 1967 from $24 to $44.

Allis-Chalmers received the tender offer on a Thursday with only a week in which to react. James Ling, at Milwaukee's Pfister Hotel, said, "We foresee a tremendous product growth in Allis-Chalmers in the decade ahead . . . We happened to have the financial resources to let us afford the luxury of paying a premium price for Allis-Chalmers stock." Ling said that he was impressed with the investment opportunities in the Middle West and particularly in Wisconsin. Ling was reassuring that he did not shake up management or disturb corporate operations after merger; he said, "We feel our biggest contribution to a community is to provide more jobs. . . . We'll make the biggest contribution (to charity) in town. But we'll let someone else run the show."

These dramatic moves built pressure for assent to the proposed tender offer by the Allis-Chalmers directors. The company was being swept into the vortex of merger, and the possibility of remaining an independent company looked slim indeed. Its only alternative to LTV's proposal seemed to be in securing a better proposal from another company.

When the first indications of a takeover attempt reached the ears of Allis-Chalmers management, steps were taken immediately to surround the company with knowledgeable attorneys in this area. It retained the highly respected law firm of Davis, Polk & Wardwell in New York and, all during its problem times, the company had the capable advice of S. Hazard Gillespie, who later became a director of Allis-Chalmers, and John J. McAtee, Jr., another attorney with the firm.

Also retained were the services of George C. Demas and Joe Flom, the latter of the New York firm of Skadden, Arps, Slate, Meagher & Flom. Both were well-known for their background in fighting takeover attempts. From 1968 through 1972 frequent meetings were held in Milwaukee and New York at which the attorneys met with company executives to plan strategies.

They were joined in September, 1970, by Richard M. Fitz-Simmons, vice president, general counsel and secretary of Allis-Chalmers, who came to the company from General Electric. From that time on he coordinated all of Allis-Chalmers legal activities against takeover attempts. It was one of the best legal teams put together for this purpose and it helped to make the Allis-Chalmers fight recognized as one of the finest business strategies ever devised.

Allis-Chalmers chairman, Robert S. Stevenson, announced that the company had been engaged in discussions with General Dynamics Corporation, investigating the merger of the two companies. Stevenson said that, while the discussions had not been conclusive, the Allis-Chalmers directors believed that an agreement would be reached and felt that the long-term interests of all shareholders would be better served by merger with General Dynamics than by the LTV proposal. The Board unanimously decided not to adopt the resolution required by LTV as a condition of its proposed tender offer.

This dramatic move by Stevenson split the vortex of merger into two possibilities, thus lessening their individual intensity. The Board's action delayed merger while Allis-Chalmers clung to its independence. The Sunday newspapers carried feature stories favorable to both LTV and General Dynamics, and they reported thoroughly the merger proposals which had been publicized. Thus the pressure for a precipitous decision was reduced.

Two days after Stevenson's announcement of merger talks with General Dynamics, Ling countered with a six-page telegram in which he proposed to acquire all the outstanding shares of Allis-Chalmers stock. The telegram outlined a proposed offer combining cash and LTV common and preferred stock. Ling said that this offer was worth between $55 and $60 per Allis-Chalmers common share and $183 to $200 per Allis-Chalmers preferred share. Reference was made to a favorable statement about LTV by *Fortune*, and Stevenson and the A-C directors were asked to obtain advice from well-known investment bankers about the future of LTV and the companies which comprised it. Ling stated, "The attitude of Allis-Chalmers directors and leading Milwaukee citizens and the fair-minded treatment of LTV by the Milwaukee press are indeed appreciated." Ling concluded by inviting them to visit his corporate headquarters in Dallas and expressing his wish to become partners with Allis-Chalmers. His second proposed tender offer was also contingent upon a favorable expression by the A-C Board within four days.

LTV's proposal, telegraphed on a Monday, was also reported nationally in the press. The reports covered the telegram fully, which gave Ling widespread coverage of his arguments and increased the momentum for assent by Allis-Chalmers. On the other hand, the views of the Allis-Chalmers management and directors were not communicated because they stated only that they were studying the offer.

Allis-Chalmers common stock had closed at $37 per share on the New York Stock Exchange the Friday preceding LTV's second offer. On Monday, the day of the second offer by LTV, Allis-Chalmers was the most actively traded stock on the exchange and closed at $39⅞, although the Dow Jones industrial average dropped 4.33 points to 916.32. The stock reached its high of $44 per common share on Thursday, August 17 and closed at $42¾, when it was again the most actively traded stock. There were obviously many buyers speculating on quick gains to be made by holding Allis-Chalmers stock at the time of merger.

The publicity attendant to the verbal exchange between Allis-Chalmers and LTV had placed great pressure on both A-C's Board and on Stevenson to accept LTV's proposal or to show an alternative by which A-C stockholders would benefit more. The pressure can be surmised by the number of transactions for the purchase of Allis-Chalmers stock. There were over 2.6 million shares of common stock exchanged between the end of July and the middle of August 1967. This volume is large in light of the 9.8 million of A-C common stock outstanding at the time. The increasing volume of transactions in A-C stock can be seen from changes in weekly volume: 109,300 shares during the week of July 24; 720,500 shares during the week of July 31; 879,600 shares during the week of August 7; and 1,323,500 during the week of August 14 (including August 18). Obviously this does not represent an equivalent turnover in ownership since some stocks may have changed hands several times during the period; but it does suggest rising expectations on the part of Allis-Chalmers stockholders for quick gains in the value of their stock.

There was yet another complication in the momentous week. A few statements of cautious concern about the rising tide of conglomeration began to be heard above the hum of excitement. Some newspaper editorials questioned the implications of conglomerates taking over companies, and one paper called companies such as LTV "wolf packs of corporate raiders." Thomas F. Nelson, Wisconsin's Commissioner of Securities, said that he would take a further look at LTV's proposal in light of Wisconsin law which required the filing of detailed information before any solicitation of stockholders.

Stevenson stood firm against the pressure, however, and on Thursday, August 17, announced that the A-C Board of Directors had unanimously refused to give favorable expression to the proposal by Ling-Temco-Vought, Inc. Stevenson said:

> "We would have grave concern as to the future health and continuity of the business under the proposed new ownership. We also question the intrinsic value of the securities being offered Allis-Chalmers shareholders in exchange, and their realizable value compared to the value claimed by LTV."

Stevenson criticized the "unusual mode of action" followed by LTV in publicizing proposals which did not require Allis-Chalmers Board action but nonetheless were contingent upon such action. He warned that the A-C Board would examine any future LTV proposal carefully and would make appropriate comments in light of the facts as they then existed.

Stevenson also filed a complaint against LTV with state Securities Commissioner Thomas Nelson:

> "Allis-Chalmers requests you to invoke your authority under SEC 189.11 Wis. Stats., to prevent further solicitation by Ling-Temco-Vought, Inc. to the shareholder of Allis-Chalmers for sales of LTV stock. We believe that the telegram of James J. Ling, Chairman of LTV, dated

August 14, 1967 and the attendant publicity given by LTV thereto may have amounted to a solicitation in violation of Wisconsin Law. In our opinion the telegram and attendant publicity appears to be misleading, had led to confusion of Allis-Chalmers stockholders and has had an unsettling effect on the market for Allis-Chalmers stock."

On the same day, Nelson prohibited LTV from further solicitation of Allis-Chalmers for its stockholders in the state. Nelson's

(Top) Low voltage switchgear like this and high voltage switchgear for industrial and utility applications are standard products from the West Allis plant. (Bottom) A new gas insulated power breaker serves at an electric utility substation.

order stated that all securities of LTV being offered in exchange for securities or assets of Allis-Chalmers must be registered before any sales or solicitations thereof were made. Such registration would have required substantial disclosure to the commissioner.

The commissioner's order heralded later action taken by the Federal administration to resist conglomerate mergers and take-overs. Stevenson's press release and letter to A-C stockholders were reported widely, as was the state commissioner's order and the reasons behind the Allis-Chalmers rejection of the LTV proposal.

One day after Allis-Chalmers rejected LTV's offer, James Ling withdrew his proposal. One can surmise that Ling-Temco-Vought might have been under pressure to complete the merger quickly or break off discussions. The company's reputation was built upon successful mergers and increasing income, and a protracted fight could have damaged its reputation and halted the growth in the price of its shares. Also, LTV was reported to have the backing of financial institutions in the first proposal and, since this type of borrowing was probably short-term, the company would have been pressured to repay the loan or release the line of credit. The market and financial community were not prepared, it would seem, to watch a long siege, but they probably expected quick gains from their investments in Ling-Temco-Vought stock.

The weeks after LTV's withdrawal were turbulent for the Allis-Chalmers Manufacturing Company, as it seemed that merger with another company was imminent. The market anticipated quick profits from A-C stock. The price fell from its high of $44 on August 17, but it remained above $38 with volume continuing abnormally high.

Allis-Chalmers could make no headway in the speculative wake following LTV's withdrawal. The company was beset with grave uncertainty of its future direction, and the momentum from its earlier turnaround was lost. Stevenson and the directors were under pressure to find a merger partner to ward off an unwelcome takeover and fulfill expectations in the financial markets. Employees and customers were raising serious questions: would the company be merged and with whom? What change in leadership would result from some future merger? The management and other employees had no sense of future direction and control. It seemed likely that merger would come, but the questions of with whom and when remained.

The market speculation about Allis-Chalmers common stock continued to make the company a prime target for takeover. The high level of trading sustained in August, 1967, continued in September. A-C common stock was either the most actively traded or among the ten most active stocks on the New York Stock Exchange during September, with its price in the mid and upper 30s.

General Dynamics and Allis-Chalmers had said that negotiations were continuing. There had been rumors that the proposed merger would provide A-C stockholders $45 per share, but General Dynamics had said the price would need to be about $32 per share of Allis-Chalmers common stock. Early in September, the two companies announced the termination of negotiations toward their merger. No reason was given for ending negotiations, but speculative trading in the market for A-C common stock would have, without doubt, raised the expectations of A-C stockholders beyond what General Dynamics was willing to pay.

In the middle of September, Kleiner, Bell & Company, a securities firm, confirmed reports that it had been a substantial buyer of A-C stock for its own account and for those of some clients. The firm was known to have been involved in gaining substantial interest in such companies as Boston & Maine Corp., Unexcelled Inc. and Studebaker Corp. Kleiner, Bell had initiated a proxy fight in connection with its investment in Studebaker stock, and its announcement quickened the winds of merger about Allis-Chalmers. As the representative of substantial stockholder interest, the securities firm was to become a catalyst for either merger or takeover.

Allis-Chalmers had no merger deal to provide quick profits for short-term investors in its stock; instead, the company announced that sales and profits were down in the first three quarters of 1967 from the same period in 1966. Earnings for the first three quarters were 82¢ per common share in 1967, compared to $2.17 per share for the first nine months of 1966, and earnings were only 1¢ per share in the third quarter.

Stevenson continued to search for a merger partner, and fruitful discussions were undertaken with Signal Oil and Gas Company. Allis-Chalmers and Signal Oil & Gas "approved in principle" a "definitive" agreement to merge which provided for a tax-free exchange of stocks. One share of Signal's Class A common stock and one share of its new $1.10 dividend cumulative convertible preferred stock would be exchanged for two shares of Allis-Chalmers common stock. Major newspapers estimated the value of the exchange from about $42 for each share of Allis-Chalmers common stock to $47. Signal said Allis-Chalmers would operate as an independent subsidiary following the planned merger.

Forrest M. Shumway, president of Signal, said that Allis-Chalmers management "felt some uncertainty and instability and felt they would have to take action." He said that the merger would be in keeping with Signal's policy of diversifying "only into well managed companies" and that he was impressed by some of the fields with "spectacular potential" where Allis-Chalmers had increased its operations.

A spokesman for Kleiner, Bell said that his firm was not the finder in the Allis-Chalmers—Signal deal. The next day, Kleiner,

Bell was reported to have expressed displeasure with the proposed terms of merger between Allis-Chalmers and Signal Oil & Gas. Stevenson responded, "I certainly can't see how Kleiner-Bell's interests as Allis-Chalmers stockholders could be served by opposing a merger of Signal Oil & Gas and Allis-Chalmers." There were recurring reports on Wall Street that other companies, still interested in acquiring Allis-Chalmers, were discussing such proposals with Kleiner, Bell. The possibility was raised by "Wall Streeters" that Kleiner, Bell was attempting to arrange a deal to get $48 per A-C share price.

A new note introduced into the public rumors possibly influenced the direction of merger and takeover. Contrary to LTV and Signal's favorable comments about A-C's performance and prospects, this new tone suggested a moribund Allis-Chalmers. The refrain was taken up by others, as one analyst said that the company sorely lacked strong management, the ability to control costs and a cohesive organization. Kleiner, Bell & Co. amplified this refrain, when Burt Kleiner said that Allis-Chalmers was a nineteenth century company with a very erratic earnings record.

Opposition to the merger intensified, and in a press release, "Signal Oil & Gas Co. and Allis-Chalmers Manufacturing Co. advised that their intended merger announced last November 28 has been called off." Burt Kleiner was reported to have been considering a proxy contest for control of the Allis-Chalmers Manufacturing Company. He said "We are presently doing a great deal of thinking about Allis. We would prefer to have a deal made but if there is no deal made, we have a fiduciary responsibility to stand up and represent our stock."

Moving quickly, Stevenson asked City Investing Company to explore merger. In March, 1968, he and City President George T. Scharffenberger announced agreement in principle to merge the two companies. Allis-Chalmers stockholders would have two alternatives under the proposed agreement: to exchange one share of Allis-Chalmers common stock for 7/10 of a share of City common stock and 1/10 of a share of City cumulative preference stock with an annual dividend rate of $6 per share and a redemption price of $100 per share, or to exchange one share of Allis-Chalmers common stock for 85/100 of a share of City common. Scharffenberger described his company's acquisition program as the conversion of wealth-potential assets into earning-generating assets since the company, the owner and operator of real estate, was selling properties to acquire other types of businesses. Allis-Chalmers was to be absorbed intact into the corporate structure of the holding company. Peter Huang, a vice president of City Investing, is quoted as having said that A-C had great assets and that City Investing had strong earning power.

This agreement had the approval of Kleiner, Bell and Company. Burt Kleiner said, "It's a very progressive step; I'm very excited about it." He stressed in a telephone interview with the *Milwaukee Sentinel* that he had played an active role in bringing the New York City holding company and the giant manufacturing firm together. Indeed, Kleiner, Bell was said to ". . . have gone so far as to threaten a proxy fight to throw present management out if it refuses or fails to find a new buyer willing and able to buy out management and stockholders alike."

The response from the stock market was not so favorable to the proposed merger. City Investing's stock fell 18.8 percent after the announcement of the agreement on March 13 until March 26, 1968. Allis-Chalmers stock, however, held in the mid 30s with heavy trading during the latter half of March. It was reported that many Wall Streeters regarded Allis-Chalmers as a "problem company" and believed its acquisition would seriously impair City Investing's earnings growth.

Despite a vigorous defense of the proposed merger by Stevenson, Kleiner and the representatives of City Investing, the financial community remained negative. Consequently, Allis-Chalmers terminated merger negotiations with City Investing.

Only seven days after the termination of these negotiations, the Allis-Chalmers management faced the annual stockholders' meeting. The market was waiting, and many stockholders were anxiously anticipating a merger or takeover of Allis-Chalmers to realize a gain on their stock purchases.

Nonetheless, the financial community was surprised when Gulf

The Allis-Chalmers HD-41 was the largest crawler tractor manufactured in the world.

& Western Industries announced its intention to tender an offer for Allis-Chalmers stock. Gulf & Western proposed to file a registration statement with the Securities & Exchange Commission covering a tender offer for up to 3 million shares of the A-C common stock. The offer was $11.50 in cash, $12.50 as the principal amount of a 6 percent subordinated twenty year non-convertible debenture, and 9/10 of a ten year warrant to purchase Gulf & Western common stock at $55 per share for each share of Allis-Chalmers common stock. Gulf & Western estimated the value of its offer at approximately $39.50 to $40 per share of Allis-Chalmers common stock. One of the dealer managers for this tender offer was Kleiner, Bell and Company. Later, a Gulf & Western prospectus showed that Kleiner, Bell and its customers owned or controlled more than 1.6 million shares of Allis-Chalmers common stock. A Gulf & Western spokesman informed Allis-Chalmers that the shares were being purchased for investment purposes only and that they would not seek representation on the Allis-Chalmers board. With earlier purchases of shares in a variety of companies, Gulf & Western had captured the attention of investors seeking market gain from conglomerate growth, and this new offer stirred additional interest.

Stevenson's contact with Gulf & Western was reportedly limited to telephone conversations on the day before the stockholders meeting. At the meeting, he said, "We were assured that the stocks were being purchased for investment, that they (Gulf & Western) weren't seeking control and wouldn't seek representation on the Board. We aren't well acquainted with them and aren't in the position to comment on the investment value of the securities they will offer." Stevenson declined to make any evaluation of the Gulf & Western offer in response to stockholder questions. He said, "Directors haven't had time to study the proposal." However, there were Wall Street reports that Gulf & Western would seek control of or eventually acquire Allis-Chalmers. The meeting was stormy, with stockholders demanding more information and expressing anger over the direction of merger activity.

Gulf & Western filed a registration statement for a tender offer with the Securities & Exchange Commission and tendered an offer for 3 million shares of Allis-Chalmers common stock. Kleiner, Bell tendered all shares it owned or controlled and recommended acceptance of the offer by its customers who held Allis-Chalmers stock.

The Federal Trade Commission had, however, advised Gulf & Western Industries, Inc. that it opposed the proposed purchase of 3 million shares of the 10.3 million outstanding. Gulf & Western responded that it disagreed with the FTC position and intended to resist vigorously any attempt to compel it to divest itself of the Allis-Chalmers stock.

The FTC complaint alleged that Gulf & Western acquisition of A-C stock constituted a violation of Section 7 of the Clayton Act and

Section 5 of the Federal Trade Commission Act. The FTC declined to comment on its complaint to Gulf & Western but announced publicly a broad study of the trend toward conglomerate mergers. The complaint against Gulf & Western, part of the widening concern of the Federal government about mergers, elaborated the refrain of the Wisconsin Department of Securities and expressed the new underpinning in Federal policy toward conglomerate mergers.

Gulf & Western Industries acquired an additional 370,000 shares in the summer of 1968, gaining almost one-third ownership of Allis-Chalmers. More than 38 percent of the shares were tendered in July, but Gulf & Western limited its purchases to the 3 million shares on a pro rata basis. By the middle of September the price of Allis-Chalmers stock had fallen below $27, although Gulf & Western had paid about $38 on its first lot of Allis-Chalmers stock. While Gulf & Western acquired sufficient stock for working control of Allis-Chalmers, its investment shrank as a consequence of the decline in the stock's market value.

The Gulf & Western management announced at a stockholder meeting its intention to seek a meeting with the Allis-Chalmers management. Their objective was to discuss A-C's future and, perhaps, Gulf & Western representation on the Allis-Chalmers Board. Later, Charles G. Bluhdorn, chairman of Gulf & Western, declined to say whether the company intended to seek representation on the Board, but he said, "We hope to help out in any way we can." David N. Judelson, president of Gulf & Western, was quoted by *Metalworking News* as saying, "I'll leave that to your best imagination" on whether his company would seek representation on the A-C Board. When asked on another occasion about an eventual merger with Allis-Chalmers, Bluhdorn declined to comment but noted that Gulf & Western had bought 370,000 additional shares since its initial tender offer.

While Gulf & Western seemed to be positioning itself for its next move, the Allis-Chalmers Manufacturing Company chose a dramatic new set of tactics. Since various mergers had been explored but had failed to produce favorable results, the company, under direction of a new executive officer, began to struggle to retain its corporate independence.

Willis G. Scholl, president of Allis-Chalmers since 1965, retired in June, 1968, and David C. Scott was elected to succeed him. Scott, taking into consideration the turmoil which Allis faced, said, "I am prepared for whatever comes." A short time later, Scott said that he did not intend to give up his future in a merger: "I didn't come into Allis to end up as the head of the Allis Division of XYZ Corp." Scott had a reputation for turning operations around; his appointment, especially with his selection from outside the company, indicated that Allis-Chalmers would not wait complacently for the next move by Gulf & Western.

Scott, fifty-two years old, had been executive vice president of Colt Industries, Inc. before coming to Allis-Chalmers. A native of Kentucky, he had formed his own engineering company after graduation from the state's university. He joined General Electric after naval service in World War II and was appointed general manager of the cathode ray tube department in 1960. GE presented him the Charles E. Coffin Award for his developmental work on the automated assembly of electronic tubes. He was appointed vice president and group executive of Colt Industries in 1963 and elected executive vice president of operations in 1965. It was at Colt that his ability to turn operations around gained widespread attention. When Scott was elected president of Allis-Chalmers, Stevenson said, "We are confident that he is particularly well qualified to take over the president's job of this great outfit."

Gulf & Western's next move after the appointment of Scott signalled an abrupt change of course. President David Judelson told security analysts that his company might dispose of its A-C shares. He said "that it was not their policy to press ahead with investments in companies where the management did not welcome their entrance or where there were legal difficulties." Allis-Chalmers Chairman Stevenson said, "We had never indicated in any way that Gulf & Western's investment in Allis-Chalmers shares was unwelcome." Gulf & Western would have, however, experienced a loss in excess of $33 million if its A-C stock were sold at the then current market price of about $27. Later, a spokesman for Gulf & Western said that Judelson's talk with financial analysts was "vastly overplayed" and that there were no plans to dump Allis-Chalmers stock.

Gulf & Western had adopted an ambivalent posture about its Allis-Chalmers stock in the financial community. The financial reports from Allis were grim, and the price of its shares had

New designs place Allis-Chalmers well in front of the agricultural equipment scene. A new AIR CHAMP planter works with a fully dual tire diesel tractor.

declined. In early 1968, Allis-Chalmers reported a substantial decline in earnings for 1967. An improvement had been anticipated, but releases in 1968 suggested an even more grim situation in that year. For the nine months ending in September, 1968, the company reported net income of only $713,000 while it had had net income of over $26 million in 1966. Further, the company reported that a study was underway involving plant relocation, rearrangement and consolidation, discontinuation of certain product lines, and an evaluation of inventory and inventory prices. This study, it was reported, could result in a very sizable loss for the full year of 1968. In October, the company said that it expected a pre-tax loss of $50 million for 1968.

After expressing its ambivalence to financial analysts in Detroit, Gulf & Western received inquiries from possible buyers of the stock. It seemed likely that a sale would occur, but people at Allis-Chalmers could not know in which direction to look for a possible takeover. This uncertainty enveloped Allis-Chalmers like a shroud of gloom, and corporate death seemed only a matter of time.

This uncertainty was dispelled by an announcement over the wire service on October 31, 1968, that White Consolidated Industries Inc. had agreed to purchase Gulf & Western's block of Allis-Chalmers common stock. Payment of $35 a share in addition to 250,000 shares of White Consolidated stock would give White ownership of about 33 percent of the total outstanding A-C common stock, enough for working control.

Allis-Chalmers declined to comment, but the financial community did not. Its expectation was that White Consolidated was about to take Allis-Chalmers over. One industry source said, "Put it this way—somebody is going to acquire Allis-Chalmers sooner or later. They (White Consolidated) now hold the tickets and they'll walk through the door." The president of White Consolidated, Edward S. Reddig, was seen as tough, shrewd and salty, with an ability to turn "ailing companies" around, "often with plenty of blood being spilled in the process."

Allis-Chalmers may have been silent about the impending takeover, but governmental agencies were not. Five days after White Consolidated announced its agreement to purchase A-C stock, Wisconsin Securities Commissioner Nelson warned that state law could be implemented were White Consolidated to make a tender offer for the remaining Allis-Chalmers stock. Any exchange of White Consolidated stock for A-C stock would require his approval "if there were a legitimate complaint about its fairness."

Within days the Federal Trade Commission again stated its interest in an independent Allis-Chalmers Manufacturing Company. Earlier the Commission had issued a notice of intent to open a formal proceeding into Gulf & Western's ownership of A-C stock. The FTC telegraphed its concern to A-C president Scott that White

Consolidated and Allis-Chalmers take no further steps to increase White Consolidated's holding or to merge the two companies:

> . . . Accordingly, it is requested that no further steps be taken by either White or Allis-Chalmers to increase White's holdings or to merge the two companies or in any other way to change the status quo until this division is furnished with complete information concerning the acquisition and can report its findings to the commission for appropriate action. May we have your prompt acceptance or rejection of this request. Thank you for your courtesy.

The FTC notice was issued after an agreement was made between White Consolidated and Gulf & Western, but before the agreement was consummated by the actual purchase. The agreement was completed on December 9 despite the FTC request to take no further action. A White Consolidated spokesman said, "It was only a request, though. It was not an order. Our attorneys studied the FTC notice carefully and decided there was no reason to hold off on consummation of the purchase."

About the time that White Consolidated was making its first moves to take over Allis-Chalmers, the Federal administration was phrasing its public arguments against conglomerate mergers. The U.S. attorney general and his assistant for anti-trust enforcement had expressed deep concern over the merger movement and were prepared to use their statutory anti-trust authority to meet all threats to competition posed by conglomeration. Industry was provided new merger guidelines. The Federal Trade Commission called for special reports from large companies on planned mergers, and the attorney general said that the Justice Department would probably oppose any mergers by the top 200 manufacturing firms.

Allis-Chalmers management and directors had said nothing about Gulf & Western's tender offer and White Consolidated Industries' purchase. They had declined to comment, refused to evaluate Gulf & Western's tender offer and asserted that they had had no conversations with White Consolidated Industries about merger. Taciturn Allis-Chalmers did speak, however, on December 19, 1968.

On that date, the Allis-Chalmers Manufacturing Company filed suit in U.S. District Court at Wilmington, Delaware, to stop a takeover attempt by White Consolidated Industries. Allis-Chalmers asked that the court enjoin White Consolidated from voting any stock it owned or controlled in any election of directors to the Allis-Chalmers Board; from acquiring or attempting to acquire any representation on the Board; from acquiring or attempting to acquire any additional shares of Allis-Chalmers stock; from soliciting, by public announcement or otherwise, in any effort to acquire more stock in Allis-Chalmers; and finally, from selling its Allis-Chalmers stock in a manner that would injure Allis-

Chalmers or restrict its vigor as a competitor. Allis-Chalmers alleged in its complaint that White's publicly declared attempt to take over the business of the plaintiff would violate U.S. anti-trust laws.

The U.S. District Court granted a temporary restraining order until it could hold a hearing on the Allis-Chalmers motion for a preliminary injunction; but on January 22, 1969, it denied the injunction request and refused to reconsider the case two days later. Allis-Chalmers expressed determination to press its case vigorously in an immediate appeal to the U.S. Circuit Court of Appeals in Philadelphia, but the Appeals Court denied a petition for a temporary restraining order while the appeal was pending. Therefore, White Consolidated was now free to take actions to gain control of Allis-Chalmers.

After the U.S. District Court had lifted its restraining order, White Consolidated had about three and one-half months before the annual meeting of Allis-Chalmers stockholders scheduled for May 14, 1969. There were about 10,700,000 shares of common stock, each with one vote, and 450,000 shares of convertible preferred stock, each with four votes. The maximum votes to be cast at this meeting, accordingly, was about 12,500,000. White Consolidated had 3,248,000 shares of common stock which gave it 26 percent of the total possible vote. One could have anticipated a vote of at least 2,175,000 to support the Allis management, about 17 percent of the total. The issue would be determined by the balance

The electric high lift order picker for warehouses carries the operator to where his load is to be selected.

of the shares of 7,750,000 votes, about 57 percent of the total possible votes. These votes were distributed among institutional investors and nearly 40,000 individual stockholders.

White Consolidated had the alternatives of negotiating a take-over with the extant A-C management, gaining the proxies of at least 3,250,000 common shares for the coming annual meeting, or making a tender offer to gain at least this number of shares. A successful tender offer would have insured a majority of stockholders favoring White Consolidated's takeover, but it would have been expensive. They had already borrowed to buy their block of stock from Gulf & Western, and the financial markets were less sympathetic than a year before to speculative lending. Also, a premium above market or book value might be necessary for a successful tender offer. Negotiations with A-C management probably did not appear fruitful to White Consolidated, since the management had indicated its intention of keeping Allis independent. In any event, White chose not to make a tender offer and instead to attempt to persuade a sufficient number of stockholders to turn over their proxies. With a majority of proxies White Consolidated could elect its own slate of directors. A victory for White at this point, however, would not necessarily have resulted in its takeover of Allis-Chalmers. As a provision of the convertible preferred shares, these holders would have had to approve the takeover in addition to the approval of the holders of common shares. Yet, the possible opposition of those with preferred shares might have been ameliorated by an A-C Board sympathetic to White

Gleaner combines are a standard of the harvesting industry. Used in contract harvesting they work from Texas to Canada.

Consolidated, along with a suit by White Consolidated to undo the purchase of the preferred shares.

White Consolidated apparently had no chosen strategy when the restraining order was lifted and it received the A-C stockholder list. White president, Edward S. Reddig, said, "We'd like to move right along as fast as we can (to acquire Allis-Chalmers) . . . Obviously we could not make any plans (during the litigation in Wilmington) . . . but we will start work on it immediately." Allis-Chalmers had appealed the District Court's lift of the temporary restraining order, but Reddig speculated that it "may take a week or so" to dispose of the Allis-Chalmers appeal.

At this critical time, Robert S. Stevenson retired as chairman of the Allis-Chalmers Board of Directors and was succeeded by David Scott. Stevenson ended a thirty-five year career with Allis-Chalmers after having had a major part in recruiting Scott and giving him full rein to reshape the company into a profitable organization. This responsibility now included development of tactics to oppose White Consolidated's imminent takeover attempt.

Edward Reddig now said, "Our first order of business will be to get that board (A-C's)," and stated that White Consolidated was preparing a proxy solicitation and ". . . may tender for A-C shares later if efforts to seat new directors are successful." Reddig expected no legal problems from the Federal Trade Commission, and he was not concerned over cautious statements by Federal agencies and the New York Stock Exchange about conglomerate merger activities.

Without an acquisition plan or tender offer, White Consolidated's case to A-C stockholders had to be negative in approach, arguing that the A-C directors and management needed to be replaced as incompetent. But White Consolidated was unable to offer A-C stockholders any assurance that their shares would increase in value if they voted for Reddig. The stockholders knew their shares were worth less than a year ago, but they were not told their future value should Allis-Chalmers be acquired. Stevenson's move to retire also weakened White's position by forcing a shift of focus from the company's recent decline in earnings and stock prices to conjecture about how the new management would perform.

Reddig finally wrote to A-C stockholders on March 28, soliciting their proxies for a slate of new directors who were also directors of White Consolidated. In his letter, he criticized the $122 million loss before taxes in 1968 and the discontinuance of dividends. He stated that the stockholders' position had been subordinated by the recent sale of 450,000 shares of convertible preferred stock by Allis-Chalmers to certain foreign and U.S. investors. Finally, he argued that White's directors could do more for A-C stockholders because White was more profitable and because the new directors, once elected, would propose an exchange of Allis-Chalmers securities for White securities.

Scott wrote in turn to solicit stockholder proxies to reelect the incumbent directors. He argued that it was in their best interest to reject White's attempted takeover, saying, "It is clear to us that White must gain control of Allis-Chalmers in order to pay off its debt incurred in buying Allis-Chalmers stock. . . . I have taken significant steps to put this company on a sound footing. The details . . . include: top management changes—streamlining of staff—plant consolidations—new acquisitions—joint ventures—new products—additional equity capital, and, finally, the facing of financial reality." He went on, "For the first two months of the current year, Allis-Chalmers had the second highest earnings for any comparable period for more than a decade—$3,900,000 before provisions for Federal income taxes. . . . We believe that the second quarter and full year results will show the positive effects of our new management." Scott warned stockholders, "We are concerned with what White might do with Allis-Chalmers if it gains control, especially in view of what White admitted with respect to its need to acquire Allis-Chalmers promptly in order to meet its pressing obligations incurred in acquiring Allis-Chalmers stock. You should ask yourself whose interest White will be serving—theirs or yours."

Scott continued his counterattack in successive letters to stockholders. He emphasized that the slate of persons nominated by White was "handpicked" by the company. He warned, "With its own nominees in control of your company, White would be in effect dealing with itself. . . . You should ask yourself whether White's nominees would be likely to get the best price for Allis-Chalmers stockholders in any transaction." Scott told stockholders that sales and earnings were up and costs were down, that 3,400 non-production employees had been removed, that excess plant capacity had been reduced and significant economies had been introduced to lower the break-even level.

Allis-Chalmers placed full page advertisements in the *Wall Street Journal* and other leading national newspapers which repeated the thrust of the appeal for proxies. The ads announced a program for profitability after "major surgery." Again, Scott wrote to the stockholders, this time shortly before the annual meeting was to take place, hitting hard at White's position and urging stockholders to endorse the management proxy.

Both Stevenson and Scott realized that any takeover attempt would require a well-organized counter-attack by Allis-Chalmers in which they would rely heavily on communications. For this job they turned to incumbent corporate vice president Charles W. Parker, Jr., who was also a former salesman and sales manager in the company's field sales operations.

He quickly organized all facets of the company to ward off any takeover attempts. Parker and his staff devoted full time, day and night, to staying on top of the situation, watching carefully for anything that was said or done that would suggest the company should put out a letter to the company's shareowners or issue a release to the news media. Company executives, including top management, attorneys, financial people, advertising and creative arts people, were kept informed of happenings so that they could react in concert to the various situations which arose.

The company, when action was taken against it, seized the initiative itself as soon as possible and issued stories in newspapers, news magazines, and other forms of communications throughout the country.

As salesman, Parker knew the value of having a ready-made Allis-Chalmers sales organization with offices in the major cities throughout the United States. Shareowner lists were broken down to geographical areas, and heavy capital equipment and farm equipment salesmen were quickly turned into shareholder relations people. They were trained to make personal calls to select shareholders in their areas, keeping them informed of the latest happenings and urging them to back the present company management.

Community people also were kept fully informed of the Allis-Chalmers story through the company's plant managers. They were organized into a communications unit contacting city, county and state governmental representatives. They also enlisted the aid of their purchasing agents to keep suppliers informed about what was happening and to obtain the backing of their resources.

For processing, metal bearing ores are ground to powder in ball and rod mills such as these.

Communications were carried on with bankers throughout the country, who would prove so important to the company. The aid of security analysts and brokerage and other financial organizations was asked for and secured. Allis-Chalmers received a vote of confidence from all areas of business and industrial life.

Special systems set up increased the communications efficiency in the writing, approval and issuing of news releases, advertisements, letters to stockholders, phone communications and personal calls enabling the company to not only react to certain situations but to seize the initiative of the situation demanded. The cooperation of printing, mailing and other services was sought and received, and these resources remained on a twenty-four hour alert during all of the takeover problems.

By the time White made its initial offer, the company had this complex yet efficient organization in operation and it was ready to carry on a communications battle against anyone who attempted to take over Allis-Chalmers.

White Consolidated countered with its own letters, advertisements and news releases. Reddig argued that the preferred stockholders now came first at Allis-Chalmers, saying that the preferred stock had been sold to "certain Rockefeller and Fiat interests." White Consolidated's continually improving performance over the last five years was detailed. In a later letter Reddig wrote to "our fellow common stockholders" at Allis-Chalmers, continuing the themes of the earlier letter and criticizing the financial arrangements pertaining to the employment of A-C President Scott. The letter was followed by others on May 1 and May 9, 1969, which continued the attack on the A-C management, directors, and Scott. They argued, again, that White had performed better than Allis-Chalmers, and that White's management could achieve good performance for Allis.

The fight for proxies was a confrontation of financial accounts as well as an exchange of words between the managements of Allis-Chalmers and White Consolidated. White Consolidated opened with President Edward Reddig telling the *Dow Jones News Service* that 1968 earnings had risen to about $27,900,000 or slightly more than $2.50 a share. Making no comment about White's opening volley, A-C President Scott acknowledged the $122 million before tax loss in 1968 and announced that International Fiat Holding S.A., Switzerland and certain Rockefeller interests had purchased 450,000 shares of convertible preferred stock for $45 million. Both companies took full page advertisements to proclaim their improved financial performance. Two Milwaukee-based securities firms evaluated Allis-Chalmers and recommended the purchase of its stock for capital appreciation, while another firm believed that White Consolidated offered investors an excellent buying opportunity.

Other actions and events were taking place around White and Allis, sweeping away speculative hopes for quick gains in share prices. Stock prices in the market were declining, and investors were taking substantial "paper losses." The price of conglomerate stocks had fallen substantially from their 1968–69 highs. For example, the price of Gulf & Western Industries' common stock fell from 66-⅛ to 34-⅝ dollars per share, City Investing from 81-⅞ to 49-¼, and White Consolidated from 42-½ to 27-¾.

While this proxy fight was underway, the Federal Trade Commission, whose earlier warning against a takeover of A-C had gone unheeded, announced its intention to issue a complaint against White Consolidated Industries, Inc. Later, the Commission made available a sixteen page complaint to be issued if the matter were not settled by a consent agreement. The proposed complaint alleged that White had violated Section VII of the Clayton Act and Section V of the Federal Trade Commission Act by acquiring approximately one-third of the common stock of Allis-Chalmers and by its plan to acquire the entire Allis-Chalmers business. The complaint was based on arguments similar to those which were being developed by the U.S. Justice Department and which were being used in the Allis-Chalmers suit against White, still pending.

The Federal Trade Commission had not officially issued a complaint, nor did the FTC news release require or permit any defense by White Consolidated. The Commission had announced its intention to issue a complaint only if there were no consent agreement. Should White Consolidated not agree to dispose of its A-C stock, the next step would be up to the FTC. The release, sent out over the wires of the *Dow Jones News Service* and published nationally, was read, no doubt, by A-C stockholders who were forming their judgments to whom to give their proxies.

Henry S. Reuss, Congressional representative from Wisconsin, "called on the Federal Trade Commission to bar a Cleveland conglomerate, White Consolidated Industries, from taking over the Allis-Chalmers Manufacturing Company in Milwaukee." He commended the FTC for its decision to issue a complaint and said that he was deeply concerned that the Commission prevent a takeover of Allis-Chalmers at its annual stockholder meeting in early May. He urged that the FTC obtain injunctive relief to bar White Consolidated from voting its stock.

The Commission gave White Consolidated four days to indicate whether it would refrain from voting its stock or seeking control of Allis-Chalmers. On the day in which White was to respond, Reddig was reported to have said, "I haven't heard a word from the FTC." Still no FTC complaint was issued, and the FTC declined to comment. Finally, William J. Boyd, Jr., chief of the FTC Merger Division, said "that the talks have been concluded temporarily and that the matter is pending before the Commission." Later, the Federal Trade Commission announced that it would withhold its proposed complaint against White Consolidated while the FTC watched the legal battle between the two firms. The public was protected, the FTC thought, since due to the appeal in the Philadelphia U.S. Court of Appeals, White Consolidated was temporarily prevented from taking over Allis-Chalmers. At the same time, the court had the opportunity to be aware of the Commission's concern.

As the May 14 annual meeting approached, both Allis-Chalmers and White Consolidated intensified their efforts to seek proxies. In the meantime, the Federal Trade Commission had proclaimed its intent to issue a complaint, and the U.S. Circuit Court had taken the Allis plea for an injunction against White Consolidated under advisement on March 28. Each day afterwards there was heightened anticipation of the court's ruling, but the court said nothing. The FTC deadline for White's response of March 10 came and passed, and the FTC, too, remained silent. The struggle was propelled toward the time of the annual meeting with doubts about what, if any, intervention would be taken by the Commission and the court. The matter was watched with great interest and concern because Allis-Chalmers had based its suit and appeal on the same arguments which the U.S. Justice Department had recently developed in its announced purpose of stopping conglomerates.

The 1969 annual meeting. (Top) President Scott at the podium and directors at the table. (Bottom) Stockholders asking questions of the president.

Only a breath away from the annual meeting, the court finally spoke. The U.S. Circuit Court of Appeals ordered the Allis-Chalmers Manufacturing Company to postpone its annual meeting until further notice because it was not feasible to rule on the appeal before that time. The enforced delay undercut White's efforts to have its slate of directors elected.

The war of words had crested and broken over the stockholders, and there was no possibility for action. Allis-Chalmers continuously rescheduled and adjourned its annual meeting while both companies waited for the court's decision. The annual meeting was scheduled and adjourned seventeen times, before it was finally held.

Two months after the original date set for the annual meeting, the U.S. Circuit Court of Appeals ended the long wait. By a vote of two to one on July 18, 1969, it reversed the Federal District Court's refusal to grant a preliminary injunction; Judge David Stahl, in stating the majority opinion of the three-judge U.S. Court of Appeals panel, wrote:

> Basically, what is at stake in the instant appeal is the life or death of Allis, a viable independent company, eager to continue as such, pitted against White, an aggressive, fast-moving acquirer of many diverse businesses, particularly in the past few years.

This opinion noted that the Federal Trade Commission had declared its intent to file a complaint against White after the earlier decision of the U.S. District Court. The Appeals Court said that the District Court should enter a preliminary injunction, permitting Allis-Chalmers to hold its stockholder meeting for the election of directors, prohibiting White Consolidated from voting its shares or taking any action calculated to give it any representation on the Board, and prohibiting White Consolidated from taking any steps calculated to increase its holdings in Allis-Chalmers.

White Consolidated asked for a rehearing before the full seven-judge panel of the U.S. Court of Appeals, but the petition was denied by a vote of four to three. White Consolidated took the matter of a temporary injunction to the Supreme Court and waited. The company waited until January of the next year, when, in a single typewritten line, the Supreme Court declined to review the Circuit Court's grant of a preliminary injunction.

The Allis-Chalmers annual meeting, postponed seventeen times, was finally held on January 22, 1970—without White Consolidated voting.

The early months of 1969 had been grim for the Allis-Chalmers Manufacturing Company. It was engaged in a desperate struggle from which it seemed to emerge only moments before being engulfed. The struggle for proxies had crested in May in anticipation of the May 14 annual meeting, then dissipated and was stilled while the courts pondered whether White Consolidated could attempt to gain control. In early June doubt was expressed in the financial community about the inevitability of a takeover, and the force of the motion toward takeover was abating.

Indications at the annual meeting were that Allis-Chalmers management and directors would have won the proxy contest with White Consolidated. There were 9,242,292 eligible votes for the annual meeting, excluding the White votes. Of the eligible votes, 7,469,671 votes were cast for the Allis slate of directors. Accordingly, over 80 percent of the eligible votes were cast for the A-C slate of directors; this would have been about 60 percent of the total vote if White had been permitted to vote.

Almost a year after the annual meeting, White Consolidated told its stockholders that it was seriously studying the possibility of selling its A-C stock. A short time later White Consolidated sold its A-C holdings to Solomon Brothers which subdivided the block among its clients. The threat of takeover, consequently was dissipated. The reported sales price was $58.5 million for stock which had originally cost $121.6 million.

The many attempts to take over Allis-Chalmers had been shattering, exacting heavy human and economic costs. The company's effectiveness was necessarily impaired as stockholders and employees were diverted by rumors and announcements. Morale and work suffered and other jobs were sought as corporate commitment was eroded by fear. Most dangerous was the diversion of key personnel from the company's problems, distracted from management by efforts to steady the helm and prevent unwanted takeovers. The great company of Edward Allis and Otto Falk seemed to flounder in the wake of raiding conglomerates.

But as the spring of 1970 began, the season seemed to hold the promise of tomorrow. It was life, not death, for the Allis-Chalmers Manufacturing Company, as its independence was assured. These years had rent the company, however, and its resurgence would need time to unfold.

The full energies of the company could be directed to its continuing reappraisal and reorganization only after the successful resistance to the takeover attempts. This struggle of the late sixties had only exacerbated the problems arising out of changes in market, competition, and technology and had slowed strategic responses aimed at their resolution.

Immediate, deliberative action was required to stop the hemorrhaging of assets and income. After record profits of $26 million in 1966, the company's net income declined to $5 million in the following year, and it had a net loss of more than $54 million in 1968. Book value dropped from thirty-six dollars for each share in 1966 to twenty-nine dollars in 1968. At the same time that income and assets were draining away, additional capital was needed to remain competitive in several of the company's markets; these capital needs seemed beyond what Allis-Chalmers could raise in the aftermath of the crisis.

Allis-Chalmers had many obsolete, inefficient plants; it was burdened with high labor costs for some products relative to its competitors, and its share of market and sales volume was insufficient in some areas to operate its plants efficiently or to invest in modern facilities. Salaried overhead was high, and the company feasted or starved in its capital goods markets. New technological developments which created sales opportunities required heavy investment and research expenditures. Finally, the company's reentry into steam turbines was devastated by new environmental and economic forces.

David Scott set about to stop the hemorrhaging of the company's income and assets, to improve the efficiency of existing operations, to invest in areas of known performance and opportunity and to extend markets in Europe, the Middle East, South America and the Soviet Union. After years of intensive command, Scott could say by 1976 that the major problems had been resolved. Net income had increased in each successive year from a low in 1971; sales expanded over the period even though the company phased out some major, unprofitable product lines.

Heroic action was taken to cauterize areas in which the company was inefficient and non-competitive. Scott reported in 1971 that total employment had been reduced by 8,000 since September, 1968; more than 6,000 were non-production employees. The greatest reduction was in corporate staff, reduced by more than 1,300 to 132 persons in 1969.

Meanwhile, in 1971 the old company name gave way to a new name—Allis-Chalmers Corporation—that symbolized the new image and new directions.

In 1968, the company had extraordinary costs, net of taxes, of $13.4 million for relocation and discontinuance of facilities and products, including employee separation costs. Costs for plant rearrangement, underutilized equipment and product discontinuance were $8.3 million in 1973; $16.4 million in 1974; and $19.0 million in 1975. At the same time (1969—75), $240.3 million was expended for capital improvements, with at least one new facility put into operation each year.

More equity and debt were required to enable Allis-Chalmers to sustain these losses and embark on a course of recovery. Although this was difficult in a time of crisis, Scott was nonetheless able to place the company in a sound financial position. He had sold $45 million of convertible preferred shares to International Fiat Holding S.A., Switzerland and certain Rockefeller interests in 1969. The sale was made at the time in which there was a $124 million writedown on assets. The move to sell the preferred shares was critical because, without it, the company would have been in violation of its covenants and probably would have been insolvent.

THE WALL STREET JOURNAL

© 1970 Dow Jones & Company, Inc. All Rights Reserved.

★ ★ ★ MIDWEST EDITION   FRIDAY, JANUARY 23, 1970

Allis-Chalmers Holds Long-Stalled Meeting, Says It Had '69 Net, May Resume Dividend

THE MILWAUKEE JOURN.

• • • • Thursday, January 22, 1970 • • • •

A-C Meeting Finally Held; Scott Expects Profit for '69

MILWAUKEE SENTIN

FRIDAY MORNING, JANUARY 23, 1970

SCOTT MAY RECOMMEND '70 DIVID

A-C Returns t Profit

In The Market

By Ross M. Dick, Journal Business Editor

White Consolidated Industries, Inc., owner of the biggest block of common stock of Allis-Chalmers, could count its shares as worth $3,248,000 more Thursday night — but it still had a paper loss of around $37 million on the price it paid for the A-C stock.

the New York Stock Exchange, A-C stock ──day to $26.50. The report given annual meeting

FORTUNE

February, 1969

Businessmen in the News

The Changing Management Roster

Shaken by take-over attempts, Allis-Chalmers Manufacturing Co. brought **David C. Scott** in from Colt Industries last September as its president. Scott came fully prepared with plans to chop out deadwood and to expand into new markets. By the time he was named chief executive a month ago at the age of 53, he had reduced the corporate staff from some 1,500 to less than 150, restructured operations, waged a str—g fight against White Consolidated ies' take-over efforts, and begun iown three construction-machinery on facilities—including a newly 000-square-foot plant that he sold cott, it's only the beginning of a hat he hopes will "turn the com- id in two years" (it was expect- ibout $25 million, including a ff, after taxes in 1968, on sales is than 1967's $822 million). 't of that program is a move s; more broadly, Scott has ners "should enter all of the neral Electric is now in and , another G.E."

Scott of Allis-Chalmers

Business Week

Reprinted with permission
from Nov. 15, 1969 BUSINESS WEEK

It's no tea party at Allis

President David Scott has his hands full trying to make Allis-Chalmers profitable while fighting a takeover bid

Newspapers and magazines throughout the United States were interested in the takeover attempt.

The acquisition of Standard Steel Corporation in the same year yielded $3.4 million net current assets, and 16,000 shares of preferred stock, valued at $1.6 million, were exchanged for Pennsylvania Electric Coil Corporation in 1972. Late in 1970, the production of industrial lift trucks was moved from the Harvey, Illinois material handling plant, to a newly constructed facility in Matteson, Illinois.

Other important steps were taken to secure the necessary financial support for the Allis-Chalmers recovery. A $41.5 million term loan having final maturity in 1977 was arranged with a group of banks in 1972. A $30.0 million loan with a group of foreign banks was also refunded in 1972 with final maturity in 1979. A $58.0 million term loan was placed with institutional investors in 1974 with maturity in 1991; these funds were used to retire term notes and short term bank debt. In 1975, the company obtained a revolving credit/term loan of $50 million from twelve domestic banks with the funds available to the company at its option over a three year period; and the debt in 1978 would be repayable over the next four years. In addition, Allis-Chalmers Credit Corporation had $242.0 million outstanding in commercial paper in 1975 to provide credit for the customers of Allis-Chalmers Corporation—with the Credit Corporation being a consistent, major contributor to the parent's earnings. The company's financial services contributed over $15.0 million to the $43.2 million of income (before taxes) in 1975, for example.

Thus Scott's strategy solidified the company's financial position and provided liquidity for the expansion of markets and for investment in areas of known performance and opportunity. Short-term debt and current maturities of long-term debt declined as a percentage of long-term debt from 105 percent in 1971 to 14 percent in 1975 for Allis-Chalmers and consolidated subsidiaries. At the same time, total debt declined as a percentage of total assets, from 33 percent in 1971 to 21 percent in 1975 for Allis-Chalmers and consolidated subsidiaries. The company, of course, has continuing obligations in connection with its credit, financial and leasing service corporations similar to other equipment manufacturers.

Scott also turned problems into opportunities through the imaginative use of acquisitions, licensing agreements and joint ventures. The company's growth had been stifled and its present position threatened by excessive risk, the high cost of technological development and great investment costs in some markets in which competitors had much greater resources. Scott sought licensing arrangements and acquisitions to gain new technology and expanded markets; he sought joint ventures to reduce risk and investment and to obtain technology and expanded markets. Sales and manufacturing effort were redoubled in international markets.

The most important area of growth during the period was in processing equipment, where major strategic moves capitalized on opportunities for profitable expansion. The Standard Steel Corporation, acquired in 1969 and renamed Stansteel Corporation, expanded markets and provided technology for the manufacture of asphalt plants, fish meal plants, small kilns, rotary driers and coolers. Stansteel's facilities and products were expanded after the acquisition and its technology applied to Allis-Chalmers Canadian subsidiary.

Also in 1969 Allis-Chalmers made a licensing agreement with Voest of Austria to market and manufacture Voest's continuous casting machinery in the United States. Another agreement with MIAG of the Federal Republic of West Germany allowed Allis to build better cement makers.

Allis-Chalmers acquired a minority interest in Svedala-Arbra AB of Sweden in 1971 and increased its interest to 95 percent by 1974. Svedala-Arbra provided facilities and opened markets for crushing and screening equipment in Scandinavia and West Germany and provided for an exchange of technology and designs between the Swedish firm and its new parent. Svedala-Arbra soon prospered and expanded, establishing a subsidiary, Svedala-Arbra GmbH, in Frankfurt.

Fabrica de Aco Paulista S.A. in Sao Paulo, Brazil was acquired in 1972 to open additional markets for crushing and screening equipment in South America. Fabrica de Aco Paulista then expanded its

The new Stansteel drum mix type asphalt plant, built by the Allis-Chalmers subsidiary, is easily transported, erected, and dismantled.

facilities by 50 percent in 1974, opening up another plant in Sorocaba, and doubled its sales volume through diversification of products and expansion of markets to other South American countries.

Stephens-Adamson, an experienced manufacturer of bulk material handling equipment, was purchased from Borg-Warner Corporation in 1974. The purchase provided Allis-Chalmers with facilities in Illinois, Mississippi, Canada, Mexico and Australia, and again added product lines and markets. In June, 1974, Allis sold its paper machinery division to a joint venture with J. M. Voith GmbH of West Germany, which provided capital for Allis and ended a drain on the company's income.

Heavy construction machinery had also been a drain upon Allis-Chalmers because its volume was generally insufficient for efficient marketing and production and for the support of developmental research. The size and resources of the competitors in heavy construction equipment made it impossible for Allis-Chalmers to increase its share of the market without great risk. Constrained by these circumstances, Scott carried through an

Capable, talented subsidiary companies abroad bring Allis-Chalmers products to important markets. Fabrica de Aco Paulista S.A. (FACO) Sao Paulo, Brazil, built these vibrating screens for an iron ore classification project, Itabira, Brazil.

imaginative strategy. He was able to conserve the company's assets in the fabrication of construction equipment and to expand markets while simultaneously reducing its risk.

In January of 1974, Allis-Chalmers completed negotiations with Fiat S.p.A. of Italy to form a joint venture, Fiat-Allis. Both firms contributed their resources for building and marketing construction machinery with production facilities in Leece, Turin and Milan, Italy; Sao Paulo and Belo Horizonte, Brazil; Essendine, England; and Deerfield and Springfield, Illinois. Initial capitalization of Fiat-Allis was approximately $313 million, with only $73 million of that amount in long-term debt. Fiat received 65 percent equity in the venture, Allis 35 percent, plus $47 million in cash. Fourteen and one-half million dollars of the $47 million were proceeds from a Fiat note, with Fiat given the option of repaying this note later with its own Allis-Chalmers stock. The A-C stock could be exchanged for the note at a stock price supported by Allis if the market price were a lesser amount.

The agreement provides Fiat with certain options to purchase the equity of Allis-Chalmers prior to 1986, although the companies could remain partners. Fiat-Allis consisted of two holding companies: Fiat-Allis B.V., a Netherlands corporation, and Fiat-Allis, Inc., incorporated in the United States. The principal North American operation is Fiat-Allis Construction Machinery, Inc., with headquarters in Illinois; the principal European operation is Fiat-Allis Macchino Movimento Terra S.p.A. with headquarters in Italy. Fiat-Allis, a world-wide operation, was established to offer

David C. Scott (left), chairman of Allis-Chalmers Corp., and Giovanni Agnelli, chairman of Fiat S.p.A., signed the agreement to set up a joint construction machinery enterprise. The signing took place in Geneva, Switzerland, on July 12, 1973.

public and private customers a full product line of high quality, to strengthen sales and service organization, to expand output and to provide increased job opportunities and to advance technical developments and foster social and economic progress.

In the aftermath of the takeover crisis, Scott also faced a wide range of products and markets which were interrelated in their manufacture and distribution. Some were profitable while others were not; Scott's strategy was to separate the potentially profitable from those which were draining the company. He invested in those of known performance and divested the unprofitable.

Allis-Chalmers sold highly diversified products and equipment to the electrical industries, competing in many markets. It sold generators and turbines for making electricity from water power as well as pumps for many other purposes. Also, steam turbines and related equipment were sold to the utilities for the generation of electricity from nuclear and fossil fuels. Finally, Allis manufactured and sold a wide range of products to regulate and control the distribution of electricity, to control and protect electrical systems and devices, and to convert electricity into mechanical power. Some of the electrical products were transformers, motors, voltage regulators, adjustable speed drives, and circuit breakers.

The company had re-entered the manufacture of steam turbines under a totally new and different concept. Scott saw electrical power requirements doubling during the 1970s with an expansion in the demand for steam turbine generating equipment. Orders had been too few for Allis-Chalmers to compete with the two manufacturing giants earlier, but the anticipated increase in demand during the 1970s was seen to permit a third profitable company in the industry.

As a first step in the company's re-entry, Scott reached an agreement to install and service steam turbines and related equipment manufactured by Siemens AG of Germany in 1968. During the next year Scott reached an agreement in principle to form a joint venture with Kraftwerk Union of Germany, owned jointly by Siemens AG and AEG Telefunken, who made turbine generating equipment.

Allis-Chalmers and Kraftwerk Union organized a jointly owned engineering and marketing company, Allis-Chalmers Power Systems, Inc. The new joint venture sold and serviced thermal electric power generating equipment in the United States. The arrangement benefited Allis by providing engineering expertise and technology for thermal electric power equipment. Allis-Chalmers also obtained licenses to manufacture and sell a wide range of electrical products under Siemens innovative technology.

The engineering knowledge and designs flowing from this venture enabled Allis to move into the manufacture of steam turbines quickly, and Allis-Chalmers reported in 1971 that it had accepted orders for five turbine generator units. Business strengthened, substantial investments were made, and the company's position was favorable.

The fortunes of Allis-Chalmers changed again, however, with the unexpected decline of demand in 1974 due to the cost of fuel, the surge of environmental problems in the thermal generation of electricity and the emergence of financial difficulties for utilities. The utilities postponed or cancelled their orders and Allis-Chalmers production proved insufficient to operate at reduced economic levels. As a consequence, Scott arranged to transfer certain production for turbine-generators from Allis-Chalmers to Kraftwerk Union to avoid significant capital expenditures. The 1975 agreement with Kraftwerk Union provided that other turbine-generator equipment would be made by Kraftwerk Union at its West German plant or at a new plant to be built in the United States. Substantial losses were incurred, but the strategic readjustment prevented continued and even greater losses.

Power transmission equipment was manufactured by Allis-Chalmers at a loss, and nine to twelve million dollars of yearly profits from 1971 to 1975 were required in other areas to cover losses from the production and sale of transformers. The company tried many ways of dealing with these losses in the late sixties and

A Stephens-Adamson bucket wheel sizer/reclaimer for high speed, high capacity handling of bulk materials.

early seventies, including modernization of plants, new technology, and the introduction of new and improved products.

Finally, Scott concluded that the company should divest itself of its overhead and underground transformer business. He wrote:

> As part of this strategy we announced, on February 20, 1975, a decision to withdraw from the electrical overhead and underground distribution transformer business because of losses and a constant cash drain on the corporation. Significant among the factors which make this move necessary are exceptionally high wage rates and benefit costs which have resulted from collective bargaining negotiations over the years at our Pittsburgh, Pennsylvania and Gadsden, Alabama plants and which are out of line with those of principal competitors. A number of efforts over the years to rectify this high labor cost situation, including repeated attempts to negotiate rates of pay and benefit levels appropriate to these businesses, have failed to produce a solution. The average wage and benefit total of the Allis-Chalmers production employee at Pittsburgh, for example, is in excess of $3 per hour more than that paid by principal competitors in this business.

The divestment of overhead and underground distribution and transformer business did not imply lessened emphasis on the company's electrical products business. Scott continued his policy of investing in electrical products of known performance. The acquisition of Pennsylvania Electric Coil Corporation in 1972 permitted Allis to enter the motor and generator service and repair business. In a continuing expansion, facilities for electrical products were built or enlarged in Jackson, Mississippi; Wichita Falls, Texas; Little Rock, Arkansas; Sanford, North Carolina; and New Orleans, Louisiana.

Power lawn and garden equipment such as this eight-horsepower Simplicity garden tractor is sold throughout the world.

The profitability of electrical products and the expansion of company sales rested largely on a major strategic move by Scott. Allis-Chalmers did not have the resources for a massive product development program to compete successfully in a broad range of electrical products, even though many were in its traditional markets. Scott took this insurmountable obstacle and turned it into an opportunity. Seeking a way to avoid development costs and yet to produce and sell such products quickly, his solution was imaginative and cut to fit the company's resources: he negotiated an agreement with Siemens AG for technical assistance and the manufacture and sales of electrical products under a Siemens license. Siemens advanced technology and superior products made Allis-Chalmers a strong competitor in American markets, and the agreement benefited Siemens, too, who avoided the large costs of initial penetration of new markets.

Scott maintained a program of investment and growth in agricultural equipment, material handling equipment and consumer products, and these areas were important contributors to the corporation's recovery. Net income attributable to these product areas grew steadily from $6.5 million in 1971 to $42.7 million in 1975. Facilities were constructed or modernized, superior products were introduced, and more extensive product lines were offered to meet a greater range of customer needs.

The Henry Corporation of Topeka, Kansas was acquired in 1968 to improve the market position in industrial tractors and equipment; the firm's headquarters for these products was moved to Topeka. Market distribution systems were improved in West Germany, Great Britain and Australia for lift trucks. Allis-Chalmers also worked out an agreement with Lancer Boss Ltd. to market its sideloading lift trucks which made it possible for Allis-Chalmers to increase dramatically its range of products.

Allis-Chalmers had acquired the Simplicity Manufacturing Company of Port Washington, Wisconsin, in 1965, and Scott continued to invest in the lawn and garden equipment manufacturer as part of his strategy for recovery. Simplicity's strength reflected both Robert Stevenson's and David Scott's understanding of how to acquire companies and manage their operations. Their approach to acquisition, licensing and joint venture demonstrated how economic and social benefits could be obtained from such programs in contrast to the devastation of the attempted forced takeover from which Allis-Chalmers had recovered.

Also strengthening the company's position as a lawn and garden manufacturer was the startup of a new plant in Lexington, North Carolina, in 1968.

A crisis had swept over Allis-Chalmers in the late 1960s which had threatened the company with extinction. One takeover raid after another broke against the Allis-Chalmers Manufacturing

Company, but the company stood firm under the leadership of Robert Stevenson and David Scott. The raids were sudden, carried out in a climate of rumor and speculation. The financial community heard outrageous and conflicting statements about Allis-Chalmers, and its imminent death had been widely reported. The emotion-charged environment provided no informed basis for rational response by stockholders to corporate raiders and the Allis-Chalmers management and directors had stood courageously against the swell of opinion.

They resisted the forced takeover attempts and, as time passed, more information became available. Agencies of the state and Federal government challenged the unwanted takeover, and the courts acted to delay forced merger. Support for the company's position developed in the financial community, and it was soon reported that Allis-Chalmers was living.

The crisis had occurred at a time of changing markets, competition and technology which had eroded the company's economic viability. At first, Robert S. Stevenson and later David C. Scott were required to address themselves to the imminent and dangerous takeover attempts. The distraction and time devoted to resisting the takeover attempts cost the company dearly in opportunities to deal with the conditions which threatened the company's economic viability. Allis-Chalmers lost great sums in the midst of the crisis. The company's competitive position was weakened by the changed conditions, and its income and assets were hemorrhaging.

These centrifugal pumps and motors are typical of those supplied to municipal pumping stations.

Newly-appointed President David Scott charted a strategy to recover from the crisis. He cauterized the hemorrhaging, improved the efficiency of existing operations, invested in areas of known performance and extended markets in Europe, South America, the Middle East and the Soviet Union. Scott turned the problems of Allis-Chalmers into opportunities by prudent divestiture decisions and by the imaginative use of licensing agreements, acquisitions and joint ventures.

Allis-Chalmers recovered from the crisis of the late sixties, and Scott reported in 1976 that their problems were behind them. The Bicentennial year, 1976, was very successful. At year end Scott reported a record net income of $58.7 million, approximately double the $29.4 million earned in 1975. Earnings per common and common equivalent share amounted to $4.51 in 1976 compared with $2.33 in the previous year. Sales were $1.5 billion in 1976 up from $1.4 billion in 1975.

PROFIT RISE 56.2% AT ALLIS-CHALMERS

THE NEW YORK TIMES, TUESDAY, MARCH 1, 1977

Net in Quarter 90 Cents a Share, Against 62 for 1975 Period — Year's Earnings Up 99.7%

MILWAUKEE SENTINEL

Business News
Tuesday, March 1, 1977

A-C Earns Record $58.7 Million

Cincinnati Post

Tuesday, March 1, 1977

Allis-Chalmers nearly doubles earnings

THE WALL STREET JOURNAL, Wednesday, Jan. 25, 1978

Allis-Chalmers Says 1977 Per-Share Profit Climbed 22% to Record

Newspapers throughout the nation chronicled the economic recovery of Allis-Chalmers.

Results for 1977 told the story best when they were compared to the previous six years in the following table:

Results for the year ending December 31:

	1971	1972	1973	1974	1975	1976	1977
Income (millions)	$ 5.2	$ 8.7	$ 16.3	$ 22.1	$ 29.4	$ 58.7	$ 67.0
Sales (millions)	$ 889	$ 960	$1,166	$1,262	$1,443	$1,519	$1,538
Earnings per share	$0.42	$ 0.70	$ 1.30	$ 1.77	$ 2.33	$ 4.51	$ 5.52
Dividends declared (per share)	$0.20	$0.204	$0.227	$ 0.26	$0.295	$0.625	$ 1.10

The company has had six years of continuous growth in assets and earnings. Allis-Chalmers had resumed its status as a great company, with multi-national markets which provided solid returns on investments and once again contributed to the nation's economic well-being.

NOTE TO EPILOGUE

1 Detailed documentation to support this chapter is contained in C. Edward Weber's, "Attempted Takeover of Allis-Chalmers Manufacturing Co." (Mimeographed paper, Milwaukee, 1976). My research utilized the extensive source materials which are housed at the Milwaukee County Historical Society. Grateful acknowledgement is made to the Allis-Chalmers Corporation and the Milwaukee County Historical Society.

APPENDIX

ALLIS-CHALMERS IN 1977

In the 116 years since Edward P. Allis purchased the Reliance Works, the Allis-Chalmers Corporation has developed into a highly diversified international corporation. The company's major manufacturing divisions have undergone radical changes over the years in response to the industrial and economic demands of the American and world markets. Today, in 1977, three of these business divisions hold central positions as core contributors to the company's total income and sales; these are the Process Equipment, Agricultural Equipment, and Electrical Groups. In addition, the Material Handling and Outdoor Power Equipment Group is developing the potential for profitable growth. Just as an examination of the sawmill, steam engine, and flour mill departments of the nineteenth century told a great deal about the company at that time, a similar look at these four business groups will help define and delineate the position of the Allis-Chalmers Corporation today.

$$* \quad * \quad *$$

The Process Equipment Groups manufacture two types of products and systems for today's economic and industrial activities: equipment for processing high-volume solids and fluids. These systems and products change the form or composition of high-volume solids, particularly metallic ores, coal, and rock; convey solid materials; move gasses and liquids; regulate fluid flow; and use water power to drive electric generators. The solids processing equipment includes crushers and grinding mills, vibrating screens and feeders, iron ore pelletizing plants, bulk conveyors, continuous casting units, copper converters, and anode furnaces, as well as systems and processes for cement, asphalt, and other non-metallic industries. The fluids processing products include hydraulic turbines, pump-turbines, water and industrial pumps, butterfly and cone valves, and a broad compressor line.

The products of the Process Equipment Groups represent a wide range of design and manufacture, from the small to the immense, from the simple to the complex. Some, like crushers and vibrating screens, have changed only slightly over recent years; others are sophisticated designs for use in improved technologies such as the direct reduction system for use in the steelmaking process and the *KilnGas* system for the gasification of coal. All these products, however, require massive moving parts which must be manufactured and fabricated with a high degree of skill.

Process equipment is sold to a broad base of companies in diverse industrial activities. These include construction, mining, grain processing, shipping, water treatment, pollution control, petrochemical refining, and cement and steel manufacturing. State and municipal governments, electric utilities, and builders of nuclear power plants are also served by the Process Equipment Groups.

In recent years, Allis-Chalmers has strengthened its market position in process equipment through acquisitions, licensing agreements, and intensified research activity. New geographic markets, many in developing countries, have been opened through acquisitions and license arrangements. New products and systems have been added through internal research activities. The work of the Allis-Chalmers Process Research and Test Center, for example, has contributed substantially to the development of pelletizing systems and direct reduction processing for steel and is now completing engineering work on a coal gasification system.

Expansion into newly industrialized nations has increased the growth rate of the Process Equipment Groups. Long term growth rates for traditional markets have tended to follow the gross national product, with a resultant rate of 9 or 10 percent per year in current dollars for the U.S. market. But Process Equipment Groups sales have increased at an average annual rate of 25 percent in the last six years. The company's movement into rapidly expanding markets in newly industrialized nations has meant significant jumps in sales and earnings.

Sales in the Process Equipment Groups have increased 211 percent over the past six years, while income was up 244 percent. The return on capital employed doubled over the same period.

PROCESS EQUIPMENT GROUPS

	1971	1972	1973	1974	1975	1976
Sales (millions)	$193	197	246	377	525	600
Income before taxes (millions)	$12.7	13.0	14.2	22.2	27.5	43.7
Capital employed* (millions)	$78.3	80.4	86.0	128.6	144.3	136.7
Return on capital employed before taxes (percent)	16.2%	16.2	16.5	17.3	19.1	32.0

*Here and subsequently, average capital employed includes dealer inventories financed by Allis-Chalmers Credit Corporation.

The Groups' range in markets served is matched by their diversity in manufacturing capability. Common requirements permit production planning at different facilities, which at present include operations in Australia, Brazil, Canada, France, Great Britain, Mexico, Spain, and Sweden, as well as the United States. These operations, along with licensing and subcontracting arrangements, enable Allis-Chalmers to produce much of its equipment in twelve countries and thus to meet international market needs and to open additional areas of credit and financing.

Several major market opportunities exist for the Process Equipment Groups in the years to come. A continued demand for oil, expected to remain the world's primary source of energy for the rest of the century, will maintain the need for compressors. Increased environmental concerns require pumps to handle cooling water and sewage, valves to control water and wastes, water control gates to enhance dam safety and flood protection, and waste processing systems for recycling and disposal of solid refuse. The market for valves and pumps is expected to be buoyed by the use of Federal funds to support municipal pollution control. New cement plants are required to replace old facilities in the United States with less energy-intensive units, as well as to supply needs in expanding foreign cement markets.

Despite current sluggishness, the long term demand for basic minerals is expected to show strong long term growth. Depletion of rich ore deposits will require the use of sophisticated processing equipment to treat crude ore. For example, the Voest-Alpine continuous caster, for which Allis-Chalmers is licensed in the United States, Canada, and Mexico, offers greater efficiency for the steel industry. Direct reduction equipment is expected to be a major market for Allis-Chalmers in five to ten years.

Increasing use of hydroelectric power, in both highly industrialized as well as less developed areas of the world, will open markets for hydraulic turbine products. The reversible pump-turbine has become attractive to balance peak and off-peak production of electricity, and bulb and *Tube* turbines will become economically feasible for large, low-level dam sites as fuel costs increase. The market opportunities for standardized units will likely be strengthened by Federal encouragement of the hydroelectric potential of smaller dams.

The gasification of coal appears to hold the major development potential for the Process Equipment Groups. It is estimated that coal gasification will supply power for 10 percent of the new electric generators by 1988, and these power plants will be 10 to 20 percent less costly than conventional coal-fired plants with sulfur-controlling stack scrubbers. Eleven electric utilities are helping to fund a pilot plant using the *KilnGas* process, in return for early purchase rights to the equipment developed.

*   *   *

The Agricultural Equipment Group, the second core business of Allis-Chalmers today, builds major farm machinery for the world production of food and fiber. The equipment includes combines, corn heads, cotton strippers, farm tractors, planters, tillage implements, diesel engines, diesel electric generator sets, and light industrial tractors.

This Allis-Chalmers agricultural equipment is sold to private farmers throughout the world, as well as to government agricultural organizations, construction contractors, and municipalities.

Agency sales complement and strengthen individual sales to farmers.

Sales in the Agricultural Equipment Group have increased 170 percent in the last six years, while income rose from a deficit in 1971 to almost $60 million. Return on capital employed has also grown from a negative return in 1971 to 19 percent in 1976.

AGRICULTURAL EQUIPMENT GROUP

	1971	1972	1973	1974	1975	1976
Sales (millions)	$204	232	294	424	511	550
Income before taxes (millions)	$(1.8)	(.3)	1.3	16.8	44.0	59.5
Capital Employed (millions)	$327.8	288.9	282.9	256.9	304.0	312.9
Return on capital employed (percent)	(.5)%	(.1)	.5	6.5	14.5	19.0

The Agricultural Equipment Group offers a broad selection of high technology equipment, without attempting to supply every possible need of the farmer. A range of high capacity machinery within established product lines is maintained, while new products are introduced consistent with ongoing field studies of market trends and the competition.

Both the production facilities and the dealership organization offer flexibility to the operation and future expansion of the Agricultural Equipment Group. Production designs utilize common components and machine tools which are applicable to a variety of operations in either long or short production runs and which reduce the costs of design changes. Capital expenditures for the improvement of production facilities have increased in the group.

A strong dealership organization is important to the success of the Agricultural Equipment Group. A dealer development program identifies specific, competitive dealer trade territories which will support a high level of activity. In line with this program, more than 100 new dealerships were established in 1977. A new parts distribution center complements the dealership organization.

Long range planning for this group must take into account recent developments in American agriculture. Industry retail sales of total tractor horsepower per 1000 acres in this country have increased from an average of 29 horsepower per acre in the 1950's to 32 horsepower in the 1960's and to 42 during the 1970's. Since the 1950's the substitution of large-capacity machinery for farm labor has expanded the size of individual farming operations and improved their productivity and earnings. More recently, large-capacity machines have been used to improve the timeliness and yield of critical field operations. Also, an overall higher level of acreage than ever before is now in production.

These developments are resulting in a need for fewer but more sophisticated machinery units. In addition, several relatively new factors are expected to influence the long term trends. Most important of these is a greater emphasis on energy efficiency in crop production and soil conservation, which will support the growth of minimum or reduced tillage practices. Already, these systems are now used in 20 percent of total U.S. planted acreage, double that of 1970. Large energy-efficient equipment is essential for timely tillage operations.

Finally, the planning for this group must consider the fluctuations which are peculiar to the agricultural economy. These factors include excess grain stocks, non-production premiums of acreage set-aside programs, export demands, weather patterns, price supports, and the livestock economy.

*   *   *

The Electrical Groups manufacture and market both steam turbine generators and a broad variety of electrical and electronic equipment that is used to power, control, protect, and regulate other machinery. This latter category includes electric motors, circuit breakers, switchgear, industrial controls, direct current devices, electronic control products and systems, power regulators, and disconnect switches—all from the Electrical Products Group. Major strategy changes in recent years have contributed to the Electrical Products Group's current position as one of the core businesses of the Allis-Chalmers Corporation.

Two major markets exist for the Electrical Products Group: industry and utilities. The domestic market for industrial electrical equipment has risen 61 percent since 1971, while the utility market rose only 21 percent, due to depressed conditions industry-wide since 1974. This utilities downturn has severely disrupted traditional growth patterns in kilowatt-hour consumption, although signs of a return to a more normal 5 to 6 percent rate are now evident. This growth, combined with purchases to meet new construction and depleted inventories, indicates a stronger utilities industry; in fact, utilities' orders in the Electrical Products Group have increased by nearly 20 percent. The group currently has approximately a 45 percent share of the market in regulators, 25 percent in switches, and 20 percent in power circuit breakers.

While the industrial operation of the Electrical Products Group also sells to the utilities market, 80 percent of its sales are to a broad spectrum of industries. These include the chemical and petro-chemical, steel, paper, textile, mining, automotive, and equipment manufacturer.

Sales for the total group have increased 85 percent since 1972; pretax income has risen from a 1973 loss to $21.5 million in 1976. Return on capital employed has grown to 28.8 percent in 1976.

ELECTRICAL PRODUCTS GROUP

	1971	1972	1973	1974	1975	1976
Sales (millions)	$114	110	126	159	182	204
Income before taxes (millions)	$4.8	.4	(2.3)	.5	2.4	21.5
Capital employed (millions)	$51.7	54.4	62.5	74.8	76.6	74.7
Return on capital employed before taxes (percent)	9.3%	.7	(3.7)	.7	3.1	28.8

Major strategic moves have increased the strength of the Electrical Products Group in recent years. New facilities which permit lower unit costs have been constructed, while the manufacture of transformers was discontinued.

The technology of the Electrical Products Group has been substantially advanced through a licensing agreement and a forthcoming joint venture with one of the world's major electrical equipment manufacturers, Siemens AG of Munich, West Germany. Siemens AG had sales of $8.8 billion, capital expenditures of $264 million, and research and development expenditures exceeding $750 million in 1976.

In 1970 a license agreement incorporating the research of Siemens AG into the product lines of Allis-Chalmers was reached. In 1978 this affiliation will become a joint venture, Siemens-Allis, Inc., comprising the businesses of the Electrical Products Group. Since 1970, Siemens technology has been applied to such projects as gas-insulated power breakers, industrial and electronic control and hydro-generators. With the formation of the joint venture, there will be additional research opportunities in products ranging from relays to instrumentation to process controls.

While the joint venture and the transfer of technology will take time to implement, it strengthens the position of the Electrical Products Group in the marketplace. The group currently serves 20 percent, or $2.6 billion, of the $13 billion domestic charter market, which is projected to grow to $24 billion by 1982. The U.S. market, in which the group competes, is expected to expand from $2.6 billion to $5 or $6 billion during the same period.

*　　*　　*

The Material Handling and Outdoor Power Equipment Group is presently undergoing a transitional growth period. It now appears that this group could, in a number of years, become a core contributor to the company's income and sales, just as the Process Equipment, Agricultural Equipment, and Electrical Products Groups are in 1977.

Sales for the Material Handling and Outdoor Power Equipment Group reflect its changing situation. While 1976 sales totaled $148 million, slightly lower than in 1975, income for the same 1976 period of $1.1 million rose from a loss of $1.3 million in 1975.

MATERIAL HANDLING AND
OUTDOOR POWER EQUIPMENT

	1971	1972	1973	1974	1975	1976
Sales (millions)	$111	127	161	186	159	148
Income before taxes (millions)	$8.3	12.2	21.6	20.7	(1.3)	1.1
Capital employed (millions)	$77.4	81.9	86.2	102.0	114.1	99.6
Return on capital employed before taxes (percent)	10.7%	14.9	25.1	20.3	(1.1)	1.1

The Material Handling segment, which makes up approximately two-thirds of the group, consists of the domestic Industrial Truck Division, a Mexican affiliate, A-C Mexicana, S.A., and Allis-Chalmers Material Handling—Europe. Its products include such lift trucks as the conventional electric, cushion- and pneumatic-tired internal combustion, narrow aisle, and walkie electric.

Sixty-five independent dealers and ten company-owned stores distribute the company's material handling products. A high degree of vertical integration combined with recent consolidations of operations and facilities have increased the division's efficiency. These strategies should allow future expansion into the food and beverage, paper and lumber industries.

The Outdoor Power Equipment segment of the group offers powered lawn and garden equipment manufactured by Simplicity Manufacturing Company. Sold under the Simplicity name and also by the Agricultural Equipment Group under the Allis-Chalmers name, these products include lawn and garden tractors, conventional and riding mowers, rotary tillers, and snow removal equipment.

The following financial data presented for 1977 is information calculated under the revised Securities and Exchange Commission regulations defining presentation of industry segment profit and loss information. Tables by groups for year 1971 through 1976 appearing on pages 412, 414, 416 and 417 have not been restated and accordingly reflect the prior year requirements of the Securities and Exchange Commission for line of business disclosure.

FINANCIAL DATA 1977

	Process Equipment	Agricultural Equipment	Electrical Products	Material Handling & Outdoor Power
Sales (Millions)	$560	$545	$226	$186
Income before taxes (millions)	55.3	40.4	26.3	8.7
Capital employed (millions)	155.9	395.3	79.7	102.5
Return on capital employed before taxes (percent)	35.5%	10.2%	33.0%	8.5%

Building on the traditions of more than a century of manufacturing, Allis-Chalmers Corporation today markets a diverse range of products. The most important of these are divided into the four central groups discussed in this Appendix: Process Equipment, Agricultural Equipment, Electrical Products, and Material Handling and Outdoor Power Equipment. The company's products have progressed far beyond the water pipes, Corliss steam engines, and band saws of the early firm; indeed, the development of modern product lines delineates the growth of the E. P. Allis Company into today's international Allis-Chalmers Corporation.

DIRECTORS 1977

William E. Buchanan, Appleton, Wisconsin
S. Hazard Gillespie, New York, New York
John L. Hanigan, Nashville, Tennessee
Ralph A. Hart, Farmington, Connecticut
Harry Holiday Jr., Middletown, Ohio
Joel Hunter, Pittsburgh, Pennsylvania
George F. Kasten, Milwaukee, Wisconsin
David C. Scott, Milwaukee, Wisconsin
Joseph W. Simpson, Jr., Milwaukee, Wisconsin
Edward E. Watson, Milwaukee, Wisconsin
Roger C. Wilkins, Hartford, Connecticut
F. Perry Wilson, New York, New York

OFFICERS 1977

David C. Scott, chairman of the board,
chief executive officer and president;
Lutgardo L. Aguilera, vice president;
Frank E. Briber, vice president;
Wendell F. Bueche, executive vice president;
Thomas L. Dineen, executive vice president;
Richard M. FitzSimmons, general counsel and
senior vice president, secretary;
Warren R. Higgins, vice president;
Gordon E. Irving, vice president;
Mitchell F. Keamy, Jr., vice president;
Paul B. Oldam, treasurer and vice president;
Charles W. Parker, Jr., vice president;
William S. Pierson, controller and vice president;
John L. Platner, vice president;
James J. Stouppe, vice president;
Jacques F. Trevillyan, senior vice president;
Roy W. Uelner, executive vice president.

INDEX

ALLIS-CHALMERS
1965

1919

1887

1949

1901

Allis-Chalmers trademark through the years.